T0291615

ATOMIC SPECTRA

In two volumes

VOLUME II

CONTENTS

Volume II. COMPLEX SPECTRA

LIST OF PLATES

CHAPTER XII

DISPLACED TERMS

1. The alkaline earths

Only about half of the bright lines of the calcium arc spectrum are accounted for by the simple terms described in a previous chapter. Most of the remaining lines, however, can be explained by the 'dashed' or 'displaced' terms introduced by Götze* in 1921, terms which Russell and Saunders† were later to attribute to the activity of two electrons (Fig. 12·2).

As an example of these terms consider the group of six bright lines occurring round 4300 A.

Terms	$2\,^3P_0'$	$2\,^3P_1'$	$2\,^3P_2'$
$2\,^3P_0$		4289·363 (40) 23306·96	
		52·20	
$2\,^3P_1$	4307·738 (45) 23207·53 47·23	4298·989 (30) 23254·76 86·77	4283·008 (30) 23341·53
		105·85	105·87
$2\,^3P_2$		4318·648 (45) 23148·91 86·75	4302·525 (60) 23235·66

Fig. 12·1. Wave-lengths and wave-numbers of the $2\,^3P' \rightarrow 2\,^3P$ multiplet of calcium. The numbers in brackets are intensities.

The intervals 52·2 and 105·9 are recognised as the intervals of the well-known $2\,^3P$ terms and these terms are therefore written in the left-hand column. The intervals 47·2 and 86·8 do not occur in the earlier analysis, but Landé's interval rule suggests that an empirical ratio of 1·84 should arise from a ^3P term, for the ideal interval ratios of ^3P and ^3D terms are 2 and 1·5 respectively.

If the combinations not observed are those forbidden by the J selection rule, then the empirical terms must have J values of 0, 1 and 2. Of the nine possible combinations, two are then forbidden because they have $\Delta J = 2$, and a third because the electron

* Götze, *AP*, 1921, **66** 285. A key to letters used in referring to periodicals is given in Appendix IV.

† Russell and Saunders, *AJ*, 1925, **61** 38. The authors write the terms $1\,^3P$ and $1\,^3P'$ instead of $2\,^3P$ and $2\,^3P'$ used here.

would have to jump from $J=0$ to $J=0$. The intersystem line $^3P_0 \rightarrow {}^1S_0$ is noticeably absent in the mercury spectrum.

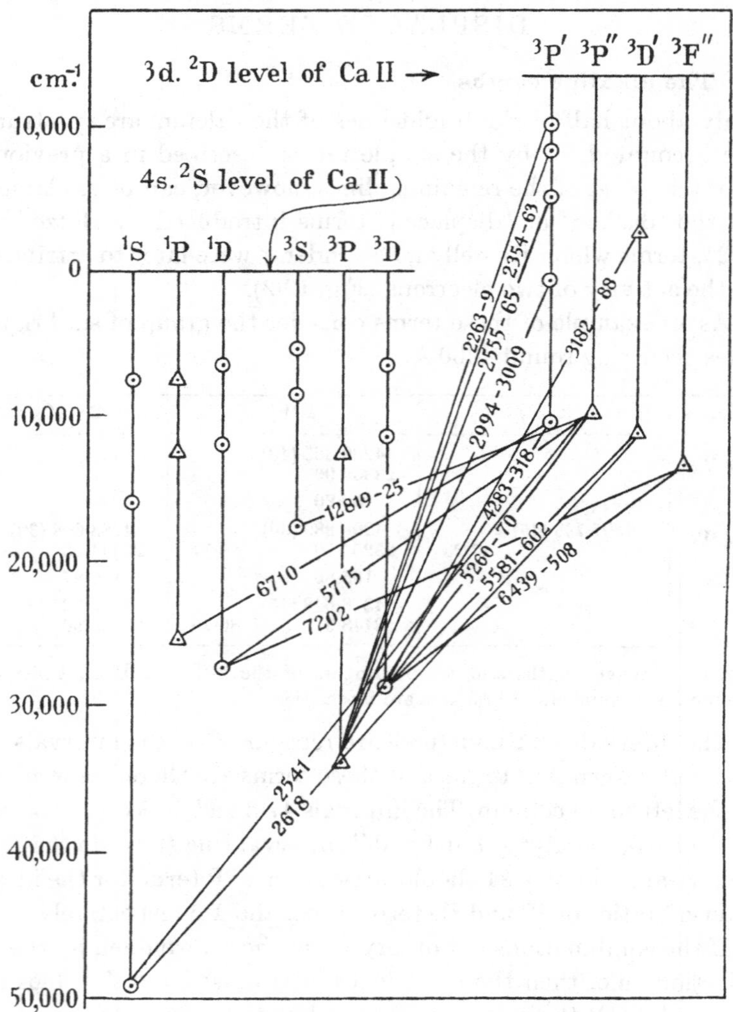

Fig. 12·2. Level diagram of calcium, extended to show the displaced terms. Even terms are shown by a circle, odd terms by a triangle.

A critic might suggest that the above argument really only determines the values of J, but the Zeeman effect has also been examined and this, connecting as it does L, J and S, serves to fix L, when J and S are known. Every indication therefore suggests that the empirical terms should be written ^{3}P, or, since they are not of the usual ^{3}P series, ^{3}P′.

So far no use has been made of the series laws, but five terms of the ^{3}P′ series have been observed, and if these are fitted to a Rydberg formula, they indicate that the limit of the series is some 13961 cm.$^{-1}$ above the common limit of the earlier known series of the calcium atom. Indeed, the 4 ^{3}P′, 5 ^{3}P′ and 6 ^{3}P′ terms are all above the limit, and contain energies of 0·68, 1·03 and 1·24 volts in excess of that required to ionise the atom. This surprisingly large energy content seems to be open to only one explanation. In the levels of this displaced series two electrons are excited, and when the atom returns to one of the normal states both electrons jump and both contribute energy to a single quantum of radiation.

This deduction may be still further refined. The second deepest term ^{2}D of the calcium ion is 13711 cm.$^{-1}$ or 1·72 volts above the basic ^{2}S term, and this is not very different from the amount by which the limit of the P′ series is raised above the normal limit. If then the two optical electrons are assumed to have orbital moments l_1, l_2, l_1 must be assigned the value 2.

The difficulty of explaining why the P′ terms combine with normal P terms is then surmounted by the hypothesis that $\mathbf{l}_1$ and $\mathbf{l}_2$ combine vectorially to give a resultant **L**. **L** combines in the usual way with **S**, itself the sum of the $\mathbf{s}_1$, $\mathbf{s}_2$ of the individual electrons, to give **J**; but the quantum transitions allowed depend on $\mathbf{l}_1$, $\mathbf{l}_2$ individually and only indirectly on **L**. The reasons already given for writing the unknown term ^{3}P′ first determine J, and then require L to be 1, but l_2 is still unassigned. If the old rule $\Delta L = \pm 1$ is to be written now $\Delta l_2 = \pm 1$, then the fact that the ^{3}P′ term combines with a P term shows that l_2 is 0 or 2; and the former is ruled out, because if $\mathbf{l}_1 + \mathbf{l}_2 = \mathbf{L}$ it is inconsistent with the previous allotment of L and l_1.

Apply the same arguments to a so-called ^{3}D′ term found by Russell and Saunders, which has the J properties of a D term but combines with ^{3}D, ^{1}S and ^{1}D. Clearly the J properties indicate $L = 2$, while the combining terms show that $l_2 = 1$; and as the ^{3}D′ series has the same displaced limit as the ^{3}P′ series, l_1 is 2.

When two electrons are excited, the atom as a whole can no longer be said to have a chief quantum number, though each electron has. Thus in the normal state of calcium, two electrons lie outside the closed shell of argon, and these occupy 4s orbits, so that the ground state is $4s^2\,{}^1S_0$. When one of these moves to a 4p orbit the configuration becomes $4s\,4p\,{}^3P$; this is the $2\,{}^3P$ term of the normal spectrum. In the lowest ${}^3P'$ term both electrons have moved to 3d orbits, so that the configuration is $3d^2\,{}^3P'$.*

Often the chief quantum number does not need to be stressed, and then $3d^2\,{}^3P'$ is abbreviated to $d^2\,{}^3P'$; or if the two d electrons have different chief quantum numbers, as in $3d.4d\,{}^3P'$, then we write $d.d\,{}^3P'$. Again as there is only one $3d.nd\,{}^3P$ series, the omission of the dash introduces no ambiguity, but in fact it is often retained as it gives a key to the transitions permitted.

What are these? Or in other words, what is the selection rule governing the combination of displaced terms? Heisenberg stated, on the basis of the quantum mechanics, that when two electrons jump simultaneously one is bound by the condition $\Delta l_1 = \pm 1$ and the second by $\Delta l_2 = 0$ or ± 2. Of this rule the transitions from the displaced to the normal terms of the alkaline earths offer a first example.

A very important consequence of this selection rule is that when an atom emits or absorbs a quantum of radiation the sum of the orbital quantum numbers of the individual electrons Σl must change by an odd number; if the sum was even it must become odd, or if it was odd it must become even. Consequently all possible terms may be divided into two groups, an even group, written in accord with the suggestion of a number of physicists† S, P, D, ..., and an odd group, written S°, P°, D°, ... A term of the one group can then combine only with a member of the other group.

Another notation adopted for several years by many spectroscopists differed somewhat from this. The even group of terms

* The $3d^2\,{}^3P'$ term is the $3\,{}^3P'$ term of this description; though the $3\,{}^3P'$, $4\,{}^3P'$ and $5\,{}^3P'$ terms arise from 3d.nd configurations, the $2\,{}^3P'$ term arises from $4p^2$; it was wrongly assigned by Russell and Saunders.

† Russell, Shenstone and Turner, *PR*, 1929, **33** 900.

was written S, P′, D, F′, ... and the odd series S′, P, D′, F, ... This notation was convenient because it was in agreement with the notation of the earlier chapters of this book and could be extended. But to-day it seems more important to be able to distinguish odd and even terms at sight, for the group to which an empirical term belongs can easily be determined, and this the dashed notation does not facilitate.

In simple spectra only one electron is excited, and for it the S, D terms have L even, while the P, F terms have L odd, and so to be quite accurate should be written P°, F°. Again, the displaced P′, D′ terms of Ca I have the sum (l_1+l_2) equal to 4 and 3 respectively; so that they are $d^2\,{}^3P$ and $p.d\,{}^3D°$.

Having dealt with two displaced series at considerable length, the other series found by Russell and Saunders, and written ${}^3P''$, ${}^3D''$ and ${}^3F''$, can be considered more briefly. All the series have the same elevated limit and consequently in every state one electron must occupy a d orbit. The double dash signified that these terms, unlike ${}^3P'$ and ${}^3D'$, do not combine with the normal terms of the same letter. Thus ${}^3P''$, though it has the J properties of a P term, combines with ${}^3S$, ${}^3D$, ${}^1S$, ${}^1D$, so that $L=1$ and $l_2=1$. ${}^3P''$ is therefore identical with $d.p\,{}^3P°$ and may be written in the latter form. Similarly, ${}^3D''$ combines with ${}^3P$, ${}^1P$ and therefore must be identical with $d^2\,{}^3D$. Combinations of ${}^3F''$ are known only with ${}^3D$ and ${}^1D$, for combinations with ${}^3S$ lie too far on the infra-red and ${}^3G$ is wholly unknown, but the energy of the term argues strongly in favour of $d.p\,{}^3F$ and against $d.f\,{}^3F$, though the evidence here adduced allows the latter.

Russell and Saunders in their analysis accounted for practically all the known lines of the three alkaline earths, though they hesitated to assign electronic states to all the terms which were suggested by the term differences, and whose energies were thus known. Wentzel,* however, has filled even this small gap, and the analysis of the visible and near ultra-violet region of calcium may be considered complete.

2. Beryllium and magnesium

In the spectra of beryllium and magnesium, and in the iso-

* Wentzel, *ZP*, 1925, **34** 730.

electronic spark spectra B II to O V and Al II to Cl VI, there occur a striking group of five lines; the separations are nearly equal, and four of the five lines are of equal intensity; the central line of the group, however, is noticeably stronger, and in some spectra it may be resolved into two close components.

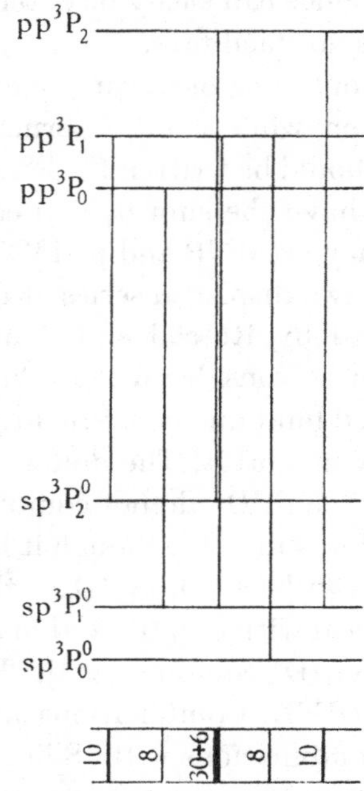

Fig. 12·3. Structure of a displaced triplet of magnesium; this arises in a $3p^2\ {}^3P \rightarrow 3s\ 3p\ {}^3P^\circ$ transition.

In 1925 Bowen and Millikan showed that these arise by the combination of the s.p $^3P^\circ$ term with a ^3P term, having so nearly he same intervals that the lines $^3P_2 \rightarrow {}^3P_2{}^\circ$ and $^3P_1 \rightarrow {}^3P_1{}^\circ$ often coincide. According to Sommerfeld's intensity rule, this central line is due to chief lines having $\Delta J = \Delta L$, while the other four lines are satellites of the first order. A more accurate theory would predict the intensities shown in Fig. 12·3. Both are in good agreement with observation.

The wave-lengths and wave-numbers of the central line in spectra isoelectronic with magnesium are shown in Fig. 12·4.*

* Bowen and Millikan, *PR*, 1925, **26** 150.

The differences show how accurately linear is the progression of frequency with atomic number; and this means that the group of lines follows the irregular doublet law, which they can do only if the transitions take place between levels having the same chief quantum number.

This argument applies as well to Be as to Mg, though figures are here adduced only for the latter. But in Be the jump ends in a 2s 2p orbit, so that it must start in a $2p^2$ orbit; no other possibility exists, for there are no 2d orbits. The close similarity of

Spectrum	λ	ν	Diff.
Mg I	2780·64	35962·96	
			20727·9
Al II	1763·95	56690·9	
			20295·5
Si III	1298·93	76986·4	
			20050·7
P IV	1030·53	97037·1	
			19948·5
S V	854·81	116985·6	
			19942·6
Cl VI	730·31	136928·2	

Fig. 12·4. The central line of displaced triplets in spectra isoelectronic with magnesium.

the lines in Mg and Be suggests that in Mg too the ^3P term arises from the $3p^2$ configuration; and this hypothesis is confirmed, when the frequency of the arc line $p^2\,{}^3P \rightarrow s\,.\,p\,{}^3P^\circ$ is compared with the spark line $p\,{}^2P \rightarrow s\,{}^2S$; the former is 35960 cm.$^{-1}$ and the latter 35760 cm.$^{-1}$, so that both must surely arise in the same electron transition $2p \rightarrow 2s$.

A comparison of these displaced terms with those found in the alkaline earths shows that in Be and Mg the lines are produced by one electron jumping, whereas in Ca, Sr and Ba two electrons jump simultaneously. But in both alike the displaced terms arise from the second lowest term of the spark spectrum.

3. Zinc, cadmium and mercury

Displaced terms have also been identified in Cd by Ruark* and in Zn and Hg by Sawyer†; but instead of the six lines to be expected four only have been found, these being interpreted as

* Ruark, *JOSA*, 1925, **11** 199. † Sawyer, *JOSA*, 1926, **13** 431.

the combination of the low s.p $^3P^\circ_{0,1,2}$ terms with the displaced terms $p^2\,{}^3P_{0,1}$ (Fig. 12·5).

Whether the $p^2\,{}^3P_2$ term is really missing or whether, as Foote, Takamine and Chenault* have suggested, it coincides with $p^2\,{}^3P_1$,

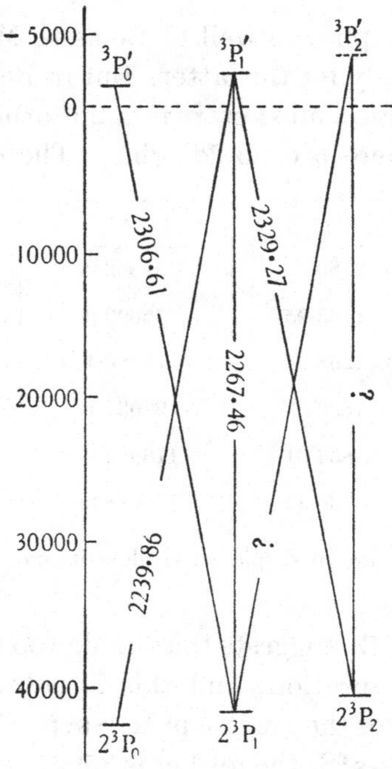

Fig. 12·5. Level diagram showing a PP′ triplet of cadmium; this arises as $5p^2\,{}^3P \rightarrow 5s\,5p\,{}^3P^\circ$.

so that in Cd the 2329 A. and 2268 A. lines are really narrow doublets, remains a question still undecided. Certainly the intensities are somewhat irregular and the 2329 A. line is much stronger than the other lines.

BIBLIOGRAPHY

These displaced spectra are fully discussed by Grotrian, *Graphische Darstellung der Spektren*, 1928, **1** 188–210.

* Foote, Takamine and Chenault, *PR*, 1925, **26** 174.

CHAPTER XIII

COMBINATION OF SEVERAL ELECTRONS

1. Combination of unlike electrons

The coupling of orbital and spin vectors, found to explain the displaced terms of the alkaline earths, gives also an adequate account of the terms of many more complex spectra.

In this coupling the orbital and spin vectors of the electrons, commonly written $\mathbf{l}_1, \mathbf{l}_2 \ldots, \mathbf{s}_1, \mathbf{s}_2 \ldots$, first combine to resultants **L** and **S** respectively, and then **L** and **S** combine to form **J**. The problem therefore is to determine all the values of L and S to which a given electron configuration can give rise; to solve this

Electron added	State of ion			
	S	P	D	F
s	S	P	D	F
p	P	SPD	PDF	DFG
d	D	PDF	SPDFG	PDFGH

Fig. 13·1. Atomic states resulting from the addition of an electron to an ion.

problem, consider the orbital and spin vectors separately. When a p and a d electron combine the **l** vectors are 1 and 2, so that their resultant must be 1, 2 or 3, values which correspond to P, D or F terms; further, each electron has a spin vector of $\frac{1}{2}$, so that the atomic spin vector due to two electrons may be 0 or 1, values which correspond to multiplicities of 1 and 3. Accordingly, a p and a d electron combine to form six terms: ^{1}P, ^{1}D, ^{1}F, ^{3}P, ^{3}D and ^{3}F.

A similar train of reasoning will determine the terms produced when an electron combines with an ion; for if an s electron is to be added to an ion in the ^{4}F state, the orbital vectors to be combined are 0 and 3, so that the resultant must be 3; while the **s** vectors are $\frac{1}{2}$ and $1\frac{1}{2}$, so that the resultant may be 1 or 2; accordingly the terms produced are ^{3}F and ^{5}F. Fig. 13·1 summarises the results obtained in this way, for it shows the terms arising when an s, p or d electron is added to an ion in the S, P, D or F state, the multiplicity being omitted since it always increases and decreases by unity.

When three or more electrons have to be combined together, the work must proceed by steps, two electrons being combined and then the third added to each of the terms produced by the two. Thus an spd configuration would give rise to the nine terms ^{2}(PDF), ^{2}(PDF) and ^{4}(PDF).

In the above discussion the word 'term' is used to denote a multiplet term, but in analysis each component of the multiplet appears as a separate empirical level or 'term'. Theory thus states precisely how many levels will arise from a given configuration,

M_L / m_l	−2	−1	0	1	2
1	−1	0	1	2	3
0	−2	−1	0	1	2
−1	−3	−2	−1	0	1
Term of atom	F	D	P		

Fig. 13·2. Addition of a p electron to an ion in the D state, showing how P, D and F states result.

and what their J values will be; and it is an outstanding achievement of the theory that the empirical terms occur in just the number predicted and have the J values assigned. Sometimes, as might be expected, some of the levels are missing, but they are nearly always levels which would produce only lines of low intensity. In the great mass of spectra so far analysed, there is only a single level which seems to be adequately substantiated and is surplus to the theory; it is found in Pd I.

Though the above method of combining two electrons is the simplest, another method due to Russell* is not without interest. Instead of combining the $\mathbf{l}_1$ and $\mathbf{l}_2$ vectors of ion and electron, combine the m_{l_1} and m_{l_2} values; all combinations are permitted,

* Russell, *PR*, 1927, 29 782.

so that a matrix of $(2l_1+1)(2l_2+1)$ values results, and this can be split into a number of sequences by Breit's alleys, exactly as in the theory of the Paschen-Back effect. Fig. 13·2 shows the calculation when a p electron is added to an ion in a D state.

The same method can be used to combine the m_{s_1} and m_{s_2} of ion and electron to obtain the multiplicity of the atomic terms.

2. Combination of equivalent electrons

Two electrons having the same values of n and l are said to be 'equivalent'.

When two equivalent electrons are to be combined, the above argument gives all the terms which might be produced, but it gives more than are produced, for the combination is restricted by Pauli's exclusion principle. Of this complication the simplest example is the combination of two electrons in the alkaline earths; thus in magnesium the two valency electrons must have $n \geqslant 3$, since the first two groups are full, and so the lowest terms may be expected to arise from the combination of two 3s electrons. s electrons, however, always have $m_l=0$, so that if Pauli's exclusion principle is to be satisfied and the two electrons are not to occupy the same orbit, m_s must be $+\frac{1}{2}$ for one and $-\frac{1}{2}$ for the other. Consequently $M_S=0$ and $S=0$, showing that the lowest term is $^1S_0$ and that a ^3S term can appear only when the two s electrons producing it have different chief quantum numbers. Thus Pauli's exclusion principle clearly explains why the $1\,^3S_1$ term is missing in all the alkaline earth spectra and the $^3S_1$ sequence begins with a $2\,^3S_1$ term.

l_1	l_2	m_{l_1}	m_{l_2}	m_{s_1}	m_{s_2}	M_L	M_S	Term
0	0	0	0	$\frac{1}{2}$	$-\frac{1}{2}$	0	0	^1S

Fig. 13·3. Combination of two equivalent s electrons, showing that only a ^1S term results.

Fig. 13·3 shows the argument summarised in a table. A similar table for two equivalent p electrons appears in Fig. 13·4.

In this table all possible combinations of m_{l_1} and m_{l_2} are included, but the two electrons are to be considered interchangeable, so that $m_{l_1}=1$ and $m_{l_2}=0$ is identical with $m_{l_1}=0$ and $m_{l_2}=1$. Further, the derivation of the terms, from the sums M_L and M_S needs to be explained. A ^{1}D term has $L=2$ and $S=0$, so that it will give rise to a series of values of M_L from 2 to -2 all

l_1 l_2	m_{l_1} m_{l_2}	m_{s_1} m_{s_2}	M_L	M_S	Terms
1 1	1 1	$\frac{1}{2}$ $-\frac{1}{2}$	2	0	^{1}D
	0	$\pm\frac{1}{2}$ $\pm\frac{1}{2}$	1	1 0 0 −1	^{3}P
	−1	$\pm\frac{1}{2}$ $\pm\frac{1}{2}$	0	1 0 0 −1	
	0 0	$\frac{1}{2}$ $-\frac{1}{2}$	0	0	^{1}S
	−1	$\pm\frac{1}{2}$ $\pm\frac{1}{2}$	−1	1 0 0 −1	
	−1 −1	$\frac{1}{2}$ $-\frac{1}{2}$	−2	0	

Fig. 13·4. Combination of two equivalent p electrons, showing how the resulting terms are calculated.

having $M_S=0$. To work from M_L and M_S to L and S therefore, start with the highest value of M_L available, namely 2 in the above figure, and cross out both a series of values of M_L from 2 to -2 and the corresponding values of M_S, namely 0. When these values have been deleted, the table of Fig. 13·4 reduces to that shown in Fig. 13·5.

M_L	M_S	Terms
1	1 0 −1	^{3}P
0	1 0 −1	
0	0	^{1}S
−1	1 0 −1	

Fig. 13·5. A step in the elucidation of the previous figure; the M_L and M_S values which remain when the ^{1}D term has been removed.

Now repeat the operation. The highest value of M_L is 1, and the greatest corresponding values of M_S also 1, so that the next term to be written in the margin must have $L=1$ and $S=1$; and the values to be deleted are $M_L=1, 0, -1$ and $M_S=1, 0, -1$. Afterwards there remains only $M_L=0$ when $M_S=0$, representing a ^{1}S term.

The calculation of terms arising from three equivalent p electrons introduces no new principles; the work, summarised in Fig. 13·6, shows that the terms to be expected are ^{2}D, ^{2}P and ^{4}S.

The combination of four equivalent p electrons is worked out in Fig. 13·7. The terms arising, be it noted, are those arising from two equivalent p electrons, and indeed this could have been predicted. For six p electrons form a closed group, in which every orbit permitted by the exclusion principle is occupied, and consequently $M_L = M_S = 0$. Therefore when there are four p electrons, the orbits left unoccupied will be precisely those

l_1 l_2 l_3	m_{l_1}	m_{l_2}	m_{l_3}	m_{s_1}	m_{s_2}	m_{s_3}	M_L	M_S	Terms
1 1 1	1	1	0	$\frac{1}{2}$	$-\frac{1}{2}$	$\pm\frac{1}{2}$	2	$\pm\frac{1}{2}$	^{2}D
			-1	$\frac{1}{2}$	$-\frac{1}{2}$	$\pm\frac{1}{2}$	1	$\pm\frac{1}{2}$	
	0	0	1	$\frac{1}{2}$	$-\frac{1}{2}$	$\pm\frac{1}{2}$	1	$\pm\frac{1}{2}$	^{2}P
			-1	$\frac{1}{2}$	$-\frac{1}{2}$	$\pm\frac{1}{2}$	-1	$\pm\frac{1}{2}$	
	-1	-1	1	$\frac{1}{2}$	$-\frac{1}{2}$	$\pm\frac{1}{2}$	-1	$\pm\frac{1}{2}$	
			0	$\frac{1}{2}$	$-\frac{1}{2}$	$\pm\frac{1}{2}$	-2	$\pm\frac{1}{2}$	
	1	0	-1	$\pm\frac{1}{2}$	$\pm\frac{1}{2}$	$\pm\frac{1}{2}$	0	$\pm\frac{1}{2}$ $\pm\frac{1}{2}$ $\pm\frac{1}{2}$ $\pm1\frac{1}{2}$	^{4}S

Fig. 13·6. Combination of three equivalent p electrons.

l_1 l_2 l_3 l_4	m_{l_1}	m_{l_2}	m_{l_3}	m_{l_4}	m_{s_1}	m_{s_2}	m_{s_3}	m_{s_4}	M_L	M_S	Terms
1 1 1 1	1	1	0	0	$\frac{1}{2}$	$-\frac{1}{2}$	$\frac{1}{2}$	$-\frac{1}{2}$	2	0	^{1}D
			0	-1	$\frac{1}{2}$	$-\frac{1}{2}$	$\pm\frac{1}{2}$	$\pm\frac{1}{2}$	1	1 0 0 -1	^{3}P
			-1	-1	$\frac{1}{2}$	$-\frac{1}{2}$	$\frac{1}{2}$	$-\frac{1}{2}$	0	0	^{1}S
	0	0	1	-1	$\frac{1}{2}$	$-\frac{1}{2}$	$\pm\frac{1}{2}$	$\pm\frac{1}{2}$	0	1 0 0 -1	
			-1	-1	$\frac{1}{2}$	$-\frac{1}{2}$	$\frac{1}{2}$	$-\frac{1}{2}$	-2	0	
	-1	-1	1	0	$\frac{1}{2}$	$-\frac{1}{2}$	$\pm\frac{1}{2}$	$\pm\frac{1}{2}$	-1	1 0 0 -1	

Fig. 13·7. Combination of four equivalent p electrons.

l_1 l_2 l_3 l_4 l_5 l_6	m_{l_1}	m_{l_2}	m_{l_3}	m_{l_4}	m_{l_5}	m_{l_6}	m_{s_1}	m_{s_2}	m_{s_3}	m_{s_4}	m_{s_5}	m_{s_6}	M_L	M_S	Term
1 1 1 1 1 1	1	1	0	0	-1	-1	$\frac{1}{2}$	$-\frac{1}{2}$	$\frac{1}{2}$	$-\frac{1}{2}$	$\frac{1}{2}$	$-\frac{1}{2}$	0	0	^{1}S

Fig. 13·8. Combination of six equivalent p electrons.

occupied by two electrons, and the values of M_L and M_S will be unaltered in magnitude. The change in sign will not affect the values of L and S or consequently the terms. For the same reasons five p electrons give rise to a ^{2}P term just as one electron does.

The terms resulting from various numbers of p electrons are collected together in Fig. 13·9, while below are similar tables for d and f electrons. Of these, the first two were given by Hund and the last by Gibbs, Wilber and White.* To save space terms of the same multiplicity are sometimes bracketed together, thus ^{4}P ^{4}F is abbreviated to ^{4}(PF); and the number of times a particular term occurs is indicated by a figure written directly below the letter.

Number of p electrons	Terms resulting			
0 or 6	^{1}S			
1 or 5		^{2}P		
2 or 4	^{1}S ^{1}D		^{3}P	
3		^{2}P ^{2}D		^{4}S

Fig. 13·9. Terms resulting from the combination of equivalent p electrons.

No. of d electrons	Terms		
0 or 10	^{1}S		
1 or 9	^{2}D		
2 or 8	^{1}(SDG)	^{3}(PF)	
3 or 7	^{2}(PDFGH) 2	^{4}(PF)	
4 or 6	^{1}(SDFGI) 2 2 2	^{3}(PDFGH) 2 2	^{5}D
5	^{2}(SPDFGHI) 3 2 2	^{4}(PDFG)	^{6}S

Fig. 13·10. Terms resulting from the combination of equivalent d electrons.

No. of f electrons	Terms			
0 or 14	^{1}S			
1 or 13	^{2}F			
2 or 12	^{1}(SDGI)	^{3}(PFH)		
3 or 11	^{2}(PDFGHIKL) 2 2 2 2	^{4}(SDFGI)		
4 or 10	^{1}(SDFGHIKLN) 2 4 4 2 3 2	^{3}(PDFGHIKLM) 3 2 4 3 4 2 2	^{5}(SDFGI)	
5 or 9	^{2}(PDFGHIKLMNO) 4 5 7 6 7 5 5 3 2	^{4}(SPDFGHIKLM) 2 3 4 4 3 3 2	^{6}(PFH)	
6 or 8	^{1}(SPDFGHIKLMNQ) 4 6 4 8 4 7 3 4 2 2	^{3}(PDFGHIKLMNO) 6 9 7 9 6 6 3 3	^{5}(SPDFGHIKL) 3 2 3 2 2	^{7}F
7	^{2}(SPDFGHIKLMNOQ) 2 5 7 10 10 9 9 7 5 4 2	^{4}(SPDFGHIKLMN) 2 2 6 5 7 5 5 3 3	^{6}(PDFGHI)	^{8}S

Fig. 13·11. Terms resulting from the combination of equivalent f electrons.

3. Deep terms of the short periods

If the terms of low energy arise from electrons of low energy then the deep terms of the short periods must be derived from

* Gibbs, Wilber and White, *PR*, 1927, **29** 790.

equivalent s and p electrons. Moreover, they can be read off from Fig. 13·9, for a closed shell contributes nothing to L or S, so that the terms which arise from an s^2p^n configuration are precisely those which would arise from p^n.

The elucidation of these deep terms is important, because a few terms often give the key to the whole spectrum, and transitions ending in deep terms yield the brightest lines.

Outer electrons			Ground term	Other deep terms	Some spectra in which these terms have been found
Total	s	p			
1	1	—	^2S	—	Na I, Cs I, Ca II, Al III
2	2	—	^1S	—	Mg I, Ca I, Al II, C III
	2	1	^2P	—	Al I, Tl I, C II, N III
4	2	2	^3P	^1D ^1S	C I, Pb I, N II, O III
5	2	3	^4S	^2D ^2P	N I, Sb I, O II, Cl III
6	2	4	^3P	^1D ^1S	O I, Se I, Cl II, A III
7	2	5	^2P	—	F I, Cl I, A II, Na III
8	2	6	^1S	—	Ne I, Kr I, Na II, Ca III

Fig. 13·12. Low terms of the short periods.

4. Two energy rules

In the spectra of the short periods the deep terms are few in number, and so those predicted by theory are easily matched with those found empirically. But of the higher terms very many will arise from a single configuration, and to know which of these normally lie low and which high is a great help to the accurate labelling of the empirical terms. This problem was first studied by Hund, who solved it in two empirical energy rules.

The first rule states that those terms lie deepest in which the electronic spin vectors are parallel to one another, and in which therefore S is a maximum. Thus in the alkaline earths, the triplet term arising from any electronic configuration lies deeper than the corresponding singlet term; while in the short periods Fig. 13·12 shows that of the terms arising from the simplest electron configuration that of highest multiplicity is the ground term.

The second rule adds that of terms which have the same multiplicity, those lie deepest in which the electronic orbital vectors are also parallel, and in which therefore L is a maximum.

Accordingly, in carbon ^1D lies lower than ^1S, and in nitrogen ^2D lies lower than ^2P.

Thus of two terms arising from a set of equivalent electrons, that which lies furthest to the right in Figs. 13·9–13·11 lies deepest in the spectrum. These rules never fail to predict the ground term of a spectrum correctly, but among higher configurations many exceptions occur.

5. Inverted terms

When a group of electrons is more than half full the deep terms arising from it are inverted. Thus the ^2P ground term of Al I is erect, but the ^2P ground term of Cl I is inverted; the ^3P ground term of silicon is erect, but that of oxygen is inverted.

This fact is best examined in the light of a third energy rule, which asserts that the deepest components of a term arise when the orbital and spin vectors of the electrons are anti-parallel. When there are no restrictions this rule ensures that in the lowest component J will be a minimum and the terms will therefore be erect; but when the group is more than half full the restrictions imposed by the exclusion principle interfere.

The ground terms of the elements of a short period may be built up graphically by adding one electron at a time. The first electron, having its **l** and **s** vectors anti-parallel, as shown in Fig. 13·13, gives rise to a $^2P_{\frac{1}{2}}$ term. On adding the second electron the first energy rule makes $\mathbf{s}_2$ parallel to $\mathbf{s}_1$, and would make $\mathbf{l}_2$ parallel to $\mathbf{l}_1$ if the exclusion principle did not interfere; but equivalent electrons may not have the same values of m_l if they already have the same values of m_s, so if m_{l_1} was -1, m_{l_2} must be 0 or 1, and of these 0 will give the lower component since it will make the vectors more nearly anti-parallel; accordingly, in Fig. 13·13 the $\mathbf{l}_2$ vector is drawn horizontal.

The addition of a third electron emphasises no new principle; but the fourth, if it is to produce the empirical term, must set its orbital vector **l** anti-parallel not to the resultant **S** but to the electronic vector $\mathbf{s}_4$. The lowest component then appears as $^3P_2$, and as the other components of the ground term are $^3P_1$ and $^3P_0$ the term appears inverted.

This theory, which can of course be applied also to d and f electrons (Fig. 13·14), explains well enough why the ground terms

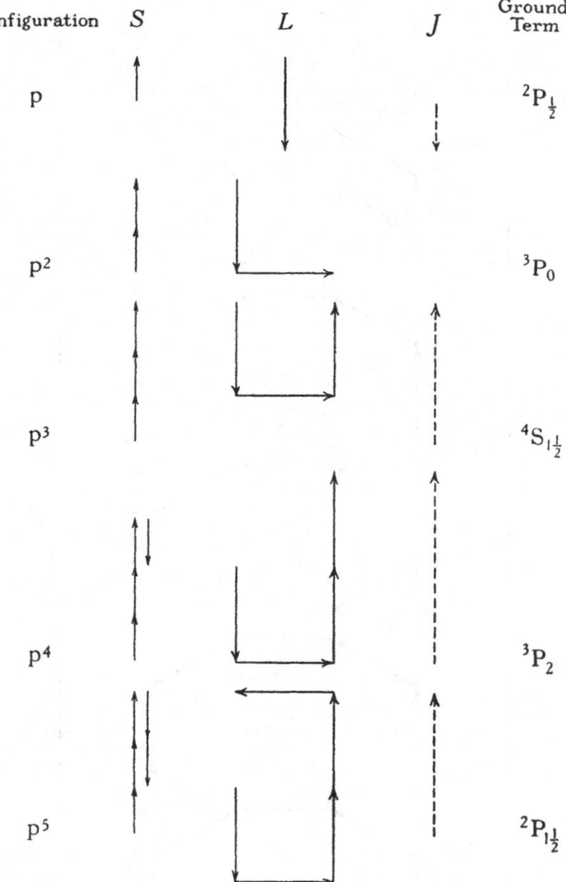

Fig. 13·13. Vector diagram showing how the ground terms of a p shell may be derived with the help of the Pauli exclusion principle.

of certain spectra are inverted, but it would not necessarily lead one to expect that many of the higher terms would also be inverted as in fact they are. A theory developed by Goudsmit,* serves however to bring out this point.

* Goudsmit, *PR*, 1928, **31** 946. Ruark and Urey, *Atoms, molecules and quanta*, 1930, 332.

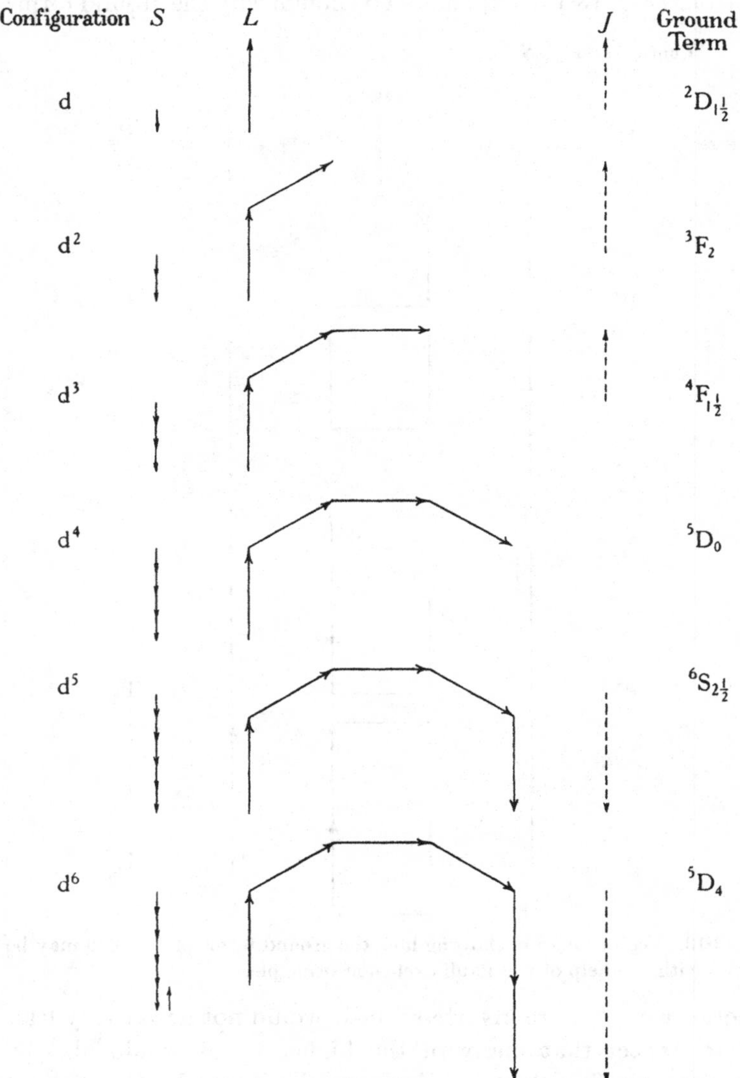

Fig. 13·14. Vector diagram showing how the ground terms of a d shell may be derived.

Previously, the wave-number of a component of a multiplet term has been written

$$E/hc = \nu_G + \Gamma, \qquad \ldots\ldots(4\cdot 1)$$

where ν_G is the wave-number of the centroid. Let us assume that when several electrons are active the atomic displacement Γ is itself the sum of electronic displacements $\gamma_1, \gamma_2, \ldots, \gamma_n$.

In the vector model the energy of an electron having orbital and spin vectors $\mathbf{l}_1$, $\mathbf{s}_1$ is assumed proportional to them and to the cosine of the angle between them, so that

$$\gamma_1 = a_1 l_1 s_1 \cos(\mathbf{l}_1 \mathbf{s}_1). \qquad \ldots\ldots(13\cdot 1)$$

Now l_1 precesses round L and s_1 round S, so that

$$\gamma_1 = a_1 l_1 \cos(\mathbf{l}_1 \mathbf{L})\, s_1 \cos(\mathbf{s}_1 \mathbf{S}) \cos(\mathbf{LS}),$$

or summing for all the electrons,

$$\Gamma = \Sigma\gamma_1 = \cos(\mathbf{LS})\, \Sigma a_1 l_1 \cos(\mathbf{l}_1 \mathbf{L})\, s_1 \cos(\mathbf{s}_1 \mathbf{S}). \ldots\ldots(13\cdot 2)$$

In order to advance beyond this equation, Slater* had to make a further postulate; the electrons are to be divided into two groups according as their spin vectors are parallel or anti-parallel to the atomic spin vector; the resultants L' and L'' of the corresponding orbital vectors are then 'action variables', which interpreted in terms of the vector model means that they must both be integral.

Consider now the ^{3}P ground term of an element, such as oxygen, with four p electrons. Three of these will have **s** parallel to **S**, while one will be anti-parallel. Clearly then $L'' = 1$, while L' might assume any value less than 3 did the exclusion principle not intervene, requiring that if all three electrons have the same value of m_s then they can none of them have the same value of m_l; consequently, when S' is $1\frac{1}{2}$, the only possible value of L' is zero. Equation (13·1) now reduces to

$$\Gamma = as\cos(\mathbf{LS})(L' - L''), \qquad \ldots\ldots(13\cdot 3)$$

for if all electrons are equivalent $a_1 = a_2 = \ldots = a_n$ and all may be written a.

This formula accounts for the empirical facts. So long as a shell is less than half full Γ will be positive and the terms erect; but

* Slater, *PR*, 1926, **28** 291.

when there are present three p, five d, or seven f electrons, $L' = L'' = 0$, and the ground term will be an S term of multiplicity one greater than the number of electrons present. When a shell is more than half full L', but not L'', is zero, so that Γ is negative and the terms are inverted.

A very pretty confirmation of this theory is found in nitrogen, most of whose terms arise from the $2p^2\,{}^3P$ configuration of N II and are consequently erect; two deep terms, however, a ${}^4P$ and a ${}^2P$, are inverted, and it seems clear that these have the electronic structure $2s\,.\,2p^4$.*

BIBLIOGRAPHY

The consequences of Russell and Saunders' paper and the Pauli exclusion principle were first developed by Hund in *Linienspektren und periodisches Systems der Elemente*, 1927; more recent accounts appear in Ruark and Urey, *Atoms, molecules and quanta*, 1930; Pauling and Goudsmit, *The structure of line spectra*, 1930.

* Compton and Boyce, *PR*, 1929, **33** 147.

CHAPTER XIV

ELEMENTS OF THE SHORT PERIODS

1. Elements to be considered

The seventh from last element of each period has a ^1S ground term, and this may be taken to mean that the configuration of each of these elements consists only of complete shells; the last six elements of each period shown in Fig. 14·1, are therefore

Column	III	IV	V	VI	VII	VIII
Configuration	p	p^2	p^3	p^4	p^5	p^6
Elements	B Al Ga In Tl	C Si Ge Sn Pb	N P As Sb Bi	O S Se Te Po	F Cl Br I	Ne A Kr Xe Nt
Ground term	^2P	^3P	^4S	^3P	^2P	^1S
Prominent multiplicities	2	1, 3	2, 4	3, 5	2, 4	1, 3

Fig. 14·1. Elements arising from configurations of p electrons.

formed by the entrance of six electrons into a new shell, the sixth electron completing the shell and forming the inert gas. As this shell must consist of p electrons in the first short period, it presumably consists of p electrons in the other periods.

If this is true, then the partly filled shell of p electrons should produce the ground term of each spectrum, the terms being ^2P, ^3P, ^4S, ^3P, ^2P, ^1S in successive columns, and in fact these are the terms found. Of the higher terms the great majority are formed by adding an electron to the ground term of the ion. Consider, for example, the spectrum of carbon in which there are only two p electrons; the ground term of C II is ^2P, and accordingly C I should consist of singlets and triplets, the terms produced being ^1P and ^3P, ^{1}(SPD) and ^{3}(SPD), or ^{1}(PDF) and ^{3}(PDF), according as the second electron moves in an s, p or d orbit. And similarly in column VI the ground term of the spark spectrum is ^4S, so that the arc spectrum should consist chiefly of the resulting triplets and quintets, though some less prominent terms may arise by the addition of an electron to the metastable states $p^3\,{}^2P$ and $p^3\,{}^2D$ of the ion.

2. Irregularities and their cause

In the early days of spectrum analysis, spectroscopists were able to order only those spectra in which series are prominent, for a series was the only regularity which they had recognised. To-day, the spectroscopist who has analysed a spectrum into a complex of terms can name those terms only if they exhibit some regularity; and the regularities on which he chiefly relies are the selection rules, which serve to determine J, the magnetic splitting factor, and the intensity and interval rules. The three last are linked together, for when a spectrum fails to obey one rule, it often fails to obey all three; and in terms of the vector model this is taken to mean that the coupling of the vectors ceases to be that postulated by Russell and Saunders.

What then are the influences which make a term irregular? Briefly they may be summarised as three; first is an increase in atomic number, carbon is more regular than lead, and neon than krypton; secondly, the column of the periodic table is significant, for the spectra of the elements on the left-hand side are more regular than those on the right, beryllium is more regular than nitrogen and nitrogen than neon; while third stands the height or energy of a term, for the higher a term lies the less likely is it to obey the simple rules, and particularly the interval rule. These influences are here stated as empirical rules; their explanation will be attempted only in a later chapter.

3. The earth metals, s^2p configuration*

The spark spectra of the earth metals, like the spark spectra of the alkalis, have a ${}^1S$ ground term, so that the arc spectra should consist of doublets; and in fact a system of doublets has been found and analysed into principal, sharp, diffuse and fundamental series. Where the terms have been resolved they have been shown

* Grotrian, *Graphische Darstellung der Spektren*, 1928, **1** 122f. and **2** 80, 96.

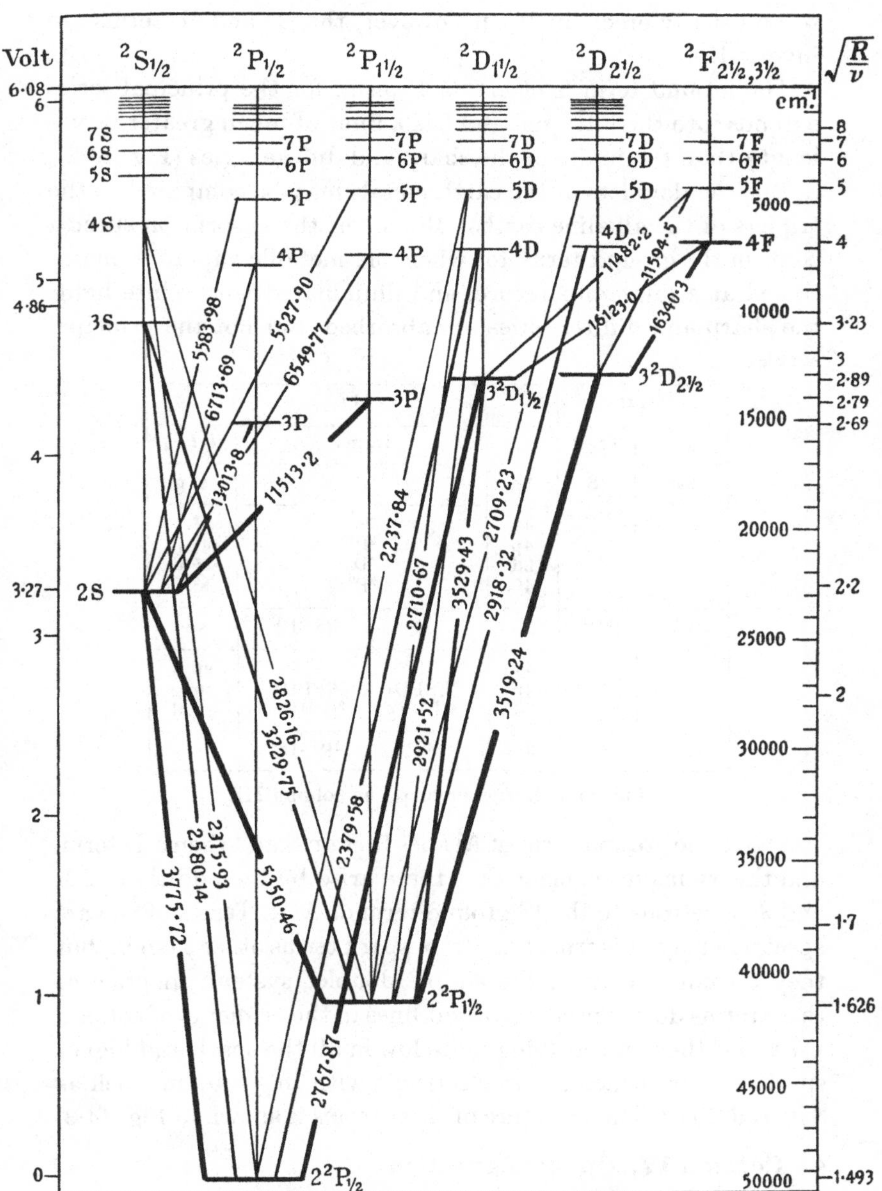

Fig. 14·2. Level diagram of thallium. (After Grotrian, *Graphische Darstellung der Spektren.*)

to be usually erect; in Pb II, however, the ^{2}D and ^{2}F terms are inverted.

The ground term is clearly a P term, for the principal series extends into the infra-red and has a limit of much greater wave-length than the limits of the sharp and diffuse series (Fig. 14·2); in this the doublets of the earth metals may be compared to the triplets of the alkaline earths. Moreover, the absorption spectra confirm the ground term, for when the metallic vapour is maintained at a low temperature, and illuminated with white light, the sharp and diffuse series are absorbed, but not the principal series.

Al II		Al I			
$l_1\,l_2$	Term	l_3	Terms		$l_1\,l_2\,l_3$
3s^2	^{1}S	3p	^{2}P$^\circ$		s^2 p
		4s	^{2}S		s^2.s
		4p	^{2}P$^\circ$		s^2.p
		3d	^{2}D		s^2.d
		4f	^{2}F$^\circ$		s^2.f
3s 3p	^{3}P$^\circ$	3p	^{4}P	^{2}(SPD)	sp^2
		4s	^{4}P$^\circ$	^{2}P$^\circ$	sp.s
		4p	^{4}(SPD)	^{2}(SPD)	sp.p
		3d	^{4}(PDF)$^\circ$	^{2}(PDF)$^\circ$	sp.d
3p^2	^{3}P	3p	^{4}S$^\circ$	^{2}P$^\circ$ ^{2}D$^\circ$	p^3

Fig. 14·3. Terms predicted in column III.

Above the ground term of Al I lies first an S and then a D term, and theory makes it clear that these arise by the addition of 4s and 3d electrons to the ^{1}S ground term of Al II. Terms of the arc spectrum may, it is true, arise from spark terms other than ^{1}S, but they do not belong to the simple doublet system; in practice these terms do not produce bright lines in the elements of column III A, but they appear lying quite low in all the spark and higher spark spectra, which are isoelectronic with this column, such as Si II and Sb III. The structure of these terms is shown in Fig. 14·3.

4. Column IV, s^{2}p^2 configuration

In the fourth column Hund's scheme predicts that the s^{2}p^2 configuration will produce five low terms $^3P_{0,1,2}$, $^1D_2$ and $^1S_0$ in that order proceeding from the ground term up; and empirically in carbon, silicon, germanium, tin and lead five low terms have been found.

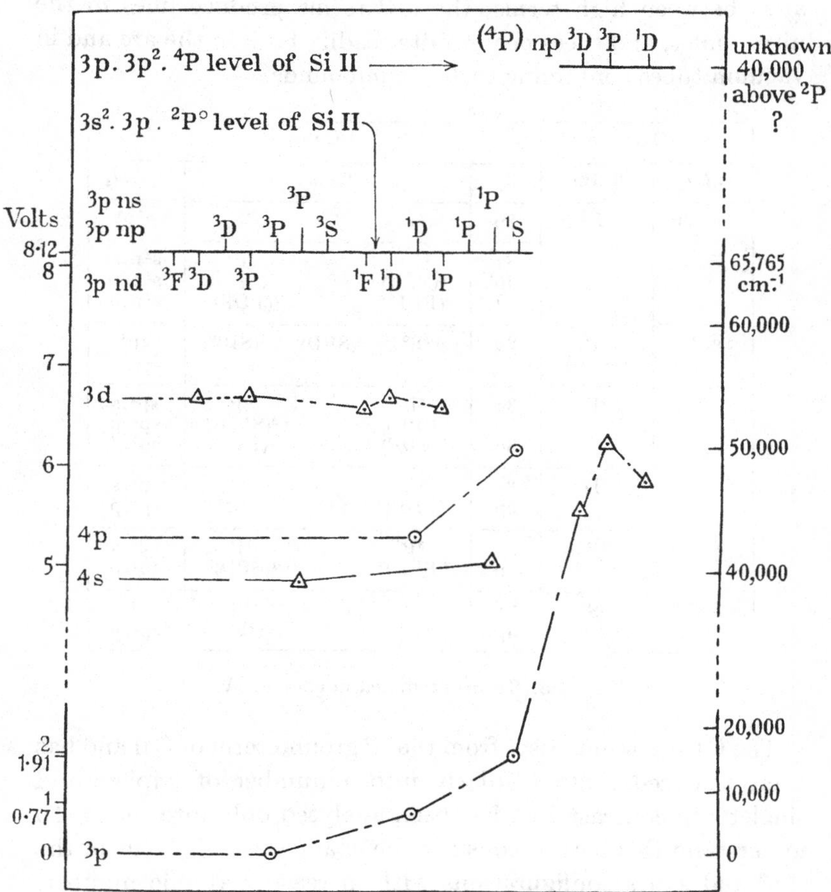

Fig. 14·4. Level diagram of silicon. In this and the following diagrams, even terms are shown by a circle and odd terms by a triangle. Terms produced by the same electron are joined by a broken line; when the electron is an s electron the line is drawn — —, when a p electron — - — and when a d electron — - - —.

In the region extending from the red end of the spectrum down to 2000 A. the carbon arc produces only one line, 2478 A.; but when a trace of carbon dioxide is added to helium at a pressure of 20 to 30 mm. many lines occur which have been attributed to the neutral carbon atom. Analysis of the spectrum shows that the reason why the visible lines are difficult to excite is that they arise between high terms; the low terms produce lines in the ultra-violet, and these occur quite readily both in the arc and in vacuum tubes containing carbon compounds.

C II		C I				
$l_1\,l_2\,l_3$	Term	l_4	Terms			$l_1 \dots l_4$
$2s^2.2p$	$^2P^\circ$	2p	^3P	^1D	^1S	s^2p^2
		3s	$^3P^\circ$		$^1P^\circ$	$s^2p.s$
		3p	$^3(SPD)$		$^1(SPD)$	$s^2p.p$
		3d	$^3(PDF)^\circ$		$^1(PDF)^\circ$	$s^2p.d$
$2s.2p^2$	^4P	2p	$^5S^\circ$	$^3(SPD)^\circ$	$^1(SD)^\circ$	sp^3
	$^2(SPD)$					
	^4P	3s	^5P		^3P	$sp^2.s$
		3p	$^5(SPD)^\circ$		$^3(SPD)^\circ$	$sp^2.p$
		3d	$^5(PDF)$		$^3(PDF)$	$sp^2.d$
	^2D	3s	^3D		^1D	$sp^2.s$
		3p	$^3(PDF)^\circ$		$^1(PDF)^\circ$	$sp^2.p$
	^2P	3s	^3P		^1P	$sp^2.s$
		3p	$^3(SPD)^\circ$		$^1(SPD)^\circ$	$sp^2.p$
	^2S	3s	^3S		^1S	$sp^2.s$
		3p	$^3P^\circ$		$^1P^\circ$	$sp^2.p$

Fig. 14·5. Terms predicted in column IV.

The C I spectrum arises from the ^2P ground term of C II and has been analysed quite regularly into a number of triplets and singlets; in contrast Pb I has been analysed only into series. To understand this better, consider the changes which occur in the s^2p^2 and $s^2p.s$ configurations with increasing atomic number. Thus in C I the ^3P ground term has nearly the ideal interval ratio of the Russell-Saunders coupling, while the extreme interval is small compared with the distance which separates the ^3P from the ^1D term. From this ideal the low terms of Si I, Ge I and Sn I fall further and further away until in lead Δ^3P_{01} is greater than Δ^3P_{12}, while the extreme triplet interval Δ^3P_{02} is as great as the separation of the $^3P_2$ and $^1D_2$ terms (Fig. 14·6).

Again examine the intervals of the $s^2p.s$ configuration which produces a ^3P and ^1P term; in carbon the extreme interval of the ^3P term is only 60 cm.$^{-1}$, while 1500 cm.$^{-1}$ separates $^3P_2$ from $^1P_1$;

	C	Si	Ge	Sn	Pb
$^3P_0$–$^3P_1$	14·8	77·1	557·1	1692	7,817
$^3P_1$–$^3P_2$	27·5	146·1	852·8	1736	2,831
$^3P_2$–$^1D_2$	10,150	6075·5	5716·0	5185	10,818
$^1D_2$–$^1S_0$	11,452	9095·4	9241·2	8550	8,000
$\Delta\,^3P_{1,2}/\Delta\,^3P_{0,1}$	1·86	1·90	1·53	1·02	0·36

Fig. 14·6. Column IV. Intervals in the ground configuration, s^2p^2.

but in tin the triplet and singlet are transformed into two diads, having intervals of only 300 and 600 cm.$^{-1}$, but separated from one another by 4000 cm.$^{-1}$ (Fig. 14·7).

	C	Si	Ge	Sn	Pb
$^3P_0$–$^3P_1$	20	77	251	274	327
$^3P_1$–$^3P_2$	40	195	1415	3714	—
$^3P_2$–$^1P_1$	1589	1037	903	628	—
$\Delta\,^3P_{0,2}$	60	272	1666	3988	—
$\Delta\,^2P_{\frac{1}{2},1\frac{1}{2}}$ of ion	64	287	1768	4253	14,071

Fig. 14·7. Column IV. Intervals in the lowest $s^2p.s$ configuration.

These changes run parallel to the increasing interval of the ^2P ground term of the spark spectrum, and the relation between the two may be explained in terms of the vector model, provided that the separation of two terms is assumed roughly proportional to the strength of the coupling producing it. This assumption is indeed implicit in the description of regular terms already given; for if, to fix ideas, the two p electrons of C I are considered, the triplet and singlet terms are found widely separated, a fact which the vector model translates to read that the spin coupling ($\mathbf{s}_1\mathbf{s}_2$) is strong; or again the two singlet terms are also widely separated, and so the orbital coupling ($\mathbf{l}_1\mathbf{l}_2$) must also be supposed strong; but the intervals of the ^3P term are small and so the coupling responsible, namely ($\mathbf{LS}$), must be weak.

Consider then the addition of an electron to an ion specified by $_i\mathbf{L}$, $_i\mathbf{S}$ and $_i\mathbf{J}$; if the coupling is Russell-Saunders, the coupling of

$_i\mathbf{L}$ and $_i\mathbf{S}$ must be split when the electron is added, and only after $_i\mathbf{L}$ has combined with $\mathbf{l}$ and $_i\mathbf{S}$ with $\mathbf{s}$ may their resultants $\mathbf{L}$ and $\mathbf{S}$ combine together to form an atomic resultant $\mathbf{J}$. In symbols, the coupling then appears as

$$\{(_i\mathbf{L}\mathbf{l})\,(_i\mathbf{S}\mathbf{s})\} = \{\mathbf{L}\mathbf{S}\} = \mathbf{J}.$$

And this theory may be expected to yield a satisfactory account of the facts, so long as the $(_i\mathbf{L}\,_i\mathbf{S})$ coupling is weak; that is, so long as the interval of the spark ground term is small, a condition satisfied in carbon but not in tin. In contrast, when $(_i\mathbf{L}\,_i\mathbf{S})$ is strong, we obtain a more satisfactory account of the term scheme by considering the ionic and electronic vectors permanently coupled to $_i\mathbf{J}$ and $\mathbf{j}$ respectively; then the splitting is produced by the strong coupling $(_i\mathbf{L}\,_i\mathbf{S})$ and the weak coupling $(_i\mathbf{J}\,\mathbf{j})$, so that in Sn I the $s^2p.s$ configuration produces not a triplet and a singlet but two diads.

This new coupling is commonly referred to as $(\mathbf{jj})$ coupling, being written symbolically as

$$(_i\mathbf{L}\,_i\mathbf{S})\,(\mathbf{ls}) = (_i\mathbf{J}\mathbf{j}) = \mathbf{J}.$$

It is characterised by strong inter-system lines; and in fact these are easily discovered in lead and can be traced greatly weakened in intensity back to silicon, but there is still a little doubt about the three lines found by Jog* in C I, for if the lines found are those sought, the singlet term values assigned by Fowler must be changed by no less than 667 cm.$^{-1}$

5. Column V, s^2p^3 configuration

Hund's theory predicts that in the elements of column V the ground term should be $^4S_{1\frac{1}{2}}$ and that above this should lie four metastable states $^2D_{2\frac{1}{2},1\frac{1}{2}}$ and $^2P_{1\frac{1}{2},\frac{1}{2}}$. And empirically these five terms have been found in the arc spectra of nitrogen, phosphorus, arsenic, antimony and bismuth.

Of these spectra only N I and P I have been fully analysed. N I was not easily observed, for many of the lines lie inconveniently far out in the infra-red or ultra-violet; moreover, when excited by an arc strong bands appear which are apt to mask the

* Jog, *N*, 1929, **123** 318.

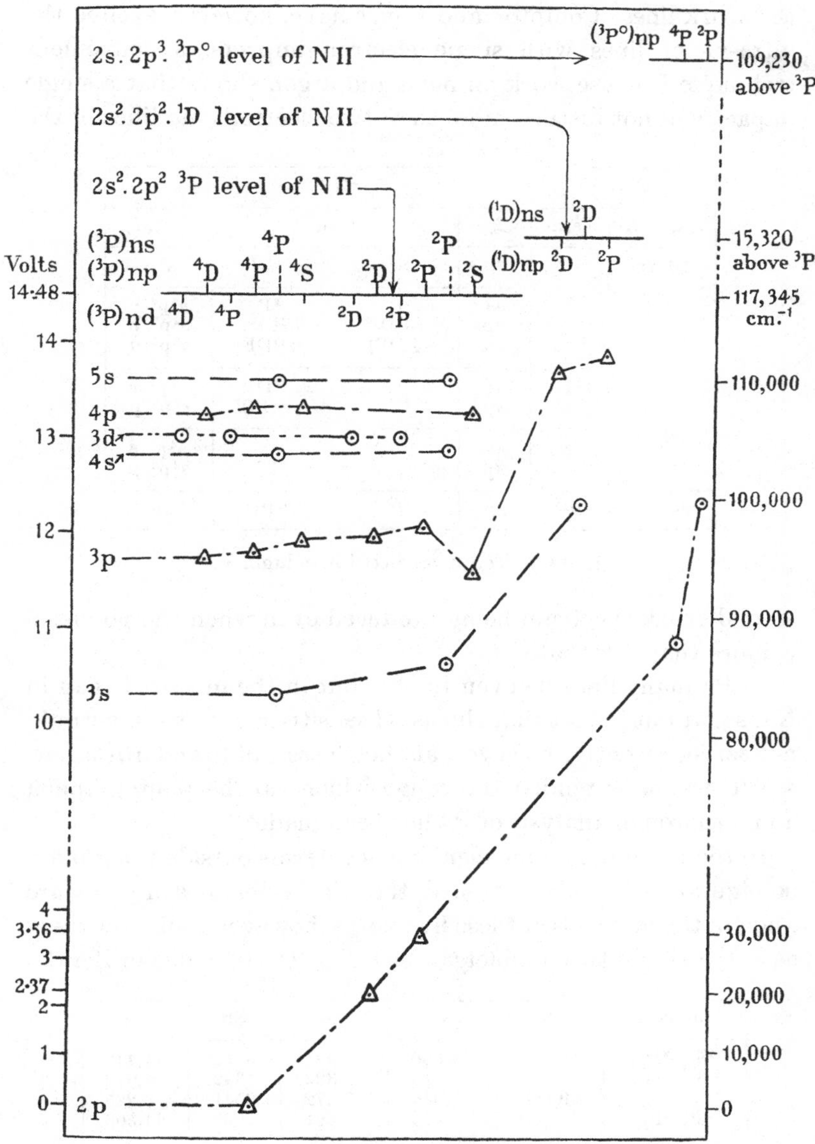

Fig. 14·8. Level diagram of nitrogen.

ultra-violet lines, while more vigorous excitation only brings out the spark lines. Compton and Boyce have, however, excited the ultra-violet lines with single electron impacts, a convenient technique because work on neon and argon shows that a single impact will not disturb more than two electrons, no lines of the

N II		N I			
$l_1 l_2 \dots l_4$	Term	l_5	Term		$l_1 \dots l_5$
$2s^2.2p^2$	^3P	2p	$^4S^\circ$	$^2D^\circ\ ^2P^\circ$	s^2p^3
		ns	^4P	^2P	$s^2p^2.s$
		np	$^4(SPD)^\circ$	$^2(SPD)^\circ$	$s^2p^2.p$
		nd	$^4(PDF)$	$^2(PDF)$	$s^2p^2.d$
	^1D	ns		^2D	$s^2p^2.s$
		np		$^2(PDF)^\circ$	$s^2p^2.p$
	^1S	ns		^2S	$s^2p^2.s$
		np		$^2P^\circ$	$s^2p^2.p$
$2s\ 2p^3$	$^3P^\circ$	2p	^4P	$^2(SPD)$	sp^4

Fig. 14·9. Terms predicted in column V.

second spark spectrum being produced even when the potential is more than adequate.

In P I many lines lie even further out in the infra-red than in N I, so far out in fact that the usual sensitiser neo-cyanine would not serve; recently, however, at the Bureau of Standards a new sensitiser has extended the range which can be photographed and a thorough analysis of P I has been made.

In the three remaining elements few terms outside the ground configuration have been named, though the energies of many are known; the intervals of these five terms, however, still show them as a singlet and two doublets in Sb I (Fig. 14·10), while in Bi I the

Interval	N	P	As	Sb	Bi
$^4S_{1\frac{1}{2}}-{}^2D_{1\frac{1}{2}}$	19,202	11,366	10,591	8512	11,418
$^2D_{1\frac{1}{2}}-{}^2D_{2\frac{1}{2}}$	−8	15	322	1342	4,019
$^2D_{2\frac{1}{2}}-{}^2P_{\frac{1}{2}}$	9,606	7,346	7,272	6541	6,223
$^2P_{\frac{1}{2}}-{}^2P_{1\frac{1}{2}}$	0	25	461	2069	11,505

Fig. 14·10. Column V. Intervals of the ground configuration, s^2p^3. Note the change from **LS** coupling in nitrogen to **jj** coupling in bismuth.

sequence of J values is still unchanged, so that the terms may be named by analogy; in an atom of high atomic weight (**LS**) coupling is not to be expected.

Of the higher terms little need be said; in N I the lowest are ^{2}P and ^{4}P from the 2p^2. 3s configuration, while very little higher lies the 2s. 2p^4 ^{4}P term. The energy relations are shown in Fig. 14·8.

Most of the term series of N I approach the ^{3}P ground term of N II as limit, but experiment shows that different series approach different components of the limit; for if two components of a multiplet tend to the same limit the interval decreases very rapidly, roughly in fact as $1/n^{*3}$; but in N I the intervals of some terms decrease only slowly or actually increase, this being true in particular of the 2p^2. ns ^{4}P and 2p^2. nd ^{2}D series.

N I					N II	
n	3	4	5	6		
$^4P_{\frac{1}{2}}$–$^4P_{1\frac{1}{2}}$	33·8	50·0	44·3	44·7	$^3P_0$–$^3P_1$	50
$^4P_{1\frac{1}{2}}$–$^4P_{2\frac{1}{2}}$	46·7	68·7	70·1	72·0	$^3P_1$–$^3P_2$	84
$^4P_{\frac{1}{2}}$–$^4P_{2\frac{1}{2}}$	80·5	118·7	114·4	116·7	$^3P_0$–$^3P_2$	134

Fig. 14·11. Intervals of the 2p^2. ns ^{4}P series in N I.

With large values of n the intervals of the 2p^2. ns ^{4}P series may reasonably be supposed to approach the values of 50 and 84 cm.$^{-1}$, which are the intervals of the ground term of N II. And if this is true then $^4P_{\frac{1}{2}}$ must tend to the lowest or $^3P_0$ limit, while $^4P_{2\frac{1}{2}}$ must approach the highest or $^3P_2$ limit. In Fig. 14·12 these facts are shown graphically, the separations being measured horizontally away from the middle terms, $^4P_{1\frac{1}{2}}$ in N I and $^3P_1$ in N II. Unfortunately in most series only two or three terms are known, and though the intervals of these may be roughly constant, they do not establish the limits beyond doubt as the above series do. Even in the 2p^2. nd ^{2}D series, of which five terms are known, one would hesitate to say whether the intervals tend to a limiting value of 50 or 84 cm.$^{-1}$, though evidently they do not tend to zero (Fig. 14·13).

The passage of the different components of a multiplet term to different limits often leads to wide departures from the interval rule, and these, like the departures due to increasing atomic number, may be described with the vector model. For consider

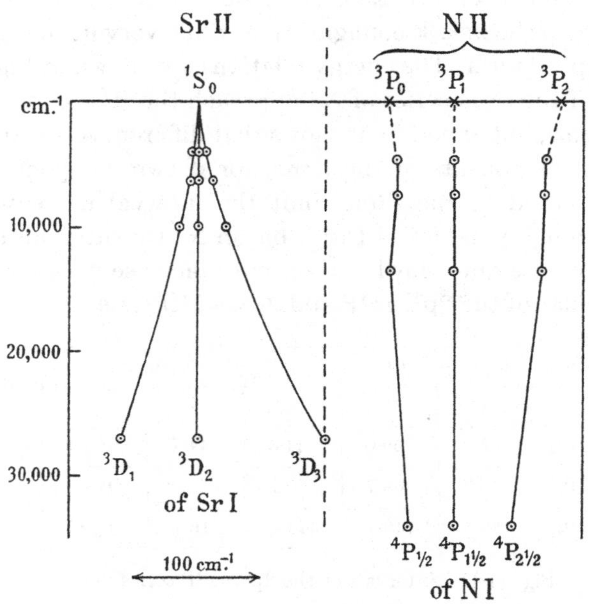

Fig. 14·12. Intervals of the 5s.*nd* ^{3}D series of Sr I contrasted with those of the 2p^2.*ns* ^{4}P series of N I; in the first the different series converge to the same limit, in the second they converge to three different limits.

n	3	4	5	6	7
$^2D_{1\frac{1}{2}}-{}^2D_{2\frac{1}{2}}$	23·5	22·2	18·3	58·3	42·6

Fig. 14·13. Intervals of the 2p^2.*nd* ^{2}D series of N I.

the addition of an electron defined by *n*, **l**, **s** to an ion defined by $_i$**L**, $_i$**S**; then in the low terms the couplings ($_i$**Ll**) and ($_i$**Ss**) are strong, while (**LS**) and (**ls**) suffice only to produce the splitting of the multiplet term into its components; but (**Ll**) and (**Ss**) decrease as the chief quantum number of the electron increases and actually vanish when *n* tends to infinity, while (**LS**) remains constant, so that when *n* is sufficiently large the latter becomes the more important and the splitting becomes that of the ionic ground term.

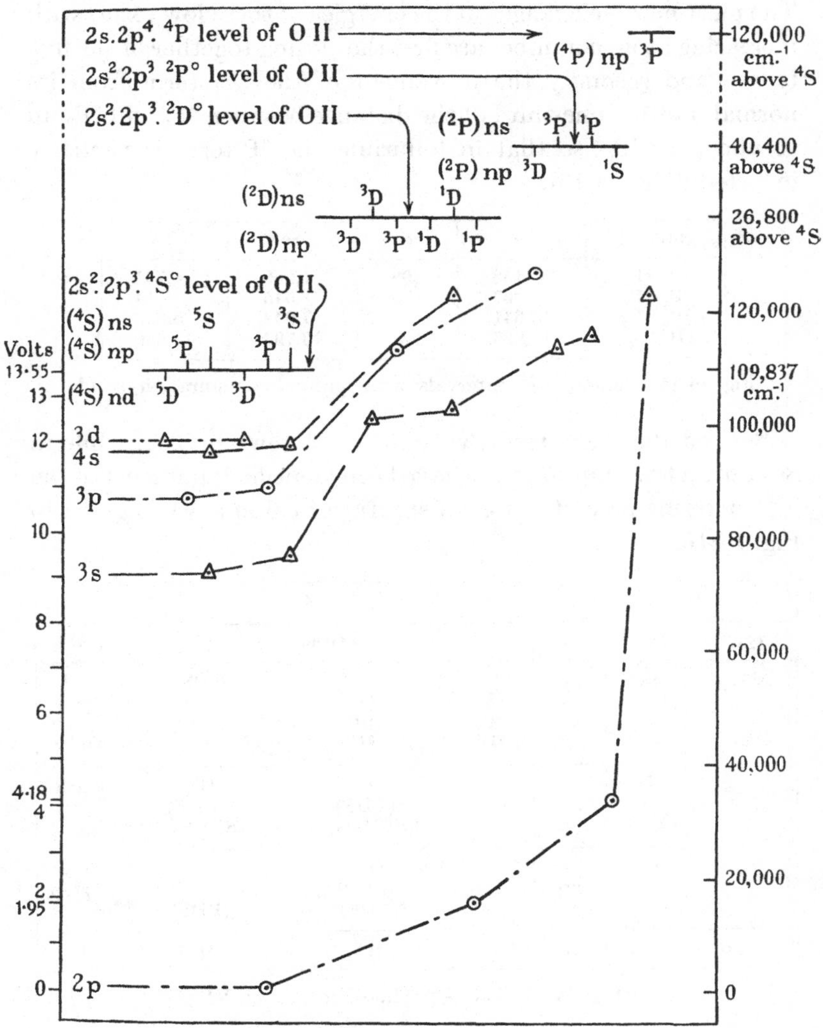

Fig. 14·14. Level diagram of oxygen.

6. Column VI, s^2p^4 configuration

Hund's theory predicts in column VI five low terms, $^3P_{2,1,0}$, $^1D_2$ and $^1S_0$, and empirically the ^3P term has been identified in the spectra of oxygen, sulphur, selenium and tellurium; while the two singlet terms are still missing only in sulphur (Fig. 14·14). The most notable changes in the energies of these low terms with increasing atomic number are first the closing together of the five terms, and secondly the movement of the $^3P_1$ term from its normal position one-third of the distance between $^3P_0$ and $^3P_2$ up to and past $^3P_0$, so that in tellurium the ^3P term is partially inverted (Fig. 14·15).

Interval	O	S	Se	Te
$^3P_2$–$^3P_1$	158	398	1,989	4,751
$^3P_1$–$^3P_0$	67	174	545	– 44
$^3P_0$–$^1D_2$	15,641	—	7,042	5,852
$^1D_2$–$^1S_0$	17,925	—	13,794	12,640

Fig. 14·15. Column VI. Intervals of the ground configuration, $s^2 . p^4$.

Beyond these low terms very little is known of Te I; while in Se I only a few quintet terms have been identified; but a list of the higher terms predicted for all spectra of column VI is given in Fig. 14·16.

O II		O I				
$l_1\, l_2 \ldots l_5$	Term	l_6	Terms			$l_1 \ldots l_6$
$2s^2 . 2p^3$	$^4S^\circ$	2p		^3P	$^1D\,^1S$	$s^2 p^4$
		*n*s	$^5S^\circ$	$^3S^\circ$		$s^2 p^3 . s$
		*n*p	^5P	^3P		$s^2 p^3 . p$
		*n*d	$^5D^\circ$	$^3D^\circ$		$s^2 p^3 . d$
	$^2D^\circ$	*n*s		$^3D^\circ$	$^1D^\circ$	$s^2 p^3 . s$
		*n*p		$^3(PDF)$	$^1(PDF)$	$s^2 p^3 . p$
		*n*d		$^3(SPDFG)^\circ$	$^1(SPDFG)^\circ$	$s^2 p^3 . d$
	$^2P^\circ$	*n*s		$^3P^\circ$	$^1P^\circ$	$s^2 p^3 . s$
		*n*p		$^3(SPD)$	$^1(SPD)$	$s^2 p^3 . p$
		*n*d		$^3(PDF)^\circ$	$^1(PDF)^\circ$	$s^2 p^3 . d$
$2s . 2p^4$		2p		$^3P^\circ$	$^1P^\circ$	sp^5

Fig. 14·16. Terms predicted in column VI.

In O I and S I series were identified by Paschen and Runge before the end of last century, and these are listed in Fowler's *Report* as singlets and triplets. Theoretically these series should be prominent, because the normal state of the ion is an S state;

but a ^4S ground term should produce systems of triplets and quintets, not of singlets and triplets. This discrepancy between experiment and modern theory was only resolved in 1923 when Hopfield identified series of lines arising from the combination of Fowler's 'singlet' S and D series with a ^3P ground term. The numerical values of the three components of the term are 109,833, 109,674 and 109,607,* showing intervals of 159 and 67 respectively and an interval ratio of 2·4 : 1. In combinations of this term the line of shortest wave-length is the most intense, so that the ^3P term must be inverted.

As inter-system lines are seldom strong, the combination of this triplet term with a 'singlet' series suggested that the supposed singlets were really triplets too narrow to be resolved. True, the ^3P ground term also combines with Fowler's 1s 'triplet' term, but the lines are much weaker, and though the $1s \rightarrow {}^3P_2$ and $1s \rightarrow {}^3P_1$ lines are quite clear, the $1s \rightarrow {}^3P_0$ line is missing in both O I and S I. If, however, the supposed triplets are really quintets the absence of this particular line is readily explained, for the 1s term is then $^5S_2$, and the combination with $^3P_0$ is forbidden, since J would have to change by two units. This evidence alone seems conclusive, but in 1923 the argument from theory was of little weight and so Paschen and Landé† felt bound to settle the question by measuring the Zeeman splitting of two quintet combinations 7771 A., $3p\,{}^5P_{123} \rightarrow 3s\,{}^5S_2$ and 3974 A., $4p\,{}^5P_{123} \rightarrow 3s\,{}^5S_2$.

Above the ^3P term thus identified the metastable states $^1D_2$ and $^1S_0$ should lie, and recently combinations of these states with various higher terms have been identified in O I; but history records that these lines were found only after McLennan‡ had shown in a brilliant research that the green auroral line 5577 A. arises in a jump between these two states. As the transition is forbidden by two selection rules the evidence requires careful scrutiny; it will be considered in a later chapter on quadripole radiation.

* Hopfield, *AJ*, 1924, **59** 114.

† Laporte, *Nw*, 1924, **12** 598, attributes this work to Paschen and Landé, but the reference given is inaccurate and the paper does not appear under either name in *Science Abstracts*, 1922–24.

‡ McLennan, *PRS*, 1928, **120** 327.

7. The halogens, s^2p^5 configuration

According to Hund's scheme the p^5 configuration produces only an inverted ^2P term, and in fact this is the normal state of all halogen atoms.

The higher terms may be divided into three groups according as they arise from the ^3P, ^1D or ^1S term of the spark spectrum. Of the predicted terms shown in Fig. 14·17, a large number have been

F II		F I			
$l_1 l_2 \dots l_6$	Term	l_7	Terms		$l_1 \dots l_7$
$2s^2.2p^4$	—	2p	$^2P^\circ$		s^2p^5
	^3P	*n*s	^4P	^2P	$s^2p^4.s$
		*n*p	$^4(SPD)^\circ$	$^2(SPD)^\circ$	$s^2p^4.p$
		*n*d	$^4(PDF)$	$^2(PDF)$	$s^2p^4.d$
	^1D	*n*s	—	^2D	$s^2p^4.s$
		*n*p	—	$^2(PDF)^\circ$	$s^2p^4.p$
		*n*d	—	$^2(SPDFG)$	$s^2p^4.d$
	^1S	*n*s	—	^2S	$s^2p^4.s$
		*n*p	—	$^2P^\circ$	$s^2p^4.p$
		*n*d	—	^2D	$s^2p^4.d$

Fig. 14·17. Terms predicted in the halogens.

found in the arc spectra of all the halogens, and in the spark spectra of neon, argon and krypton. In particular, Bakker, De Bruin and Zeeman* have made an extensive magnetic analysis of A II and have studied the spectrum very thoroughly; it provides much material for the study of irregular g values and of series limits, and will be referred to again in that context; the terms obey the normal multiplet laws badly, but the Rydberg formulae rather well.† In the lighter elements the empirical terms are easily named, for they approximate to the Russell-Saunders laws, but only the J values of the higher terms of I I are known.

Directly above the ground term come even terms from the $p^4.s$ and $p^4.d$ configurations. Of these the $p^4(^3P)s$ configuration should produce five terms, named according to the Russell-Saunders scheme $^4P_{2\frac{1}{2},1\frac{1}{2},\frac{1}{2}}$ and $^2P_{1\frac{1}{2},\frac{1}{2}}$; and in fact in F I the extreme

* Bakker, De Bruin and Zeeman, *K. Akad. Amsterdam, Proc.*, 1928, **31** 780.
† Rosenthal, *AP*, 1930, **4** 80.

intervals of the two multiplets are less than 450 and 350 respectively, while they are separated by nearly 2000 cm.$^{-1}$ In I I, on

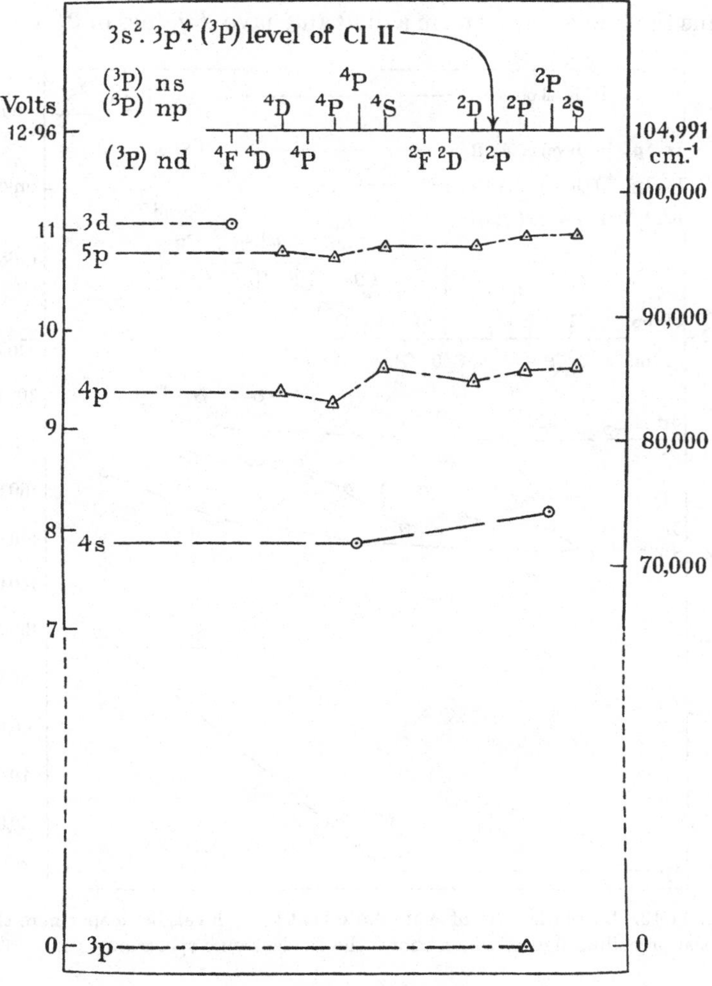

Fig. 14·18. Level diagram of chlorine.

the other hand, the division of the five terms into multiplets is not justified by their energies; as the sequence of J values is the same as in chlorine, the empirical terms may be named by analogy, but if this is done $\Delta^4P_{1\frac{1}{2},\frac{1}{2}}$ is 4800 cm.$^{-1}$ while $^2P_{1\frac{1}{2}}$–$^4P_{\frac{1}{2}}$ is only 900 cm.$^{-1}$ (Fig. 14·20). If the five terms are divided by their energies, they form two diads below and a monad above; thus the lowest terms form a diad and have J values of $2\frac{1}{2}$ and $1\frac{1}{2}$,

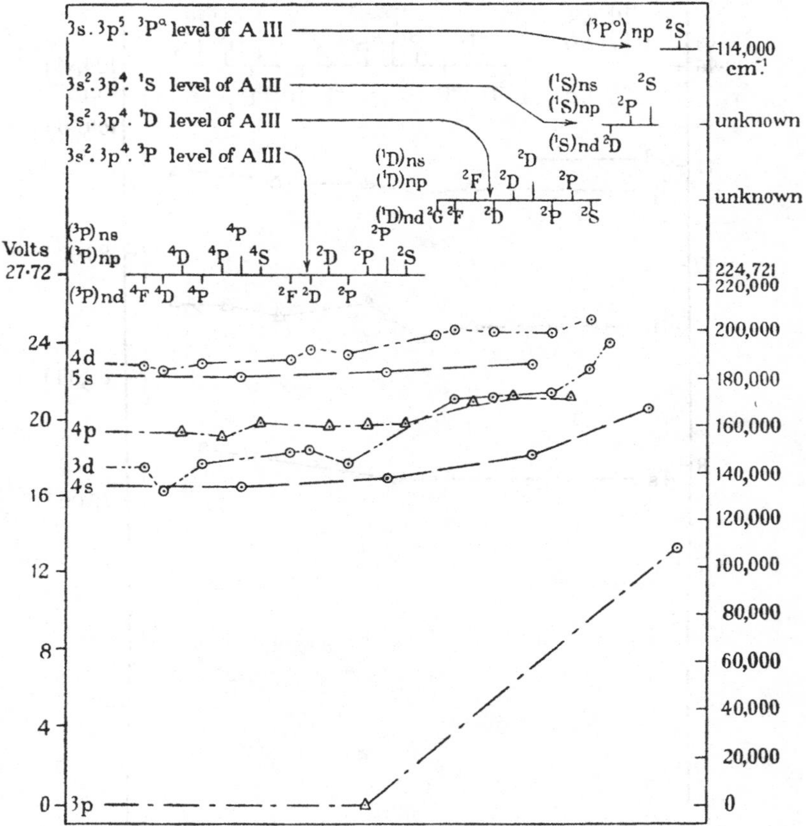

Fig. 14·19. Level diagram of A II. Note that the d levels lie deeper here than in the preceding figure, which shows the isoelectronic spectrum Cl I.

being formed presumably by the addition of an s electron to the $^3P_2$ spark term; similarly, the next diad has J values of $1\frac{1}{2}$ and $\frac{1}{2}$, and so may be assumed to arise from the addition of the s electron to the $^3P_1$ spark term; and highest of all lies the single level with a J value of $\frac{1}{2}$, arising as $(^3P_0)$ s. These energy values clearly justify us in saying that the coupling in F I is Russell-Saunders, but in I I is (jj).

A comparison of the isoelectronic spectra Cl I and A II brings out one common effect of increasing the nuclear charge; in Cl I the succession of the configurations is apparently 4s, 4p, 5s, 5p, 3d,

	F	Cl	Br	I
$^4P_{2\frac{1}{2}}-^4P_{1\frac{1}{2}}$	275	530	1471	1459
$^4P_{1\frac{1}{2}}-^4P_{\frac{1}{2}}$	160	338	1977	4803
$^4P_{\frac{1}{2}}-^2P_{1\frac{1}{2}}$	1892	1398	300	924
$^2P_{1\frac{1}{2}}-^2P_{\frac{1}{2}}$	325	640	1787	4530
Δ^3P_{02} of ion	491	991	—	—

Fig. 14·20. Intervals of the p^4 (^3P).s configuration, showing the change from **LS** to jj coupling.

terms arising from all the configurations except 5s having been identified; but in A II the succession is 4s, 3d, 4p, 5s, 4d, the increased nuclear charge having thus moved the d terms down relative to the s and p terms. Of this movement much more will be heard in the long periods.

8. The inert gases, s^2p^6 configuration

The p^6 group, being a complete shell, produces only a ^1S term, and empirically this always lies so much deeper than any of the odd terms that the lines which result lie in the far ultra-violet. The ground term and the lowest configuration of odd terms have been identified in the arc spectra of neon, argon, krypton and xenon and also in the spark spectra Na II, K II, Rb II and Cs II; of these neon was the subject of such a thorough study at a time when the structure of complex spectra was very little understood that it still deserves pride of place (Figs. 14·21, 22).

The spectrum of neon consists of two parts, one in the visible analysed into series by Paschen in 1918,* and a few ultra-violet lines unknown to Paschen, but discovered by Lyman and Saunders in 1925† and attributed at once to the deep ^1S ground term. The large number of series discovered by Paschen may be

* Paschen, *AP*, 1919, **60** 405; 1920, **63** 201.

† Lyman and Saunders, *PR*, 1925, **25** 886*a*.

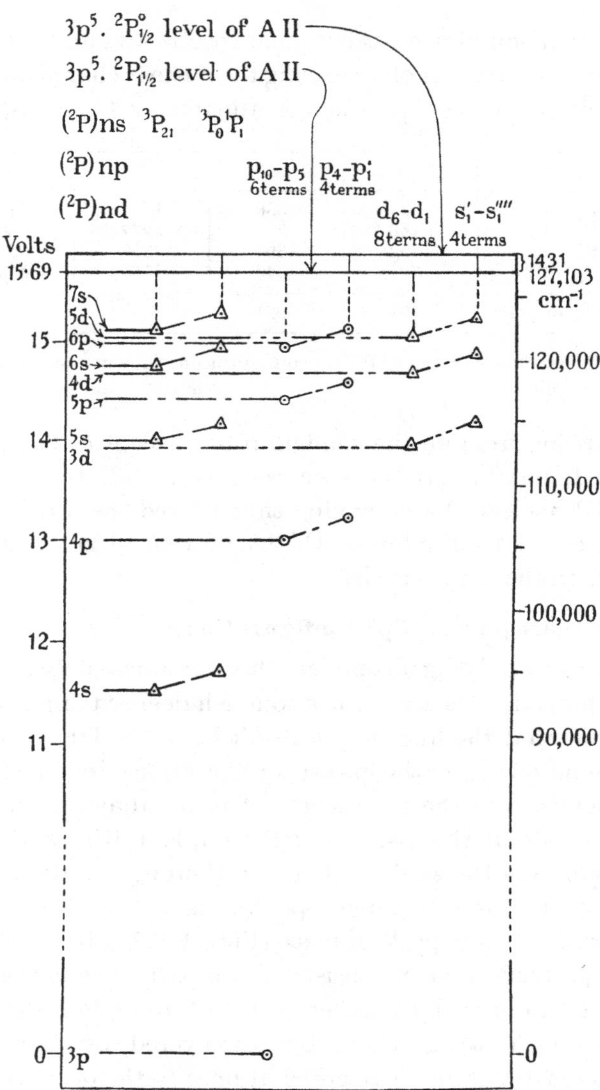

Fig. 14·21. Level diagram of argon; this differs from the level diagram of neon only in the energies of the terms.

divided by the energies of their lowest terms into three groups; of these the lowest contains four terms, named from below up s_5, s_4, s_3 and s_2, and identified in modern theory with the $2p^5\,(^2P)\,3s$ configuration. Above these lie ten terms called by Paschen p terms, and written from below up p_{10} to p_1; these combine with

Ne II		Ne I			
$l_1\,l_2 \ldots l_7$	Terms	l_8	Terms	Paschen's terms	$l_1 \ldots l_8$
$2s^2.2p^5$	$^2P^o$	2p	^1S	—	s^2p^6
		*n*s	$^3P^o$ $^1P^o$	s_2, s_3, s_4, s_5	$s^2p^5.s$
		*n*p	$^3(SPD)$ $^1(SPD)$	$p_1 \cdots p_{10}$	$s^2p^5.p$
		*n*d	$^3(PDF)^o$ $^1(PDF)^o$	s_1 and d	$s^2p^5.d$

Fig. 14·22. Terms predicted in the inert gases.

the s terms and clearly arise from the $p^5\,(^2P)\,3p$ configuration. Higher still appear twelve terms written some of them d and some s_1; these arise from the $p^5\,(^2P)\,3d$ configuration. All these terms were assigned *J* values by Landé* as the result of a magnetic analysis (Fig. 14·23), and these *J* values agree precisely with those required by theory.

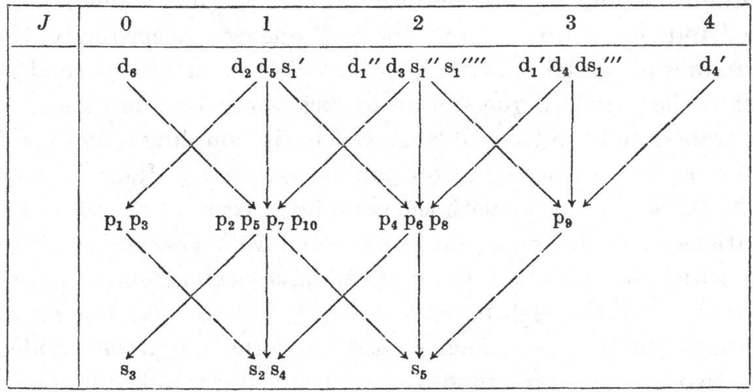

Fig. 14·23. Permitted combinations and *J* values of the empirical terms of neon.

Though thus far agreement is so satisfactory the s terms alone come near to obeying the multiplet laws. By Hund's energy rules

* Landé, *PZ*, 1921, 22 417.

these four terms should be $^3P_{2,1,0}$ and $^1P_1$ in order of increasing energy (Fig. 14·23); the empirical J values establish the $^3P_2$ and $^3P_0$ terms in the positions indicated; while for s_4 and s_2 Back* obtained g values of 1·46 and 1·03, in satisfactory agreement with the theoretical values of 1·5 for $^3P_1$ and 1·0 for $^1P_1$. Moreover, this allotment is confirmed by the ultra-violet combinations with the $^1S_0$ ground term, for the line of lower frequency $s_4 \rightarrow {}^1S_0$ is weaker than the line of higher frequency $s_2 \rightarrow {}^1S$†; and this agrees with the general rule that inter-system lines are weaker than combinations between terms of the same system.

Terms		g factor		Limit	
Emp.	Theor.	Emp.	Theor.	Emp.	Theor.
s_2	$^1P_1$	1·04	1	Upper	$^2P_{\frac{1}{2}}$
s_3	$^3P_0$	$\frac{0}{0}$	$\frac{0}{0}$	Upper	$^2P_{\frac{1}{2}}$
s_4	$^3P_1$	1·46	1·5	Lower	$^2P_{1\frac{1}{2}}$
s_5	$^3P_2$	1·50	1·5	Lower	$^2P_{1\frac{1}{2}}$

Fig. 14·24. The $2p^5.ns$ terms of neon; the empirical g values are those of the first or 3s term of the series.

The p terms present much greater difficulties; a very brief examination shows that the interval rule is quite useless, while the Landé g formula is none too well obeyed. Accordingly, the matching proposed by Hund‡ (Fig. 14·24) must be regarded at best as the reasoned guess of an expert; for a thorough study of the transition from Russell-Saunders to (jj) coupling is necessary before a satisfactory solution can be reached.§ Judging from H. N. Russell's success with the complex spectra of the iron row, a satisfactory solution might be expected from a careful study of intensities, but this does not seem to have been attempted yet.

In ordering the high terms to series based on these low terms Paschen found clear evidence that the series might be divided into two groups, some tending to a lower limit and others to a limit 780 cm.$^{-1}$ higher; these two limits appear in modern theory

* Back, *AP*, 1925, **76** 317.
† Shenstone, *N*, 1928, **121** 619.
‡ Hund, *ZP*, 1929, **52** 601.
§ Pogány, *ZP*, 1935, **93** 376 and chapter XVIII, § 7.

as the two components of the ^{2}P ground term of Ne II, now known to have an interval of 782 cm.$^{-1}$ Of the s terms s_2 and s_3 tend to the upper limit, s_4 and s_5 to the lower limit; the limits of the p series are shown in Fig. 14·25.

Empirical				Theoretical	
Term	J	g	Limit	Term[a]	g
p_1	0	$\frac{0}{0}$	Upper	$^3P_0$	$\frac{0}{0}$
p_3	0	$\frac{0}{0}$	Lower	$^1S_0$	$\frac{0}{0}$
p_2	1	1·340	Upper	$^3P_1$	1·5
p_5	1	0·999	Upper	$^1P_1$	1·0
p_7	1	0·699	Lower	$^3D_1$	0·5
p_{10}	1	1·984	Lower	$^3S_1$	2·0
p_4	2	1·301	Upper	$^3P_2$	1·5
p_6	2	1·229	Lower	$^3D_2$	1·167
p_8	2	1·137	Lower	$^1D_2$	1·0
p_9	3	1·329	Lower	$^3D_3$	1·333

Fig. 14·25. The $2p^5.3p$ terms of neon.

[a] The naming of the terms is based on the most exiguous evidence.

Interval	Ne	A	Kr	Xe
$^3P_2$–$^3P_1$	417	607	945	978
$^3P_1$–$^3P_0$	359	803	4275	8152
$^3P_0$–$^1P_1$	1070	846	655	988

Fig. 14·26. Intervals of the p^5 (^{2}P).s configuration in the inert gases.

The spectra of the heavier inert gases are very similar to neon, but as the interval of the ground term of the spark spectrum increases, its influence on the arc spectrum becomes increasingly evident; it appears, for instance, in the intervals of the four (^{2}P) s terms, shown in Fig. 14·26; while in Kr I the p terms divide themselves into two groups with an interval of no less than 5200 cm.$^{-1}$ between them; the upper group consists of four terms and the lower group of six just as theory will be shown to require, but the chance of fitting the Russell-Saunders notation to these terms is more remote even than in neon.

Indeed, the heavy inert gases centre in themselves three influences which all conspire to break up the simple Russell-

Saunders coupling; first, increasing atomic number makes for loss of regularity; this was discussed particularly in columns IV and V, but it is visible in every column; secondly, the increasing separation of the ground term of the ion tends to divide the terms by limits rather than by multiplets; and lastly spectra grow less regular as one passes from left to right across the periodic table, neon is less regular than carbon and carbon less regular than sodium.

BIBLIOGRAPHY

The first systematic account was by Hund in *Linienspektren und periodisches System der Elemente,* 1927, and this is still standard.

Term values are largely taken from Bacher and Goudsmit, *Atomic Energy States,* 1932. References to particular elements can be traced in the select bibliography of Appendix VI.

CHAPTER XV

LONG PERIODS

1. The ground terms

In the short periods the s electron carries appreciably less energy than the p electron, and very little study suffices to show that the first two electrons enter an s shell and the last six a p. In the long periods however, while an s electron still carries less energy than a p, the s and d electrons carry roughly the same energy, so that a first glance shows one group of twelve elements instead of two groups of two and ten.

1	2	3	4	5	6	7	8	9	10	11	12
K	Ca	Sc	Ti	V	Cr	Mn	Fe	Co	Ni	Cu	Zn
Rb	Sr	Y	Zr	Cb	Mo	Ma	Ru	Rh	Pd	Ag	Cd
Cs	Ba	La–Lu	Hf	Ta	W	Re	Os	Ir	Pt	Au	Hg

Bohr's work on the periodic system showed that in the first two of these twelve elements the electron enters an s orbit, but that thereafter as the nuclear charge increases the d orbits grow more stable and the s less, until in the spark spectrum of a system of ten electrons all ten occupy d orbits, and produce the $^1S_0$ ground term characteristic of a complete shell. The eleventh and twelfth electrons then re-enter the s shell producing copper, zinc and their homologues. Thus in only eight of the twelve elements are the configurations still in doubt; these eight were enclosed by Bohr in a frame, and will be described in successive periods as the elements of the iron frame, the palladium frame and the platinum frame.

Since only two electrons can enter an s shell, an atom with n outer electrons has the choice of only three configurations, d^n, $d^{n-1}.s$ and $d^{n-2}.s^2$; the chief quantum numbers will usually be omitted, in order to simplify the discussion, for the orbits are always 3d and 4s in the iron frame, 4d and 5s in the palladium frame and 5d and 6s in the platinum frame.

The terms, which any configuration produces, can be calculated; and if the lowest term of each configuration is that with largest spin and orbital vectors, a list of possible configurations and the ground terms which they produce can be compiled, as Fig. 15·1 shows. But the ground term is the easiest of all terms to determine empirically, so that the argument which produced this table may be reversed, and the ground term used to decide which configuration produces the lowest term (Figs. 15·2–4). When

No. of electrons	Configuration		
	d^n	$d^{n-1}.s$	$d^{n-2}.s^2$
0	^1S		
1	^2D	^2S	
2	^3F	^3D	^1S
3	^4F	^4F	^2D
4	^5D	^5F	^3F
5	^6S	^6D	^4F
6	^5D	^7S	^5D
7	^4F	^6D	^6S
8	^3F	^5F	^5D
9	^2D	^4F	^4F
10	^1S	^3D	^3F
11		^2S	^2D
12			^1S

Fig. 15·1. Ground terms of three configurations.

No. of electrons	Arc spectra			Spark spectra			No. of electrons
	Atom	Ground term	Configuration	Atom	Ground term	Configuration	
1	K	^2S	s	Ca^+	^2S	s	1
2	Ca	^1S	s^2	Sc^+	^3D	d.s	2
3	Sc	^2D	$d.s^2$	Ti^+	^4F	$(d^2.s)$	3
4	Ti	^3F	$d^2.s^2$	V^+	^5D	d^4	4
5	V	^4F	$d^3.s^2$	Cr^+	^6S	d^5	5
6	Cr	^7S	$d^5.s$	Mn^+	^7S	$d^5.s$	6
7	Mn	^6S	$d^5.s^2$	Fe^+	^6D	$d^6.s$	7
8	Fe	^5D	$d^6.s^2$	Co^+	^3F	d^8	8
9	Co	^4F	$(d^7.s^2)$	Ni^+	^2D	d^9	9
10	Ni	^3F	$d^8.s^2$	Cu^+	^1S	d^{10}	10
11	Cu	^2S	$d^{10}.s$	Zn^+	^2S	$d^{10}.s$	11
12	Zn	^1S	$d^{10}.s^2$	Ga^+	^1S	$d^{10}.s^2$	12

Fig. 15·2. Ground terms of the iron frame elements.

In this and the two following figures, the ground term is given only when it is known empirically; a ? following the term shows that it is still open to doubt. A configuration enclosed in brackets cannot be deduced from the ground term alone, but is known from further study of the spectrum; where a configuration is given in the absence of the ground term, reliance has been placed on the argument of §4 of this chapter.

there are 3, 6 or 9 electrons, two configurations produce the same ground term, and then further study is necessary to distinguish between them; where a configuration has been obtained in this way, it is enclosed in brackets.

No. of electrons	Arc spectra			Spark spectra			No. of electrons
	Atom	Ground term	Configuration	Atom	Ground term	Configuration	
1	Rb	^2S	s	Sr^+	^2S	s	1
2	Sr	^1S	s^2	Y	^1S	s^2	2
3	Y	^2D	$d.s^2$	Zr^+	^4F	$d^2.s$	3
4	Zr	^3F	$d^2.s^2$	Cb^+	^5D	d^4	4
5	Cb	^6D	$d^4.s$	Mo^+	^6S?	d^5	5
6	Mo	^7S	$d^5.s$	Ma^+	—	$d^5.s$	6
7	Ma	—	$d^5.s^2$	Ru^+	^4F	d^7	7
8	Ru	^5F	$d^7.s$	Rh^+	—	d^8	8
9	Rh	^4F	$(d^8.s)$	Pd^+	^2D	d^9	9
10	Pd	^1S	d^{10}	Ag^+	^1S	d^{10}	10
11	Ag	^2S	$d^{10}.s$	Cd^+	^2S	$d^{10}.s$	11
12	Cd	^1S	$d^{10}.s^2$	In^+	^1S	$d^{10}.s^2$	12

Fig. 15·3. Ground terms of the palladium frame elements.

No. of electrons	Arc spectra			Spark spectra			No. of electrons
	Atom	Ground term	Configuration	Atom	Ground term	Configuration	
1	Cs	^2S	s	Ba^+	^2S	s	1
2	Ba	^1S	s^2	La^+	^3F	d^2	2
				Lu^+	^1S	s^2	
3	La	^2D	$d.s^2$	Hf^+	^2D	$d.s^2$	3
	Lu	^2D	$d.s^2$				
4	Hf	^3F	$d^2.s^2$	Ta^+	—	—	4
5	Ta	—	$d^3.s^2$	W^+	—	—	5
6	W	^5D	$(d^4.s^2)$	Re^+	—	—	6
7	Re	^6S	$d^5.s^2$	Os^+	—	—	7
8	Os	—	$d^6.s^2$	Ir^+	—	—	8
9	Ir	^2D?	d^9?	Pt^+	—	—	9
10	Pt	^3D	$d^9.s$	Au^+	^1S	d^{10}	10
11	Au	^2S	$d^{10}.s$	Hg^+	^2S	$d^{10}.s$	11
12	Hg	^1S	$d^{10}.s^2$	Tl^+	^1S	$d^{10}.s^2$	12

Fig. 15·4. Ground terms of the platinum frame elements.

With only one exception the arc spectra of the iron frame (Fig. 15·2) have $d^{n-2}.s^2$ as their ground configuration, while the spark spectra have $d^{n-1}.s$ or d^n, a contrast which shows how general is the tendency of the d electrons to sink relative to the s electrons as the nuclear charge increases; of this tendency an oft-quoted example is the descent of the s electron of K I and Ca II into the d orbit of Sc III. In the palladium frame (Fig. 15·3) the tendency is again in evidence; the d^n configuration produces the ground terms of six spark spectra, but only one arc. Comparison of the iron and palladium frames shows that the d orbits are more stable in the latter; of the platinum frame (Fig. 15·4) not much is yet known, but there is some reason to think that in the early elements the d orbits are more stable still, but that after the intrusion of the f shell between lanthanum and lutecium, the d shell is less stable even than in the iron frame. Thus La II has a $d^2\,{}^3F$ ground term in contrast to the ds ${}^3D$ of Sc II and the $s^2\,{}^1S$ of Y II, but the ground states of Lu II, Lu I, Hf I and Re I all seem to contain two s electrons, besides varying numbers of d electrons.

2. Configurations and analysis

In all three frames the s and d electrons have roughly the same energy, and this determines the general form of the spectra. A configuration which contains only s and d electrons outside the last closed shell will consist of even terms, and accordingly the low terms of all elements are even; they will usually arise from two different configurations, and as no combinations are permitted, the higher of the two is metastable.

Above these low terms come a group of odd terms containing a p electron, and above those again a second group of even terms containing only s and d electrons. Theory allows many higher terms, but in practice the lines involving these terms are weak or absent, so that series consist of only two or at most three terms.

This means that the method of analysis differs from that applied to simple spectra*; the spectroscopist tries to find not series, but multiplets. As in other spectra constant differences between the frequencies of pairs of lines indicate terms, and these may be separated into odd and even, because an odd term combines only with an even. The ground term can usually be picked out because it gives rise to the *raies ultimes* and most of the absorption lines; the spark produced under water is also useful,

* For an account of methods of analysis, see Russell, *AJ*, 1927, **66** 348.

for the spectrum consists of a few lines all ending in the ground term.

The approximate energy of a term can be determined by examining the temperature class of the lines to which it gives rise. If it is of low energy content, then it will appear bright at a low temperature and in King's classification will belong to a low class. This and the division into odd and even terms will often determine the configuration to which a term belongs.

Finally, the terms of a configuration may be divided into multiplets by examining the intensities of the lines to which they give rise, for when the coupling is not normal the intensities are far less seriously disturbed than either Landé's interval ratios or the magnetic splitting factors.

To suit these new methods of analysis, new ways of specifying the empirical terms have had to be developed; thus the analysis may succeed in finding terms, but be unable to divide them into multiplets; the terms are then numbered 1, 2, 3, ... beginning from the term of lowest energy, or if the J value is also known then this may be added as a suffix $1_{2\frac{1}{2}}$. Or again the terms may have been worked out, and then each term is specified by a small letter placed before its term symbol; in this notation the letters a to e are reserved for the low terms, z, y, x, ... for the middle or odd terms, and f, g, h, ... for the high terms; in each group the terms are lettered from low to high energy.

3. Individual spectra

The spectra of the long periods are most naturally classified by the number of electrons outside the last inert gas shell, and in this order they will be reviewed.

The energies of all the low terms and a varying number of odd terms are tabulated; the energy given for any multiplet is that of the component with greatest J, for this will produce the strongest lines; the energies are measured up from the ground term, and the ground term itself is in clarendon type. To avoid any misunderstanding the chief quantum number is added for spectra of the iron frame; to apply to the palladium and platinum frames it should be increased by one and two respectively.

One electron. K I, Rb I, Cs I; Ca II, Sr II, Ba II, Ra II; Sc III, Y III, La III, Lu III; Ti IV, Zr IV, Ce IV.

In the arc spectra of the alkalis and the spark spectra of the alkaline earths the ground term is always ^{2}S, the single electron occupying a 4s, 5s or 6s orbit. With further increase in the nuclear charge, however, the electron falls into a d orbit, for Sc III, Y III, Zr IV, and Ce IV are all known to have a ^{2}D ground term.

The introduction of a group of f electrons produces, however, a surprising change; Lu III, which differs from La III only in having this group, has a ^{2}S ground term, so that the 6s orbit must have grown more stable than the 5d.

Low terms

	K	Rb	Cs	Ca II	Sr II	Ba II	Ra II
4s $^2S_{\frac{1}{2}}$	**0**	**0**	**0**	**0**	**0**	**0**	**0**
3d $^2D_{2\frac{1}{2}}$	21,539	19,355	14,597	13,711	16,837	5,675	26,209

	Sc III	Y III	La III	Lu III	Ti IV	Zr IV	Ce IV
4s $^2S_{\frac{1}{2}}$	25,537	7466	13,590	**0**	80,379	38,258	5152
3d $^2D_{2\frac{1}{2}}$	**198**	**725**	**1,604**	8648	**348**	**1,250**	**3305**

Fig. 15·5. One electron. Energies of the low terms.

Two electrons. Ca I, Sr I, Ba I; Sc II, Y II, La II, Lu II; Ti III, Zr III.

The ground terms of the arc spectra of the alkaline earths and the second spark spectra of column IV are respectively $^1S_0$ and $^3F_2$, the first arising from the s^2 configuration and the second from the d^2; and this is natural enough, since the d orbit normally sinks relative to the s as the nuclear charge increases. Between these two extremes the singly ionized earth metals should provide a natural transition, and in some measure they do, for the ground term of Sc II is d.s ^{3}D and of La II d^2 ^{3}F; while if the f shell screens the outer electrons from the nucleus the s^2 ^{1}S ground term of Lu II does not go unexplained. But that the ground term of Y II should be s^2 ^{1}S is a striking anomaly, for the above scheme would predict it as arising either from d.s or d^2; moreover, the ground term of nearly every spectrum is derived by the addition of an electron to the ground term of the next higher spark spectrum, but Y II forms an exception to this rule also, for the single electron of Y III occupies a d orbit.

The other low terms which theory dictates are shown in Fig. 15·7; and just these terms are actually found in the spark

Low terms

		Ca	Sr	Ba	Sc II	Y II	La II	Lu II	Ti III	Zr III
$4s^2$	$^1S_0$	**0**	**0**	**0**	11,736	**0**	7,395	**0**	—	—
3d.4s	$^3D_3$	20,371	18,320	9,596	**178**	1,450	3,250	14,199	38,425	19,533
	$^1D_2$	21,849	20,149	11,395	2,541	3,296	1,395	17,332	41,704	16,122
$3d^2$	$^3F_4$	—	—	—	4,988	8,743	**1,971**	32,504	**422**	**1,487**
	$^3P_2$	48,564	32,625	23,919	12,154	14,098	6,227	—	10,721	8,840
	$^1G_4$	—	—	—	14,261	15,683	7,473	—	14,398	2,534
	$^1D_2$	—	—	—	10,945	14,833	10,095	—	8,473	3,392
	$^1S_0$	—	—	—	—	—	—	—	14,053	3,835

Fig. 15·6. Two electrons. Energies of the low terms.

Low terms

Configuration	Terms	
	Triplets	Singlets
$4s^2$	—	S
3d.4s	D	D
$3d^2$	PF	SDG

Middle terms

Ion		Atom		
l_1	Term	l_2	Terms	$l_1\,l_2$
3d	^2D	4p	$^3(PDF)^\circ$ $^1(PDF)^\circ$	dp
4s	^2S	4p	$^3P^\circ$ $^1P^\circ$	sp

High terms

Ion		Atom		
l_1	Term	l_2	Terms	$l_1\,l_2$
3d	^2D	*n*s	^3D ^1D	d.s
		*n*d	$^{3,1}(SPDFG)$	d.d
4s	^2S	*n*s	^3S ^1S	s.s
		*n*d	^3D ^1D	s.d

Fig. 15·7. Terms predicted from two electrons.

spectra; $d^2\,{}^1S$ has been a little difficult to locate, but it has been found all right in Ti III and Zr III. These low terms are in satisfactory agreement with Hund's energy rules; in the d.s configuration of Sc II, for example, the ${}^3D$ term lies below ${}^1D$, while in the d^2 configuration ${}^3F$ lies below ${}^3P$; among the singlets, however, $b\,{}^1D$ is lower than $a\,{}^1G$. Deviations such as this are common in all spectra.

The middle and high terms have also been collected in Fig. 15·7; and the general relations of a typical spectrum are shown in

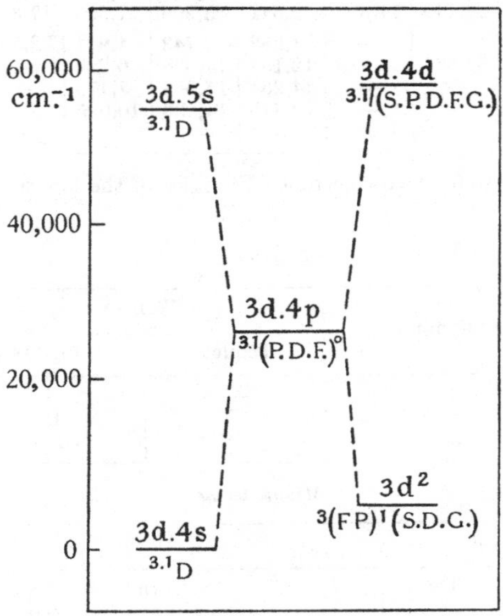

Fig. 15·8. Simplified level diagram of Sc II.

Fig. 15·8, where the four groups of even terms, arising from the configurations $3d^2$, 3d.4s, 3d.4d and 3d.5s, all combine with the central group of odd terms, but not at all with each other. This means, of course, that all terms of the $3d^2$ configuration are metastable, for no direct return to the ground state is possible.

In the high 3d.4d configuration of Sc II the agreement with Hund's energy rules is much less satisfactory than among the low terms; among the triplets ${}^3F$ and ${}^3P$ lie close together, but some 4000 cm.$^{-1}$ above the ${}^3G$, ${}^3D$ and ${}^3S$ levels; while among the singlets, on the other hand, ^1F and ^1P lie some 4000 cm.$^{-1}$ below the ^1G, ^1D and ^1S levels, each group being fairly compact. This curious alternating arrangement of energy levels finds a parallel in pentads of similar origin in other spectra, but it is quite contrary to Hund's energy rules and has only recently received any theoretical explanation.

All the triplet terms of Sc II are erect, and most of them obey the interval rule well (Fig. 15·9).

Configuration	Interval		Interval quotient
3d.4s	$\Delta^3D_{1,2}$	67·7	33·8
	$\Delta^3D_{2,3}$	109·9	36·6
$3d^2$	$\Delta^3F_{2,3}$	80·7	26·9
	$\Delta^3F_{3,4}$	104·2	26·0
$3d^2$	$\Delta^3P_{0,1}$	27·4	27·4
	$\Delta^3P_{1,2}$	52·9	26·4

Fig. 15·9. Interval rule in the low terms of Sc II.

Three electrons. Sc I, Y I, La I, Lu I; Ti II, Zr II, Ce II, Hf II.

The ground terms of the arc spectra of scandium, yttrium and lutecium are d.s^2 ^2D, while the usual subsidence of the d relative to the s orbit produces d^2.s ^3F in the spark spectra of titanium and

Low terms

		Sc	Y	La	Lu	Ti II	Zr II	Ce II	Hf II
$3d.4s^2$	$^2D_{1\frac{1}{2}}$	**168**	**530**	**1053**	**1994**	25,193	18,397	**420**	—
$3d^2.4s$	$^4F_{1\frac{1}{2}}$	11,677	11,532	4122	—	**393**	**1,323**	1923	?8362
	$^4P_{2\frac{1}{2}}$	—	15,477	—	—	10,025	8,058	1870	—
	$^2G_{4\frac{1}{2}}$	20,237	18,499	—	—	15,258	14,190	—	—
	$^2F_{3\frac{1}{2}}$	15,042	15,864	—	—	4,898	6,468	—	—
	$^2D_{2\frac{1}{2}}$	17,013	16,159	—	—	8,744	4,505	—	**?3051**
	$^2P_{1\frac{1}{2}}$	—	19,406	—	—	16,625	6,112	—	—
	$^2S_{\frac{1}{2}}$	—	—	—	—	21,338	25,202	—	—
$3d^3$	$^4F_{4\frac{1}{2}}$	33,906	29,843	—	—	1,216	3,758	—	—
	$^4P_{2\frac{1}{2}}$	36,573	32,366	—	—	9,518	9,969	—	—
	$^2H_{5\frac{1}{2}}$	—	—	—	—	12,775	12,360	—	—
	$^2G_{4\frac{1}{2}}$	—	—	—	—	9,118	8,153	—	—
	$^2F_{3\frac{1}{2}}$	—	—	—	—	20,892	19,433	—	—
	$^2D_{2\frac{1}{2}}$	36,330	—	—	—	12,758	14,733	—	—
	$^2D_{2\frac{1}{2}}$	—	—	—	—	—	14,163	—	—
	$^2P_{1\frac{1}{2}}$	—	—	—	—	9,976	20,080	—	—

Fig. 15·10. Three electrons. Energies of the low terms.

zirconium. It has been suggested that the ground term of Hf II is $d^2.s\,^2D$, but this irregularity still awaits confirmation.

The low terms predicted by theory are collected in Fig. 15·10; every one of these terms has been found in Zr II, while all the odd terms have been found in Y I; thus once again the great power of Hund's theory is demonstrated. The greatly increased complexity of Sc I and Y I compared with Sc II and Y II is perhaps so obvious that it hardly deserves mention.

Low terms

Configuration	Terms	
	Quartets	Doublets
$3d.4s^2$	—	D
$3d^2.4s$	PF	SPDGF
$3d^3$	PF	PDFGH 2

Middle terms

Ion		Atom		
$l_1\,l_2$	Term	l_3	Terms	$l_1\,l_2\,l_3$
$4s^2$	^1S	4p	^2P	s^2p
3d.4s	^3D	4p	$^4(PDF)$ $^2(PDF)$	d.sp
	^1D	4p	$^2(PDF)$	
$3d^2$	^3F	4p	$^4(DFG)$ $^2(DFG)$	$d^2.p$
	^3P	4p	$^4(SPD)$ $^2(SPD)$	
	^1G	4p	$^2(FGH)$	
	^1D	4p	$^2(PDF)$	
	^1S	4p	^2P	

Fig. 15·11. Terms predicted from three electrons.

The lines due to transitions from high to middle terms are, in the spectrum of Y I, so much weaker when the terms are doublet than when quartet, that very few high doublet terms have been identified.

Four electrons. Ti I, Zr I, Hf I; V II, Cb II.

The arc spectra of titanium, zirconium and hafnium all have as ground term $^3F_2$ of the $d^2.s^2$ configuration; but the d orbit evidently falls abruptly relative to the s at this point, for the spark spectra seem to have $d^4\,^5D$ as ground term.

The spectra of titanium and zirconium have been very thoroughly analysed at the Bureau of Standards by Russell and Kiess, and almost all the low terms from the $d^2.s^2$ and $d^3.s$ configurations have been identified. In their analysis of hafnium

Low terms

		Ti	Zr	Hf	V II	Cb II
$3d^2.4s^2$	$^3F_4$	**387**	**1,241**	**4568**	—	—
	$^3P_2$	8,602	4,186	8984	—	—
	$^1G_4$	12,118	8,057	—	—	—
	$^1D_2$	7,255	5,101	5639	—	—
	$^1S_0$	15,167	13,142	—	—	—
$3d^3.4s$	$^5F_5$	6,843	5,889	—	3,163	4147
	$^3F_4$	11,777	12,342	—	9,097	—
	$^5P_3$	14,106	11,258	—	13,741	—
	$^3P_2$	18,912	15,932	—	—	—
	$^3H_6$	18,193	15,120	—	20,363	—
	$^1H_5$	20,796	18,739	—	—	—
	$^3G_5$	15,220	12,773	—	14,655	—
	$^1G_4$	18,288	17,753	—	—	—
	$^3F_4$	—	15,700	—	—	—
	$^1F_3$	29,818	—	—	—	—
	$^3D_3$	17,540	14,697	—	—	—
	$^1D_2$	20,210	17,228	—	—	—
	$^3D_3$	—	—	—	—	—
	$^1D_2$	—	—	—	—	—
	$^3P_2$	18,145	17,143	—	—	—
	$^1P_1$	20,063	—	—	—	—
$3d^4$	$^5D_4$	—	22,398	—	**339**	**1225**
	$^3H_6$	—	—	—	—	—
	$^3G_5$	36,201	—	—	16,532	—
	$^3F_4$	—	—	—	13,608	—
	$^3F_4$	—	—	—	—	—
	$^3D_3$	—	—	—	—	—
	$^3P_2$	—	—	—	11,908	—
	$^3P_2$	—	—	—	—	—

Fig. 15·12. Four electrons. Energies of the low terms.

Meggers and Scribner had to rely on wave-lengths and furnace intensities only, but this has not prevented them from identifying seven even and 63 odd terms. From a study of the observed combinations, J values have been assigned and the suggestion put forward that the lowest three terms are $d^2.s^2\,{}^3F$. The J values would allow the next terms to be $^1D_2$ and ^3P, and this allotment would make the low terms very similar to those of Ti I; but further discussion must wait on the promised measurements of the Zeeman effect.

Among the odd terms those due to the configuration $d^2.sp$ lie lowest in the arc spectra, but terms from the two other configurations have been recognised.

Numerous combinations have been found between both singlet and triplet terms, and between triplet and quintet terms, but very

Low terms

Configuration	Terms		
	Quintet	Triplet	Singlet
$3d^2.4s^2$	—	PF	SDG
$3d^3.4s$	PF	PDFGH 2 2 2	PDFGH 2
$3d^4$	D	PDFGH 2 2	SDFGJ 2 2 2

Middle terms

Ion		Atom		
$l_1 \dots l_3$	Term	l_4	Terms	$l_1 \dots l_4$
$3d^2.4s$	^{4}F	4p	^{5,3}(DFG)	$d^2.sp$
	^{4}P		^{5,3}(SPD)	
	^{2}G		^{3,1}(FGH)	
	^{2}F		^{3,1}(DFG)	
	^{2}D		^{3,1}(PDF)	
	^{2}P		^{3,1}(SPD)	
	^{2}S		^{3,1}P	
$3d^3$	^{4}F	4p	^{5,3}(DFG)	$d^3.p$
	^{4}P		^{5,3}(SPD)	
	^{2}H		^{3,1}(GHI)	
	^{2}G		^{3,1}(FGH)	
	^{2}F		^{3,1}(DFG)	
	^{2}D		^{3,1}(PDF)	
	^{2}D		^{3,1}(PDF)	
	^{2}P		^{3,1}(SPD)	
$3d.4s^2$	^{2}D	4p	^{3,1}(PDF)	$d.s^2p$

Fig. 15·13. Terms predicted from four electrons.

few between singlets and quintets, and even these few are all susceptible of the same explanation. When two terms of the same configuration and the same J have nearly the same energy, they share both their intensities and their g values. Thus $d^3.p\,{}^1D_2^\circ$ combines with the low even ${}^5D_3$ term because the former lies very near to the ${}^5P_2^\circ$ term of the same configuration; the empirical g values of the $4d^3.5p\,{}^1D_2^\circ$ and ${}^5P_2^\circ$ terms of Zr I are both 1·42, whereas the Landé values are 1·000 and 1·833 respectively; their respective energies are 34,850 and 34,761 cm.$^{-1}$

Five electrons. V I, Cb I; Cr II, Mo II.

The ground term of these spectra starts as ^4F of the $d^3.s^2$ configuration in the iron row, changes to ^6D from $d^4.s$ in the palladium row, and settles down as $d^5\ ^6S$ in both the spark spectra; these changes clearly conform to the general type.

Low terms

		V I	Cb I	Cr II	Mo II
$3d^3.4s^2$	$^4F_{4\frac{1}{2}}$	553	—	—	—
	$^4P_{2\frac{1}{2}}$	9,825	—	—	—
$3d^4.4s$	$^6D_{4\frac{1}{2}}$	2,425	1050	12,498	known
	$^4H_{6\frac{1}{2}}$	15,063	—	30,393	—
	$^4G_{5\frac{1}{2}}$	17,242	—	33,696	—
	$^4F_{4\frac{1}{2}}$	15,770	—	31,221	—
	$^4D_{3\frac{1}{2}}$	8,716	—	20,025	known
	$^4P_{2\frac{1}{2}}$	15,572	—	—	—
	$^2P_{1\frac{1}{2}}$	—	—	17,593	—
$3d^5$	$^6S_{2\frac{1}{2}}$	—	—	0	ground?
	$^4G_{5\frac{1}{2}}$	—	—	20,514	—
	$^4F_{4\frac{1}{2}}$	—	—	32,856	—
	$^4D_{3\frac{1}{2}}$	—	—	25,035	—
	$^4P_{2\frac{1}{2}}$	—	—	—	—
	$^2D_{2\frac{1}{2}}$	—	—	21,824	—

Fig. 15·14. Five electrons. Energies of the low terms.

Of the other low terms little is known outside the Cr II spectrum, in which many quartet but only two doublet terms have been found; these doublets are conspicuous in breaking Hund's energy rules, for they lie among the lower of the quartet terms.

The odd terms are best developed by the addition of a p electron to the lowest terms of the parent spectrum; in this way a number of triads are formed, and these agree very satisfactorily with the terms found empirically. Thus in V I there occur a triad of sextet terms at 18,000 cm.$^{-1}$ and a triad of quartet terms at about 22,000 cm.$^{-1}$; the J values show that in fact these are the two triads $^6(DFG)^\circ$ and $^4(DFG)^\circ$ which should arise by the addition of a p electron to the low $d^3.s\ ^5F$ term of V II.

Of all these five electron spectra Cr II has been the most thoroughly analysed, and in it nearly all the terms both obey the interval rule reasonably well and have normal g values. The $^6D^\circ$ and $^4P^\circ$ terms of the $d^4.p$ configuration are irregular judged by either criterion; but a term which deviates from one rule, often deviates from the second as well.

Low terms

Configuration	Terms		
	Sextet	Quartet	Doublet
$3d^3.4s^2$	—	PF	PD_2FGH
$3d^4.4s$	D	$P_2D_2F_2GH$	$S_2D_2FG_2J$
$3d^5$	S	PDFG	$SPD_3F_2G_2HJ$

Middle terms

Ion		Atom		
$l_1 \ldots l_4$	Term	l_5	Terms	$l_1 \ldots l_5$
d^4	^5D	p	$^{6,4}(PDF)$	$d^4.p$
$d^3.s$	^5F	p	$^{6,4}(DFG)$	$d^3.sp$

Fig. 15·15. Terms predicted from five electrons.

Six electrons. Cr I, Mo I, W I; Mn II.

The analysis of these spectra has hardly extended beyond the septet and quintet systems, only a few triplets having been found, and no singlets though these undoubtedly occur. The arc spectra of chromium and molybdenum and the spark spectrum of manganese all have $d^5.s\ ^7S$ as ground term; in contrast the ground term of tungsten is probably ^5D, but the analysis is so little advanced

Low terms

		Cr I	Mo I	W I	Mn II
$3d^4.4s^2$	$^5D_4$	8,307	12,346	**6219**	—
$3d^5.4s$	$^7S_3$	**0**	**0**	2951	**0**
	$^5S_2$	7,593	10,768	—	9,473
	$^5G_6$	20,519	16,828	—	—
	$^5F_5$	—	—	—	—
	$^5D_4$	24,282	—	—	—
	$^5P_3$	21,841	18,229	—	—
$3d^6$	$^5D_4$	—	—	—	14,324

Fig. 15·16. Six electrons. Energies of the low terms.

that this cannot be surely assigned to the $d^4.s^2$ as against the d^6 configuration.

Low terms

Configuration	Terms			
	Septet	Quintet	Triplet	Singlet
$3d^4.4s^2$	D	D	PDFGH 2 2	SDFGJ 2 2 2
$3d^5.4s$	S	SPDFG	SPDFGHJ 2 4 3 3	SPDFGHJ 3 2 2
$3d^6$	—	D	PDFGH 2 2	SDFGJ 2 2 2

Middle terms

Ion		Atom		
$l_1 \dots l_5$	Term	l_6	Terms	$l_1 \dots l_6$
$3d^3.4s^2$	^4F	4p	$^{5,3}(DFG)$	$d^3.s^2p$
$3d^4.4s$	^6D	4p	$^{7,5}(PDF)$	$d^4.sp$
$3d^5$	^6S	4p	^7P ^5P	$d^5.p$

Fig. 15·17. Terms predicted from six electrons.

The list of low terms is complete, but the middle terms are developed from the low terms of the spark spectrum; a complete list would be so long as to give no guidance at all.

Seven electrons. Mn I, Re I; Fe II, Ru II.

The ground term of the arc spectra of manganese and rhenium is $d^5.s^2\,{}^6S$, but it changes first to $d^6.s\,{}^6D$ and then to $d^7\,{}^4F$ in the spark spectra of iron and ruthenium respectively; thus once again the d orbit is seen to sink relative to the s orbit, when we pass from one period to the next or from an arc to a spark spectrum.

Low terms

		Mn	Re	Fe II	Ru II
$3d^5.4s^2$	^6S	**0**	**0**	23,318	—
$3d^6.4s$	$^6D_{4\frac{1}{2}}$	17,052	11,754	**0**	9,151
	$^4G_{5\frac{1}{2}}$	—	—	25,429	—
	$^4F_{4\frac{1}{2}}$	—	—	22,637	—
	$^4D_{3\frac{1}{2}}$	23,297	—	7,955	19,379
	$^4P_{2\frac{1}{2}}$	—	—	20,830	—
$3d^7$	$^4F_{4\frac{1}{2}}$	—	—	1,873	**0**
	$^4P_{2\frac{1}{2}}$	—	—	13,474	8,257
	$^2F_{3\frac{1}{2}}$	—	—	—	22,289

Fig. 15·18. Seven electrons. Energies of the low terms.

The ground term of Mn I has been recently confirmed by showing that all the resonance lines arise from jumps ending in the ^6S state.*

Ion		Atom		
$l_1 \dots l_6$	Term	l_7	Terms	$l_1 \dots l_7$
$d^5.s$	^7S	s d	^6S ^6D	$d^5.s^2$ $d^6.s$
	^5S	s d	^6S (as above) ^6D (as above) ^4D	$d^5.s^2$ $d^6.s$
$d^4.s^2$	^5D	d	^6S (as above) ^{4}(PDFG)	$d^5.s^2$

Fig. 15·19. Low terms of Mn I developed from the low terms of Mn II.

Configuration	Terms		
	Sextets	Quartets	Doublets
$3d^5.4s^2$	S	PDFG	SPDFGHJ 3 2 2
$3d^6.4s$	D	PDFGH 2 2 2	SDFGJ 2 2 2
$3d^7$	—	PF	PDFGH 2

Middle terms

Ion		Atom		
$l_1 \dots l_6$	Term	l_7	Terms	$l_1 \dots l_7$
$d^5.s$	^7S ^5S	p p	^8P ^6P ^6P ^4P	$d^5.s.p$
$d^4.s^2$	^5D	p	^{6,4}(PDF)	$d^4.s^2.p$
d^6	^5D	p	^{6,4}(PDF)	$d^6.p$

Fig. 15·20. Terms predicted from seven electrons.

With the increasing number of electrons, even the low terms become so complex that they are not easily identified; to meet this difficulty the low terms may be developed from the lowest terms of the spark spectrum, just as the odd terms have been developed. Thus in Fig. 15·19 there are only a few terms where in Fig. 15·20 there is a huge mass.

* Fridrichson, *ZP*, 1930, **64** 43.

Since the ground term of the spark spectra of Mn and Re is ^{7}S, series are to be expected in the arc spectra; and in fact Catalan* did first analyse Mn I into a number of series. The lowest terms of the odd series are shown in Fig. 15·20.

The terms of these spectra are nearly all inverted; while those of Ru II show very regular g values.

Eight electrons. Fe I, Ru I; Co II.

The ground term of the arc spectrum of iron is d^6.s^2 ^{5}D, but the ground term changes in the normal way to d^7.s ^{5}F and then to d^8 ^{3}F in the arc spectrum of ruthenium and the spark spectrum of cobalt respectively.

Low terms

		Fe I	Ru I	Co II
3d^6.4s^2	$^5D_4$	0	? 7,483	—
3d^7.4s	$^5F_5$	6,928	0	3,350
	$^5P_3$	17,550	? 8,771	17,771
	$^3F_4$	11,976	6,545	9,813
	$^3P_2$	22,838	—	—
3d^8	$^3F_4$	—	? 9,120	0
	$^3P_2$	—	?10,623	—

Fig. 15·21. Eight electrons. Energies of the low terms.

Ion		Atom		
$l_1 \dots l_7$	Term	l_8	Terms	$l_1 \dots l_8$
d_6.s	^{6}D	s	^{5}D	d^6.s^2
		d	^{5}(PF)	d^7.s
	^{4}D	s	^{5}D (as above) ^{3}(PDFG)	d^6.s^2
		d	^{5}(PF) (as above) ^{3}(PDFG)	d^7.s
d^7	^{4}F	s	^{5}F ^{3}F	d^7.s
		d	^{3}(PF)	d^8
	^{4}P	s	^{5}P ^{3}P	d^7.s
		d	^{3}(PF) (as above)	d^8

Fig. 15·22. Low terms of Fe I developed from the low terms of Fe II.

Hund's theory predicts for the d^6.s^2 configuration one quintet and seven triplet levels besides a number of singlets. In the iron

* Catalan, *Phil. Trans. R.S.*, 1922, **223** 127.

arc the ^5D term is inverted and follows the interval rule fairly well, as the following table shows (Fig. 15·24).

Low terms

Configuration	Terms		
	Quintet	Triplet	Singlet
$3d^6.4s^2$	D	PDFGH 2 2	SDFGJ 2 2 2
$3d^7.4s$	PF	PDFGH 2 2 2	PDFGH 2
$3d^8$	—	PF	SDG

Middle terms

Ion		Atom		
$l_1 \dots l_7$	Term	l_8	Terms	$l_1 \dots l_8$
$d^6.s$	^6D	p	^{7,5}(PDF)	$d^6.s.p$
	^4D	p	^{5,3}(PDF)	$d^6.s.p$
d^7	^4F	p	^{5,3}(DFG)	$d^7.p$
	^4P	p	^{5,3}(SPD)	—

Fig. 15·23. Terms predicted from eight electrons.

Term	Interval $\Delta E/ch$	A
$^5D_0$		
	89·9	89·9
$^5D_1$		
	184·1	92·1
$^5D_2$		
	288·1	96·0
$^5D_3$		
	415·9	104·0
$^5D_4$		

Fig. 15·24. Intervals of the $d^6.s^2$ ^5D term of Fe I.

Of the triplet levels, only four have been found; these are all inverted and obey the interval rule about as well as the ^5D term. Goudsmit's theory of the displacement sum predicts that the ^3D term will be erect, but the term is still unidentified.

The other low configuration of the iron arc is $d^7.s$, the four terms so far identified being clearly produced by an s electron adding itself to the d^7 ^4F and ^4P terms, for the derivation is in-

PLATE V

1. Fluorescent spectrum of chromium in Al_2O_3 at $-186°$ C. At the right-hand or blue end of the spectrum a strong principal doublet appears, a thousand times over-exposed. Next to these are the subsidiary lines, weaker but sharp, while on the left are still weaker bands. The line near the centre is the 7032 A. of neon. (After Deutschbein, *PZ*, 1932, **33** 875.)

2. DF septet from the iron arc. This multiplet arises as $3d^6 . 4s\,(^6D)\,5s\,^7D \rightarrow (^6D)\,4p\,^7F°$. (Lent by Prof. H. Dingle.)

Plate V

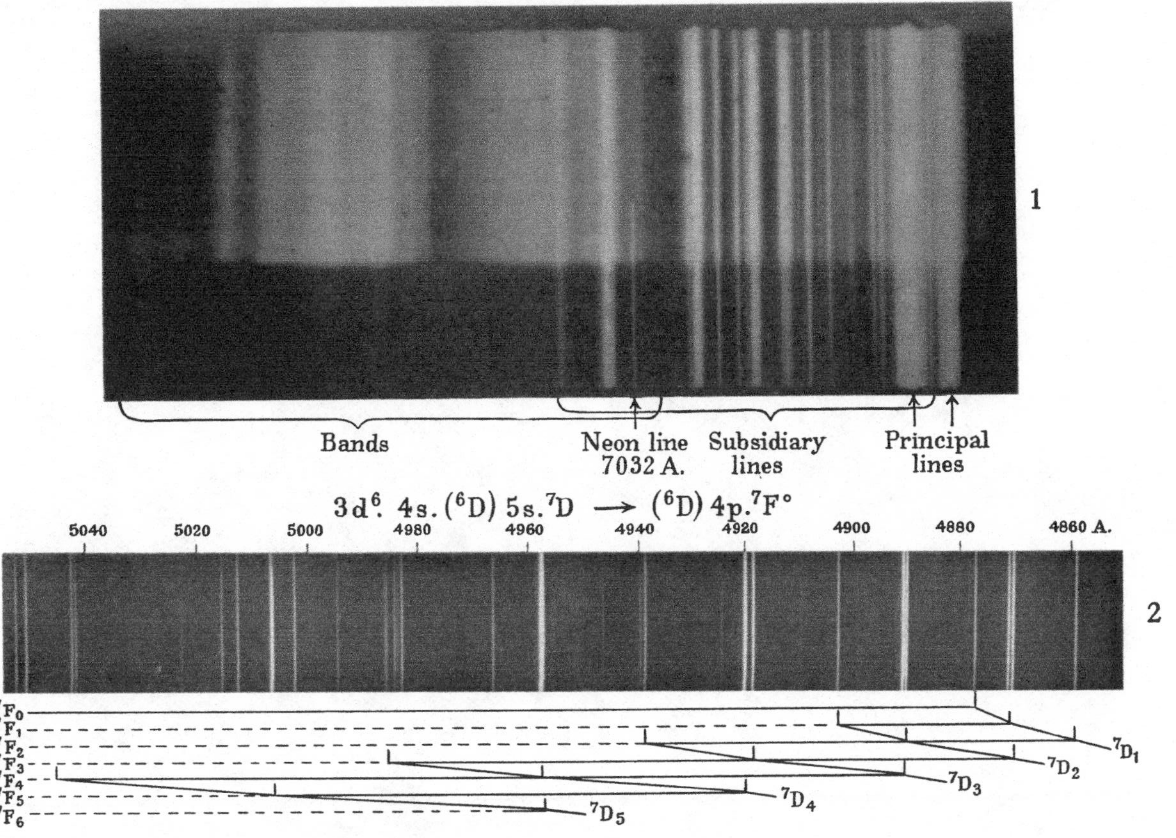
1
Bands
Neon line
7032 A.
Subsidiary
lines
Principal
lines
2
$3d^6.\ 4s.\ (^6D)\ 5s.\ ^7D \longrightarrow (^6D)\ 4p.\ ^7F^\circ$
5040
5020
5000
4980
4960
4940
4920
4900
4880
4860 A.
$^7F_0$
$^7F_1$
$^7F_2$
$^7F_3$
$^7F_4$
$^7F_5$
$^7F_6$
$^7D_1$
$^7D_2$
$^7D_3$
$^7D_4$
$^7D_5$

dicated both by the energies and the intervals of the terms. In the spark ^4F lies 11,000 cm.$^{-1}$ below ^4P, and in the arc ^5F and ^3F lie about the same distance below the corresponding ^5P and ^3P terms. Again, ^4F obeys the interval rule well, while the derived ^5F and ^3F terms do not deviate much (Fig. 15·25). But the ionic ^4P term is quite irregular and so are the derived terms of the arc spectrum; even the irregularities are similar (Fig. 15·26).

Term	$\Delta E/ch$	A	Term	$\Delta E/ch$	A	Term	$\Delta E/ch$	A
$^5F_1$								
	168·9	84·5				$^4F_{1\frac{1}{2}}$		
$^5F_2$			$^3F_2$				279·6	111·8
	257·7	85·9		407·6	135·9	$^4F_{2\frac{1}{2}}$		
$^5F_3$			$^3F_3$				407·7	116·5
	351·3	87·8		584·7	146·2	$^4F_{3\frac{1}{2}}$		
$^5F_4$			$^3F_4$				557·6	123·9
	488·5	97·7				$^4F_{4\frac{1}{2}}$		
$^5F_5$								

Fig. 15·25. Intervals of the d^7 (^4F).s $^{5,3}F$ terms of Fe I and the d^7 ^4F term of Fe II.

Term	$\Delta E/ch$	A	Term	$\Delta E/ch$	A	Term	$\Delta E/ch$	A
$^5P_1$			$^3P_0$			$^4P_{\frac{1}{2}}$		
	200·4	100·2		104·9	104·9		231·7	154·5
$^5P_2$			$^3P_1$			$^4P_{1\frac{1}{2}}$		
	176·8	58·9		108·5	54·2		198·7	79·5
$^5P_3$			$^3P_2$			$^4P_{2\frac{1}{2}}$		

Fig. 15·26. Intervals of the d^7 (^4P).s $^{5,3}P$ terms of Fe I and the d^7 ^4P term of Fe II.

A very large number of odd terms have been identified and named in the spectra of Fe I and Co II, for the terms satisfy the simple laws; their intervals are regular, and almost without exception they are inverted. But the terms of Ru I are much less regular, so that a thorough Zeeman analysis was necessary before the terms could be named, and even now only a J value has been assigned to very many; the interval law is only roughly satisfied.

Nine electrons. Co I, Rh I; Ni II, Pd II.

The ground term starts as $d^7.s^2\,^4F$ in the arc spectrum of cobalt, changes to $d^8.s\,^4F$ in the arc spectrum of rhodium and finishes as $d^9\,^2D$ in the spark spectra of nickel and palladium;

Low terms

		Co I	Rh I	Ni II	Pd II
$3d^7.4s^2$	$^4F_{4\frac{1}{2}}$	0	12,733	—	—
	$^4P_{2\frac{1}{2}}$	15,184	—	—	—
	$^2F_{3\frac{1}{2}}$	—	—	—	—
	$^2P_{1\frac{1}{2}}$	20,501	—	—	—
$3d^8.(^3F)\ 4s$	$^4F_{4\frac{1}{2}}$	3,483	0	8,393	25,081
	$^2F_{3\frac{1}{2}}$	7,442	5,691	13,549	32,278
$(^3P)\ 4s$	$^4P_{2\frac{1}{2}}$	13,795	9,221	25,035	36,281
	$^2P_{1\frac{1}{2}}$	18,390	11,968	29,069	43,648
$(^1G)\ 4s$	$^2G_{4\frac{1}{2}}$	16,468	16,018	32,498	44,506
$(^1D)\ 4s$	$^2D_{2\frac{1}{2}}$	16,778	13,521	23,107	41,198
$(^1S)\ 4s$	$^2S_{\frac{1}{2}}$	—	—	24,825	—
$3d^9$	$^2D_{2\frac{1}{2}}$	21,920	3,310	0	0

Middle terms

		Co I	Rh I	Ni II	Pd II
$3d^7.4s\ (^5F).4p$	$^6G_{6\frac{1}{2}}$	25,138	—	—	—
	$^6F_{5\frac{1}{2}}$	23,612	—	—	—
	$^6D_{4\frac{1}{2}}$	24,628	—	—	—
	$^4G_{5\frac{1}{2}}$	28,845	—	—	—
	$^4F_{4\frac{1}{2}}$	28,346	—	—	—
	$^4D_{3\frac{1}{2}}$	29,294	—	—	—
$(^3F)\ 4p$	$^4G_{5\frac{1}{2}}$	41,528	—	—	—
	$^4F_{4\frac{1}{2}}$	41,225	—	91,797	—
	$^4D_{3\frac{1}{2}}$	39,649	—	—	—
	$^2G_{4\frac{1}{2}}$	31,700	—	—	—
	$^2F_{3\frac{1}{2}}$	31,871	—	93,525	—
	$^2D_{2\frac{1}{2}}$	33,463	—	—	—
$3d^8.(^3F)\ 4p$	$^4G_{5\frac{1}{2}}$	32,431	29,105	53,495	68,611
	$^4F_{4\frac{1}{2}}$	32,842	29,431	54,556	69,878
	$^4D_{3\frac{1}{2}}$	32,027	27,076	51,557	65,247
	$^2G_{4\frac{1}{2}}$	33,440	31,614	55,299	72,285
	$^2F_{3\frac{1}{2}}$	35,451	32,004	57,079	73,327
	$^2D_{2\frac{1}{2}}$	36,092	32,046	57,419	72,733
$(^3P)\ 4p$	$^4D_{3\frac{1}{2}}$	—	36,787	70,776	83,056
	$^4P_{2\frac{1}{2}}$	—	35,334	66,569	76,767
	$^4S_{1\frac{1}{2}}$	—	—	74,299	86,280
	$^2D_{2\frac{1}{2}}$	—	—	71,770	83,802
	$^2P_{1\frac{1}{2}}$	—	—	72,984	85,151
	$^2S_{\frac{1}{2}}$	—	—	74,282	85,071
$(^1G)\ 4p$	$^2H_{5\frac{1}{2}}$	—	—	75,719	85,593
	$^2G_{4\frac{1}{2}}$	—	—	79,923	89,982
	$^2F_{3\frac{1}{2}}$	—	—	?75,916	86,043
$(^1D)\ 4p$	$^2F_{3\frac{1}{2}}$	—	—	67,693	79,708
	$^2D_{2\frac{1}{2}}$	—	—	68,634	82,057
	$^2P_{1\frac{1}{2}}$	—	—	68,864	80,956
$(^1S)\ 4p$	$^2P_{1\frac{1}{2}}$	—	—	60,502	—

Fig. 15·27. Nine electrons. Energies of the low and middle terms.

moreover, whereas the ground ^{2}D term is only 8000 cm.$^{-1}$ below the next lowest term in Ni II it is 25,000 cm.$^{-1}$ in Pd II. These changes are in the usual order.

The energies of the low and middle terms are always related, but no group of electrons shows these relations more clearly than the group of nine here considered. In Co I the ground term is of the 3d^7.4s^2 configuration and the lowest odd terms are derived

Low terms

Configuration	Terms	
	Quartet	Doublet
3d^7.4s^2	PF	PDFGH (2)
3d^8.4s	PF	SPDFG
3d^9	—	D

Middle terms

Ion		Atom		
$l_1 \ldots l_8$	Term	l_9	Terms	$l_1 \ldots l_9$
d^8	^{3}F	p	^{4,2}(DFG)	d^8.p
	^{3}P	p	^{4,2}(SPD)	
d^7.s	^{5}F	p	^{6,4}(DFG)	d^8.sp
	^{5}P	p	^{6,4}(SPD)	

Fig. 15·28. Terms predicted from nine electrons.

from 3d^7.4s.4p; on the other hand, when the ground term is from d^8.s as it is in Rh I or from d^9 as in Ni II and Pd II, practically all the known odd terms are from d^8.p. Again, in Ni II the energies of the even and odd terms run strikingly parallel; of the 3d^8.4s configuration ^{2}D and ^{2}S lie exceptionally low, and this is matched among the odd terms by the anomalous positions of ^{2}(PDF) and ^{2}P from 3d^8.p. Presumably these two irregularities have a common cause in the d^8 ^{1}D and d^8 ^{1}S terms of Ni III, but unfortunately the latter spectrum has not yet been analysed.

In Ni II all but one of the low even terms are inverted, but of the odd d^8.p configuration eight terms are erect or only partially inverted. In Pd II the intervals are rather irregular.

Ten electrons. Ni I, Pd I, Pt I; Cu II, Ag II, Au II; Cd III, Hg III; In IV, Tl IV.

The ground term of all the spark and higher spark spectra is $d^{10}\,{}^1S$; but the arc spectra exhibit two irregularities. Thus the ground term of nickel is $3d^8.4s^2\,{}^3F$, although the ground term of Ni II is $3d^9\,{}^2D$ and the ground term of an arc spectrum is usually obtained by adding an electron to the ground term of the spark; the only other exception to this rule seems to be Y II. Again, if the ground terms of nickel and platinum are respectively $d^8.s^2\,{}^4F$ and $d^9.s\,{}^4F$, the ground term of palladium might be expected to be one or the other of these, whereas in fact it is $d^{10}\,{}^1S$.

Low terms

		Ni I	Pd I	Pt I	Cu II	Ag II	Au II
$3d^8.4s^2$	$^3F_4$	0	25,101	824	—	—	—
	$^3P_2$	15,610	37,952	6,567	—	—	—
	$^1G_4$	22,102	—	21,967	—	—	—
	$^1D_2$	13,521	—	26,639	—	—	—
	$^1S_0$	—	—	—	—	—	—
$3d^9.4s$	$^3D_3$	204	6,564	0	21,925	39,164	15,036
	$^1D_2$	3,410	11,722	13,496	26,261	46,045	29,618
$3d^{10}$	$^1S_0$	14,729	0	6,140	0	0	0

Fig. 15·29. Ten electrons. Energies of the low terms.

Low terms

Configuration	Terms	
	Triplet	Singlet
$d^8.s^2$	PF	SDG
$d^9.s$	D	D
d^{10}	—	S

Middle terms

Ion		Atom		
$l_1 \dots l_9$	Term	l_{10}	Terms	$l_1 \dots l_{10}$
d^9	^2D	p	$^{3,1}(PDF)$	$d^9.p$
$d^8.s$	^4F	p	$^{5,3}(DFG)$	$d^8.sp$
	^2F	p	$^{3,1}(DFG)$	—

Fig. 15·30. Terms predicted from ten electrons.

The changes in the arc spectra in passing from row to row are also instructive; in Ni I the term separations are wide and the multiplets overlap; but, though complex, the spectrum is admirably regular; all the even terms and nearly all the odd terms are inverted, and the intervals conform to Landé's rule; so simple was the analysis indeed that the Zeeman effect has not been studied. In the Pd I spectrum the wide interval, 3512 cm.$^{-1}$, of the ^{2}D ground term of Pd II begins to exert an effect; as some triplet series tend to the lower and some to the higher limit, the interval ratios change rapidly with the serial number. And the terms of Pt I are even less regular than those of Pd I; in analysis the interval rule is useless, and the intensity rules can be treated only as approximations; the measurements of g, too, depart from Landé's values, but they suffice with the J selection rule to determine the J value of all the empirical terms. The determination of the orbital and spin vectors is, however, far more difficult, for the failure of the simple rules is a sure sign that the coupling is no longer Russell-Saunders. However, all the low terms predicted from $d^8.s^2$, $d^9.s$ and d^{10} seem to have been identified, save only $d^8.s^2\,{}^1S$, which has not yet been found in any spectrum. Names have also been assigned to some of the odd terms arising from the $d^8.sp$ and $d^9.p$ configurations.

In Cu II and Ag II the terms are generally inverted, but the interval rule is very badly satisfied; in Ag II irregular g values are also indicated.

In the analysis of Pd I there is a surplus level, known as k_1, which is of interest chiefly as being the only level which fails to fit into the Hund scheme. The level is determined by five exact combinations, but the lines due to k_1 are all listed as diffuse and differ in this from all other lines of the palladium spectrum; the level can hardly be a hyperfine component, for no other levels show a similar structure, and the interval separating it from the nearest normal level is over 3 cm.$^{-1}$

Eleven electrons. Cu I, Ag I, Au I; Zn II, Cd II, Hg II.

As the d level can hold only ten electrons, the elements copper, silver and gold, which have eleven electrons outside the last inert gas shell, should exhibit the simple alkali spectra provided only that d^{10} is firmly bound. In fact all three elements do exhibit doublet series and they all have a ^2S ground term. But besides

Low terms

	Cu I	Ag I	Au I	Zn II	Cd II	Hg II
$3d^9.4s^2\ ^2D_{2\frac{1}{2}}$	11,203	—	9161	62,721	69,259	35,514
$3d^{10}.4s\ ^2S_{\frac{1}{2}}$	0	0	0	0	0	0

Fig. 15·31. Eleven electrons. Energies of the low terms.

this simple system, which was discovered in very early days, the last decade has revealed in copper and gold other quartet and doublet terms; in silver these terms have not been found, though all the lines of Ag I have been classified.*

Cu^+		Cu		
$l_1 \dots l_{10}$	Term	l_{11}	Terms	
$3d^{10}$	^1S	s	^2S	Alkali-like systems
		p	$^2P^o$	
		d	^2D	
$3d^9.4s$	$^3D\ ^1D$	4s	^2D	Deep even term
	^3D	4p	$^4(PDF)$ $^2(PDF)$	Middle group of odd terms
	^1D	4p	$^2(PDF)$	
	^3D	*n*s	$^4D\ \ ^2D$	High even terms
		*n*d	$^4(SPDFG)$ $^2(SPDFG)$	
	^1D	*n*s	^2D	
		*n*d	$^2(SPDFG)$	

Fig. 15·32. Terms predicted in copper.

The lowest term of Cu II is $d^{10}\,^1S$; and from it arise by the addition of s, p and d electrons the alkali-like system; from the ^3D and higher ^1D terms of Cu II arise all the other known terms; those predicted are shown in Fig. 15·32, while those actually found are

* Blair, *PR*, 1930, **36** 1531.

arranged to their proper ionic limits in Fig. 15·33. In both the (^{3}D) ns ^{2}D and the (^{1}D) ns ^{2}D series two terms are known, and the

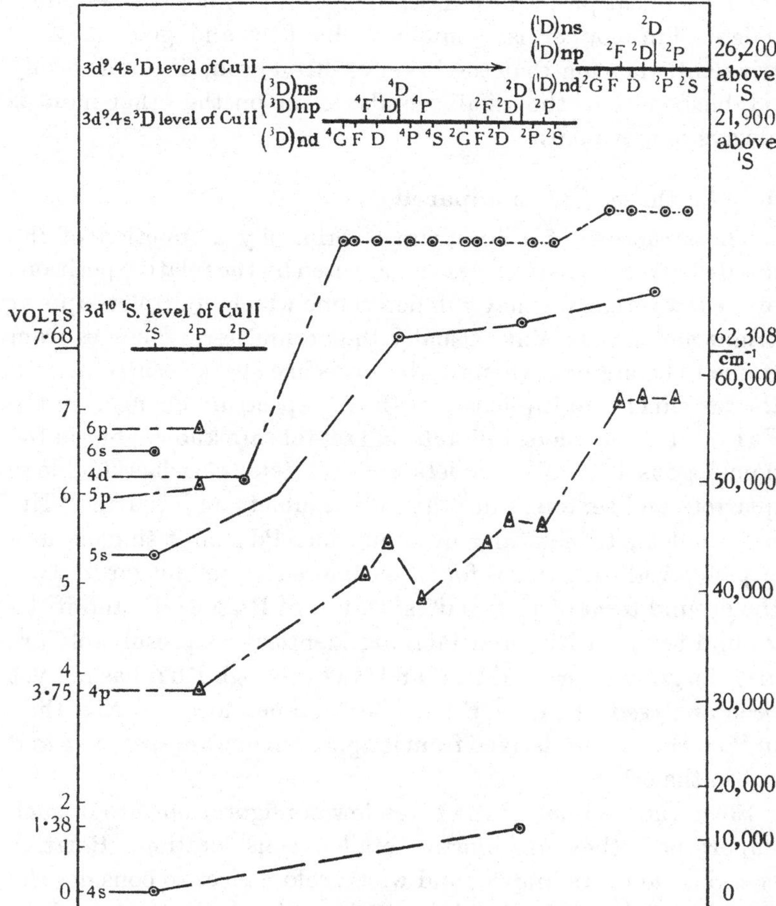

Fig. 15·33. Level diagram of copper. The even terms are shown by circles, the odd terms by triangles.

Rydberg formula then determines the height of the ^{3}D series limits above the ^{1}S limit as 22,200 cm.$^{-1}$

In gold the deep ^{2}D term has been found and also some higher terms, but as the coupling is roughly of the (jj) type only the five lowest terms can be named.

The contrast between copper and gold on the one hand and silver on the other appears not only in their spectra but also in their chemical properties. Both copper and gold can be monovalent, but copper is commonly divalent and gold may be trivalent, showing that the lower electron group is not as firmly established as in the alkali metals; silver on the other hand is always monovalent.

4. The three rows compared

The structure of a spectrum is primarily a function of the number of electrons; but it is also affected by the relative positions of the low terms, for they will determine which multiplets appear bright and which faint. Usually the ground term changes from frame to frame, and when this happens in a spark spectrum, even the prominent multiplicities of the arc spectrum change. In the Fe I spectrum triplets, quintets and septets are known, but in the homologous Ru I only triplets and quintets; Co I has doublets, quartets and sextets, but Rh I only doublets and quartets; Ni I has singlets, triplets and quintets, but Pd I only singlets and triplets. And the reasons for these differences are not far to seek; the ground term of Fe II is $d^6.s\,{}^6D$, but of Ru II $d^7\,{}^4F$; and if the ground terms of Ni II and Pd II are identical, as presumably are also the ground terms of Co II and Rh II, though Rh II has not yet been analysed, the $d^8.s\,{}^4F$ term lies so much lower in Ni II than in Pd II that terms derived from it appear in one arc spectrum and not in the other.

Since the energies of the three low configurations are of such importance, they are worth detailed consideration. Relative energies alone are known, and we therefore elect to consider the energies of the $s^2.d^{n-2}$ and d^n configurations relative to $s.d^{n-1}$, since the last has been identified in more spectra than either of the other two. Many figures are still missing from the tables, but a fair sequence of $s^2.d^{n-2}$ is available in the arc spectra of the iron and palladium rows, while rather less complete sequences for d^n are found in the spark spectra of the same rows; accordingly $(s^2.d^{n-2}-s.d^{n-1})$ and $(s.d^{n-1}-d^n)$ are plotted against n for the arc (Fig. 15·34) and spark (Fig. 15·35) spectra respectively. That

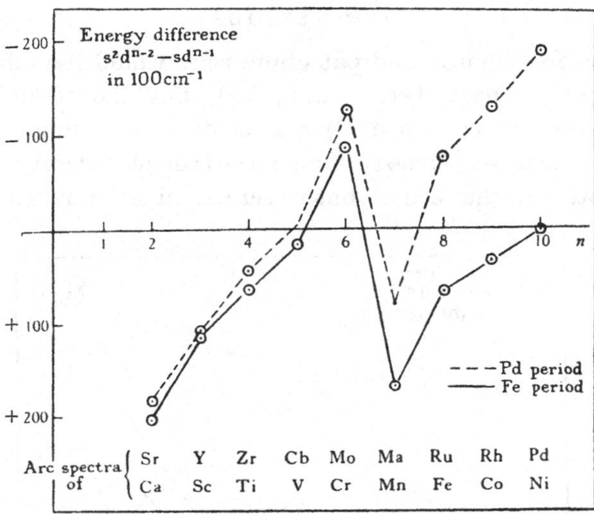

Fig. 15·34. The energy difference of the $s^2.d^{n-2}$ and $s.d^{n-1}$ configurations in the arc spectra of the iron and palladium frames. n is the number of electrons outside the inert gas shell. Due to a slip in drafting the energy scales of this and the two succeeding figures read down instead of up.

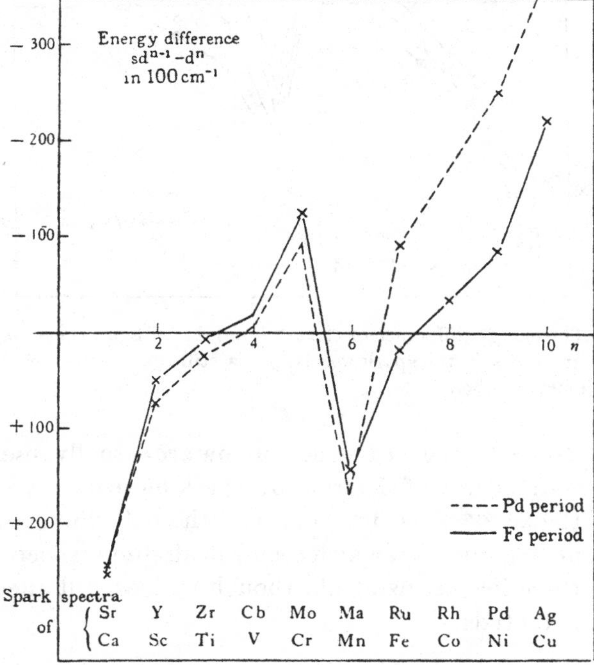

Fig. 15·35. The energy difference of the $s.d^{n-1}$ and d^n configurations in the spark spectra of the iron and palladium frames.

the curves for the iron and palladium rows would resemble one another might have been anticipated, but the resemblance between the $(s^2.d^{n-2}-s.d^{n-1})$ and $(s.d^{n-1}-d^n)$ differences is altogether surprising; true the first curve is displaced a step to the right relative to the second, but closer examination reveals that

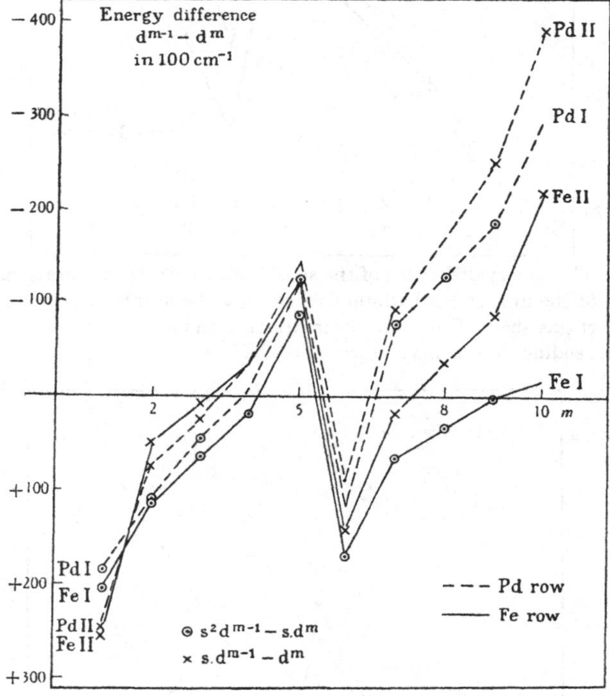

Fig. 15·36. The energy difference of the d^{m-1} and d^m configurations. m is here $n-1$ in the arc and n in the spark spectra; n is still the number of electrons outside the inert gas shell.

the $(s^2.d^{n-2}-s.d^{n-1})$ curve for the iron row arc actually resembles the $(s.d^{n-1}-d^n)$ curve of the iron row spark more closely than it resembles the $(s^2.d^{n-2}-s.d^{n-1})$ curve of the palladium row arc. The fourfold magnification which occurs in all simple spectra does not affect these low terms at all, though it does still affect the ionisation potentials.

To make these facts stand out, write $m=(n-1)$ in the $(s^2.d^{n-2}-s.d^{n-1})$ difference and $m=n$ in $(s.d^{n-1}-d^n)$, and then plot both the differences $(s^2.d^{m-1}-s.d^m)$ and $(s.d^{m-1}-d^m)$ against m (Fig. 15·36). That the four curves are so very similar can only mean that they all measure essentially the difference between the configurations d^{m-1} and d^m, the number of s electrons being of little moment.

The regularity thus revealed enables us to fill up certain gaps in our knowledge; thus in the palladium row, there now seems little doubt that the ground terms of Ma I, Ma II and Rh II—the

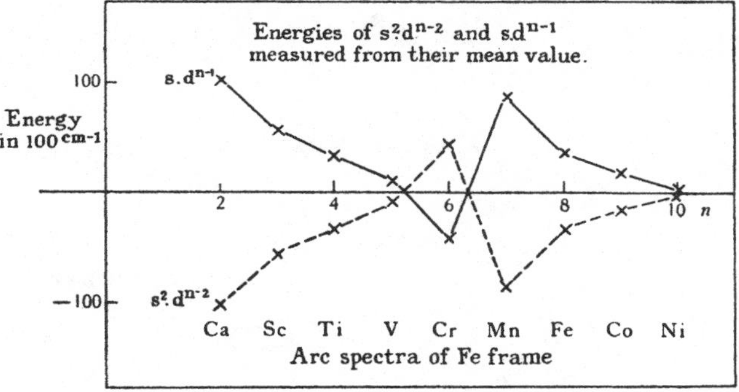

Fig. 15·37. Arc spectra of the iron frame. The energies of the $s^2.d^{n-2}$ and $s.d^{n-1}$ configurations measured from their mean value.

only spectra whose ground terms are not known—must be $d^5.s^2\,{}^6S$, $d^5.s\,{}^7S$ and $d^8\,{}^3F$ respectively; while if in the platinum row the ground term of W I is ${}^5D$ as has usually been supposed, then it must arise from the $d^4.s^2$ configuration and probably the ground configuration of every arc spectrum from Lu I to Os I is of the $d^{n-2}.s^2$ type. The supposed ${}^2D$ ground term of iridium must be accepted with extreme reserve, for it would necessitate a d^9 configuration in the normal atom.

The similarity of the four $(d^{m-1}-d^m)$ curves further excites a desire to explain at least their more striking features; and in fact Hund has already done this, for if the energies of the $s^2.d^{n-2}$ and $s.d^{n-1}$ configurations are measured from their mean value (Fig. 15·37), the sharp rise of the $(d^{m-1}-d^m)$ curves at $m=5$ and the precipitate drop at $m = 6$ are revealed as both due to the low energy of the d^5 configuration compared with either d^4 or d^6.

BIBLIOGRAPHY

As for the short periods, the only systematic account seems to be by Hund in *Linienspektren und periodisches System der Elemente*, 1927.

The values of energy levels are taken largely from Bacher and Goudsmit, *Atomic Energy States*, 1932. Other statements about particular elements are based on the select bibliography appearing in Appendix VI.

CHAPTER XVI

THE RARE EARTHS

1. In the periodic system*

Bohr's theory explains the intrusion of fourteen rare earths into the sixth period as due to the filling of a shell of 4f electrons. Now Bohr's theory also explains the increase in metallic properties which occurs in descending a column of the table as due to the valency electrons occupying orbits of successively higher quantum numbers; while the decrease in metallic properties which occurs in passing across the table from left to right he ascribed to the firmer binding of the electrons as the nuclear charge increases. If these predictions are general, they should be valid in the rare earth elements. Yttrium should be more metallic than scandium, and lanthanum than yttrium, but in passing through the rare earths from lanthanum to lutecium the elements should grow steadily less metallic.

The firmness with which the valency electron is bound is best measured by the ionisation potential; but for most of the rare earths the potential has only been estimated from the conductivities of the oxides in a flame,† and many have felt that the values are not so sure as those obtained by more direct methods. In the last two years this view has been confirmed by the analysis of the spectra of lanthanum and cerium; the ionisation potentials thus obtained are 5·59 and 6·54, which compare with the flame values of 5·49 and 6·91 volts.

For confirmation turn first to the molecular volume of homologous compounds, and then to two chemical reactions. The more firmly the valency electrons are bound, the smaller should be the volume of the compound, and in fact this prediction is fulfilled in the sesqui-oxides and sulphates. The sesqui-oxides of the rare earths exist in three crystalline forms, which Goldschmidt named *A*, *B* and *C*. *A* is stable at high temperatures, *C* at low, but the

* Von Hevesy, *Die seltenen Erden*, 1926, 21f.

† Rolla and Piccardi, *PM*, 1929, 7 286, and Fig. 10·13 of Volume I.

transition temperature changes from element to element, rising from lanthanum to lutecium; so that at room temperature the form *A* is stable in lanthanum and the form *C* in lutecium. The hexagonal crystal *A* has been measured in four elements between lanthanum and neodymium, the pseudo-trigonal crystal *B* in four between neodymium and gadolinium, and the regular crystal *C* in scandium, yttrium and in all the elements which succeed samarium (Fig. 16·1). These measurements show that in the sixth

		Sc	Y	La	Ce	Pr	Nd	Il	Sm	Eu
Sesquioxide	*A*	—	—	50·28	47·89	46·65	46·55	—	—	—
	B	—	—	—	—	—	*c.* 51 ?	—	46·9	46·5
	C	35·53	45·13	*c.* 57	—	—	—	—	48·38	48·28
Sulphate octahydrate		—	240·8	—	—	253·9	252·4	—	247·9	247·3
Element		—	—	22·6	21·0	21·8	20·7	—	—	—

		Gd	Tb	Dy	Ho	Er	Tu	Yb	Lu
Sesquioxide	*A*	—	—	—	—	—	—	—	—
	B	*c.* 43	—	—	—	—	—	—	—
	C	47·58	46·38	45·49	44·89	44·38	44·11	42·5	42·25
Sulphate octahydrate		246·4	—	242·8	241·1	239·3	—	235·1	234·7
Element		—	—	—	—	—	—	19·8	—

Fig. 16·1. Molecular volumes of some rare earth compounds.

period the molecular volume of all three crystal forms decreases as the atomic number increases, while in travelling from scandium through yttrium to lanthanum there is a considerable expansion. These two variations are in the directions predicted by Bohr; together they bring yttrium out with very nearly the same atomic volume as holmium, an interesting coincidence, as a comparison of chemical properties would assign yttrium the same place in the rare earth sequence; the four elements which succeed holmium are thus actually less metallic than yttrium, though still much more metallic than scandium.

When the rare earth sulphates crystallise as the octahydrates, all except cerium are isomorphous. Their densities, which have been measured by Auer von Welsbach, show that once again the molecular volume decreases in passing from praseodymium to lutecium, and once again yttrium appears next to holmium.

There are also chemical methods by which the decreasing basicity may be demonstrated. If one measures the iodine liberated in the reaction

$$E_2(SO_4)_3 + 5KI + KIO_3 + 3H_2O = 2E(OH)_3 + 3K_2SO_4 + 3I_2,$$

increase in the iodine is a sign of decreasing basicity. The order in which this reaction places the elements is identical with that obtained from the molecular volumes, while the difference between samarium and europium is abnormally small here as there.

Another method is to warm a solution of the sulphate with an exactly equivalent weight of sodium carbonate, and to measure the rate at which carbonic acid is liberated. The order obtained is again the same as that of the molecular volumes: Pr, Nd, Sm, Eu, Gd, Tb, Dy, Y, Tu, Yb.

2. Valency

If the periodic system was based solely on chemical grounds, there would be no choice but to crowd all the fifteen elements between lanthanum and lutecium into column III; elsewhere in the periodic system the valency changes by one when the atomic number changes by one, but all fifteen rare earths are trivalent

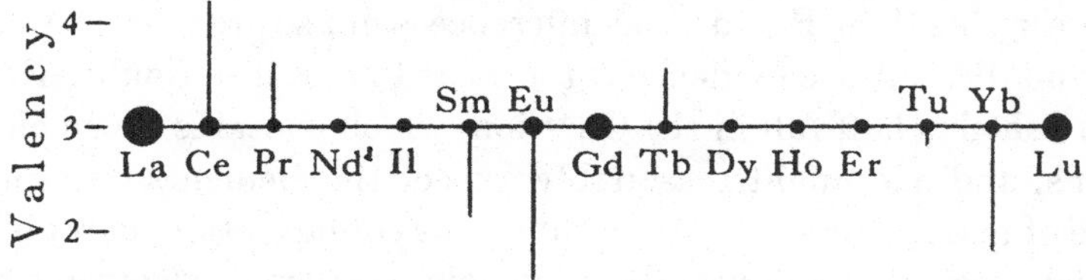

Fig. 16·2. Valencies of the rare earths; all are trivalent, and the size of the point shows the relative stability of the ion, Me^{3+}; the appearance of quadrivalent compounds is shown by a line running up, and of divalent compounds by a line running down; the lengths of these lines give a rough measure of stability. (After Jantsch and Klemm, *Z. f. anorg. u. allg. Chem.* 1933, **216** 80.)

and so similar in other ways that chemists have been able to separate them from one another only by such laborious methods as fractional crystallisation of their salts and fractional decomposition of their nitrates. Six of the rare earths however form compounds, in which they exhibit a second valency. Cerium, praseodymium and terbium can all be quadrivalent; samarium, europium and ytterbium divalent (Fig. 16·2).

Five of these deviations can be linked up with the theories of atomic structure developed in earlier chapters. The more electrons a shell contains the less willing is it to part with one of them; sodium is more reactive than magnesium, and aluminium than silicon; if then the 4f shell obeys the same laws, it should be easier to remove an electron from cerium in which only one is present than from succeeding elements in which there are several. And in fact the first two elements are quadrivalent, while cerium becomes quadrivalent much more readily than praseodymium, for compounds of the latter readily oxidise cerous compounds to ceric; further, only one quadrivalent compound of praseodymium has been isolated pure, whereas a whole series of ceric compounds are known.

The divalency of europium and the quadrivalency of terbium are due to quite another cause. Both the p and d shells show that they are more stable when they are just half full than when they contain one electron more or less; nitrogen with three p electrons has a higher ionisation potential than either carbon or oxygen, while the examination of the low terms of the frame elements, carried out in the last chapter, shows that the d^5 configuration is more stable than d^4 or d^6; the difference is not so great as between d^9 and d^{10}, but the evidence is too clear to admit of doubt. Now the f shell is half full in the Gd^{3+} ion, which contains seven electrons, and one might reasonably expect the elements on either side of gadolinium to try to assume this configuration, europium which precedes gadolinium by keeping an extra electron and being divalent instead of trivalent, and terbium which succeeds gadolinium by parting with an extra electron and becoming quadrivalent.

The divalency of ytterbium exhibits the tendency to form the closed shell of fourteen f electrons; but the divalency of samarium remains unexplained, for there is no reason to think the f^6 configuration much more stable than the f^5; in the d shell d^4 is more stable, but not much more stable, than d^3. But we must not make too much of this failure, for $SmCl_2$ is definitely less stable than $EuCl_2$.

3. Arc and spark spectra

The paramagnetic susceptibilities of the rare earths have already been cited as strong evidence that in the trivalent ions, which occur in crystals and in solution, all electrons outside the xenon core occupy f orbits. But to obtain evidence of the structure of the elements themselves, appeal must be had to the arc and spark spectra. These exhibit an exceptional number of lines; in the spectrum of dysprosium, for example, over 3000 have been measured. This naturally makes the identification of small amounts of the rare earths very difficult, for if one finds the weak yttrium line 3468·0 A., one cannot distinguish it from four other lines, 3467·4 A. of Gd, 3467·8 A. of Cd, 3468·2 A. of Tb or 3468·4 A. of Th, unless one can measure the wave-length to a few tenths of an angstrom. To surmount this difficulty spectroscopists have been driven to use the 'residual lines', that is, the lines which are the last to fade when the proportion of the element in the mixture is steadily reduced. For example, if the line measured as 3468·0 A. is really an yttrium line, then the strongest yttrium line 3710·3 A. must appear much stronger than 3468·0 A. on the same plate; moreover, as 3710·3 A. is a residual line, it must be the last to vanish as the material examined is diluted.

As long ago as 1922 Bohr* stated that the atom in which an f electron first appears is cerium, and that the f shell is full in ytterbium; this statement he based on chemical properties and on a mathematical comparison of the stabilities of alternative orbits. Since then many spectra of elements lying just before and just after the rare earths have been analysed; while within the rare earth frame the low terms of Ce I, Sm I, Eu I and Gd I are now known.†

The fifty-fifth electron occupies the 6s shell in Cs I and Ba II, and sinks into a 5d orbit in La III; on Bohr's authority it was commonly expected to sink into a 4f orbit in Ce IV, but when the

* Bohr, *Theory of Spectra and Atomic Constitution*, 1922, 110.

† Recent analyses of rare earth spectra are: Ce I, Karlson, *ZP*, 1933, **85** 482; Ce III, Kalia, *Indian Journ. Phys.*, 1933, **8** 137; Eu II, Albertson, *PR*, 1934, **45** 499*a*; Eu I, Russell and King, *PR*, 1934, **46** 1023; Sm I and Gd I, Albertson, *PR*, 1935, **47** 370.

Ce IV spectrum was analysed the ground term was found to be ^2D.* Thereafter it seemed rather improbable that any of the three electrons required to produce neutral cerium would enter a 4f orbit; but apparently Bohr was correct after all, for the ground term of Ce I is ^3H and this arises in the $4f.5d.6s^2$ configuration.

No. of electrons	Atom	Ground term	Configuration
1	Cs I Ba II La III Ce IV	^2S ^2S ^2D ^2F	s s d f
2	Ba I La II Ce III	^1S ^3F ^3F	s^2 d^2 d^2
3	La I	^2D	$d.s^2$
4	Ce I	^3H	$f.d.s^2$
8 9 10	Sm I Eu II Eu I Gd I	$^7F_0$ $^9S_4$ $^8S_{3\frac{1}{2}}$ $^9D_2$	$f^6.s^2$ $f^7(^8S).s$ $f^7.s^2$ $f^7.d.s^2$
15	Tu I Yb II Lu III	 ^2S	 $f^{14}.s$
16	Yb I Lu II	 ^1S	 $f^{14}.s^2$
17	Lu I Hf II	^2D ^2D	$f^{14}.d.s^2$ $f^{14}.d.s^2$

Fig. 16·3. Ground terms of the arc and spark spectra with the configurations in which they arise.

At the other end of the rare earth frame the ground terms of Lu III and Lu II are ^2S and ^1S respectively, showing that the 4f shell is complete when the nuclear charge is 71, for these terms must arise from the configurations $f^{14}.s$ and $f^{14}.s^2$; but as the f shell is bound to grow more stable, when the nuclear charge increases, like the d shell of preceding periods, these results do not prove that the f shell is complete in ytterbium. Thus the spectra which have been analysed since 1922 are not inconsistent with Bohr's hypothesis, but they have not yet banished doubt.

* Gibbs and White, *PR*, 1929, **33** 157. Lang has since found a lower ^2F term. *PR*, 1936, **49** 552*a*.

4. Absorption spectra

As there are no specific chemical tests for any of the rare earths except cerium, the chemist relies on optical tests instead. Of these at least six are available; the arc and spark spectra, the absorption and phosphorescent spectra of a crystal or solution, and X-ray lines and absorption edges. Any of these can be used to identify an element, but of them all the absorption spectrum is often the simplest, for many of the trivalent ions are brightly

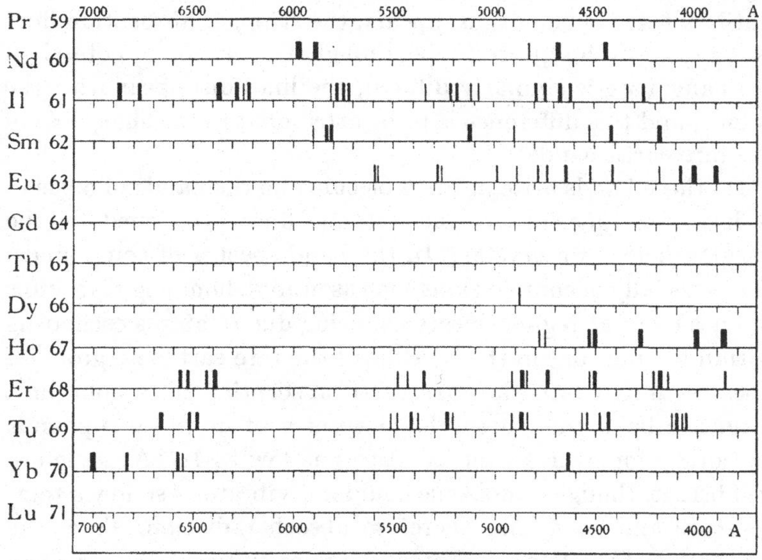

Fig. 16·4. Absorption spectra of the rare earth ions. (After Hevesy, *Die seltenen Erden*, 1926, 39.)

coloured. Praseodymium is green, neodymium red-violet, samarium yellow and element 61 probably yellowish green; dysprosium and holmium are both yellow, erbium is rose and thulium green; of the rest, four, cerium, gadolinium, ytterbium and lutecium, are quite colourless, while europium and terbium show little trace (Fig. 16·4). Thus the elements near the ends of the block, and those round gadolinium, show little or no colour; this can be explained as depending on the depth of the ground term. Thus the ^8S ground term of the gadolinium ion, calculated by

Hund and confirmed by the paramagnetic susceptibility, should be peculiarly stable, since it arises in an f^7 configuration; and in fact the absorption bands of gadolinium lie in the ultra-violet.

When the absorption spectrum is more carefully examined, it is found to consist of surprisingly narrow bands; many indeed are only 1 or 2 A. wide, whereas the absorption bands of the coloured ions of earlier periods often cover 100 A. or more; indeed the rare earth bands are better described as 'lines more or less diffuse' than as bands. Thus if one dilutes the solution of a rare earth salt until all trace of colour disappears, the stronger absorption lines still appear in the spectroscope, though if a solution of potassium permanganate is similarly diluted, the lines disappear with the colour; and this difference is to be attributed to the sharpness of the rare earth bands.

Do these bands arise in the molecule like the bands of a gas in a discharge tube, or are they atomic lines broadened by the varying fields of a crystal? In the band spectra of compounds, and especially of complex ions such as uranyl, homologous groups of lines recur at regular intervals, being due to atoms oscillating within the ion; but in the crystals of the rare earths no group of lines recurs. Again, the rare earth absorption lines split in a magnetic field, though the band spectra of uranyl compounds do not.* Moreover, at temperatures as low as 1·7° A. some are still bright, though a molecule could not vibrate at so low a temperature and could not therefore absorb radiation. True, the intensities of the lines change as the temperature falls, for while some remain bright, others fade; but the lines which fade are easily explained as arising from levels above the ground state, levels in which the Boltzmann distribution allows very few atoms at low temperatures.†

The positions of the bands are largely independent of the anion and are the same in the solid as in solution. This alone is strong evidence that the lines are atomic in origin, and arise in a shell

* Becquerel, *le Radium*, 1907, **4** 328; *K. Akad. Amsterdam, Proc.* 1929, **32** 749.

† Becquerel, *Livre jubiliare de Kamerlingh Onnes*, 1922, 288; Ehrenfest, *ibid.* 362.

which is well screened from the forces of other ions, as the 4f shell may be supposed screened by the 5s and 5p shells.

If the rare earth lines arise within the atom, it should be possible to identify the levels by the methods which have been so successful in arc spectra. The magnetic susceptibilities of the rare earths are strong evidence that the ground terms of the trivalent ions all arise in an f^n configuration and are those predicted by Hund's energy rules. Do the absorption spectra confirm this hypothesis?

As long ago as 1907 Becquerel showed that the absorption bands, narrow at room temperature, become narrower still as the temperature is reduced, until at the temperature of liquid hydrogen some bands are almost as sharp as the lines of gaseous spectra.* Becquerel and his co-workers however used minerals, which contained several rare earths, and perhaps also other ions in solid solution; these irregularities in the crystal lattice would produce strains which might well blur the absorption lines. Further advance waited until 1929, when Freed and Spedding† resorted to synthetic crystals; while the following year Saha‡ suggested that the principal lines of a crystal may arise in forbidden transitions of the ion. Of this suggestion Deutschbein and Tomaschek have made full use, though thus far work has been largely confined to gadolinium and samarium.§

5. Gadolinium

The gadolinium ion exhibits the simplest spectrum of all the rare earths, and produces sharp lines even at room temperature; both the multiplets and their components are well separated from one another. The basic level ^8S is known to be very little disturbed in electric fields,|| and as experimental evidence shows that all lines arise from it, the lines constitute in effect an energy level diagram. The only misfortune is that so many lines lie in the ultra-violet.

* Becquerel, *le Radium*, 1907, **4** 328; *K. Akad. Amsterdam, Proc.* 1929, **32** 749.

† Freed and Spedding, *PR*, 1929, **34** 946. ‡ Saha, *N*, 1930, **125** 163.

§ See chapter XXII on Fluorescent crystals.

|| Spedding, *PR*, 1931, **38** 2080*a*.

Freed and Spedding* worked first with the chloride, $GdCl_3.6H_2O$, which crystallises in colourless plates. The spectra were taken along the principal axis, that is, perpendicular to the faces, but tests made perpendicular to the principal axis showed that within the limits of measurement the spectrum is independent of the direction in which the light passes. Moreover, in solutions of varying concentration, photographs show that the positions and general spacing of the multiplets are similar to that found in the crystal, except that the lines are rather more blurred and the multiplets are shifted slightly towards higher frequencies.

When the absorption spectrum of the chloride is compared with that of other gadolinium salts, the negative ion is seen to produce little change in the positions of the multiplets, though it changes the number of lines and their positions within the multiplet. Accordingly, the multiplets must arise from electronic transitions of the Gd^{3+} ion, but the splitting of these levels must depend on the forces exerted by the surrounding atoms. These forces seem to depend more on the crystal symmetry than on the negative radical; the spectra of monoclinic $Gd_2(SO_4)_3.8H_2O$, $GdCl_3.6H_2O$† and $GdBr_3.6H_2O$‡ are almost identical, but different from hexagonal $Gd(C_2H_5SO_4)_3.9H_2O$ and $Gd(BrO_3)_3.9H_2O$; while work on the triclinic acetate § suggests still a third type, but only one band of the acetate has been examined, so it is perhaps unwise to generalise.

As the temperature is reduced little change appears in the spectrum; true the multiplet intervals increase slightly, and the whole spectrum shifts slightly towards the red, but this movement averages only some 4 $cm.^{-1}$ in a change from the laboratory to liquid hydrogen. These two changes are primarily due to the contraction of the crystal, which brings the ions closer together, and therefore makes the electric field more intense.

* Freed and Spedding, *PR*, 1929, **34** 945. $GdCl_3.6H_2O$; $Gd_2(SO_4)_3.8H_2O$.

† Spedding and Nutting, *PR*, 1931, **38** 2294*a*.

‡ Spedding and Nutting, *Am. Chem. Soc. J.* 1930, **52** 3747. $GdBr_3.6H_2O$.

§ Spedding and Nutting, *Am. Chem. Soc. J.* 1933, **55** 503. $Gd(C_2H_5SO_4)_3.9H_2O$ and $Gd(BrO_3)_3.9H_2O$.

6. Samarium

At room temperature the absorption spectrum of samarium consists of diffuse lines and bands, lying chiefly between 3000 and 5000 A. As the temperature is reduced the lines sharpen, until when the crystal is in liquid hydrogen the lines are very fine; the lines also change in intensity, and this change is much more striking in samarium than in gadolinium; at low temperatures some lines disappear, while new lines make their appearance*; and this is to be expected, since Hund's theory makes the ground term $f^5\,{}^6H$, and at low temperature lines arising in the higher components of this term must be very weak.

The Boltzmann distribution law indeed makes it certain that any absorption lines which appear below room temperature must arise in a level lying less than 500 cm.$^{-1}$ above the ground level. The visible spectra of crystals of chloride and bromate thus arise between half a dozen levels lying below 500 cm.$^{-1}$ and other levels lying between 17,000 and 27,000 cm.$^{-1}$ In order to follow the changes in intensity more closely, the absorption spectrum of the chloride, $SmCl_3.6H_2O$, was photographed at five temperatures between 293° and 20° A., first with a single crystal and then with powdered crystals†; the latter method brings out the weak lines, though it blurs the strong, as multiple internal reflection lengthens the path. Comparison of these photographs suggests the division of the lines into two groups; one consists of lines which increase in intensity as the temperature is lowered, many only appearing when the temperature has already fallen to − 195° C., while in the other group the lines decrease in intensity as the temperature falls, many being absent at 20° A. These two groups are conveniently referred to as 'low temperature lines' and 'high temperature lines'. Many of the fainter low temperature lines appear on the violet side of a multiplet, while the high temperature lines seem to congregate on the red side.

The intensities of both groups of lines depend in part on the populations of the lower levels, and this in turn is governed by

* Freed and Spedding, *N*, 1929, **123** 526.

† Spedding and Bear, *PR*, 1932, **42** 58, 76. $SmCl_3.6H_2O$; single crystal and powdered crystal.

Boltzmann's law; if three low levels exist, separated by intervals of 150 cm.$^{-1}$, and all are of equal weight, the numbers of atoms in these levels at 20° and 78° A. are those given in Fig. 16·5. Thus in

Level / Temp.	0 cm.$^{-1}$	150 cm.$^{-1}$	300 cm.$^{-1}$
20° A.	1	$8{\cdot}3.10^{-6}$	$6{\cdot}9.10^{-11}$
78° A.	1	$2{\cdot}4.10^{-2}$	$5{\cdot}7.10^{-4}$

Fig. 16·5. Normal fraction of atoms existing at any time in three low levels.

liquid hydrogen the number of ions lying in levels above the ground level is small, and there seems little doubt that the low temperature lines may be attributed to the ground level, while the high temperature lines arise in a group of levels lying between 100 and 300 cm.$^{-1}$

These predictions are very satisfactorily confirmed by a search for constant intervals between lines of the low and high temperature groups (Fig. 16·6). In the crystals of $SmCl_3.6H_2O$

Low temperature line		High temperature line	$\Delta\nu$ cm.$^{-1}$
λ A.	ν cm.$^{-1}$	ν cm.$^{-1}$	
5592·6	17875·8	17730·6	145·2
5582·5	17908·0	17763·1	144·9
4988·2	20042·5	19897·0	145·5
4899·6	20404·5	20259·0	145·5
4513·2	22151·8	22006·0	145·8
5592·6	17875·8	17716·3	159·5
5582·5	17908·0	17748·6	159·4
4988·2	20042·5	19882·7	159·8
4899·6	20404·5	20246·0	158·5
4866·0	20544·8	20386·0	158·8

Fig. 16·6. Doublet intervals of 145 and 159 cm.$^{-1}$ found in the absorption spectra of $SmCl_3.6H_2O$ at low temperatures.

these differences reveal levels at 145, 160, 204, 217 and 300 cm.$^{-1}$ above the ground level. Some of these levels are probably complex, the 300 cm.$^{-1}$ level in particular consisting perhaps of components at 295 and 315 cm.$^{-1}$, for the spread varies with the

PLATE VI

1. Single crystal absorption spectrum of $SmCl_3 . 6H_2O$ at four different temperatures. The photographs were taken at the temperatures shown on the left, these being the boiling points of the substances shown on the right. All the lines grow sharper as the temperature is reduced, but the intensity may increase or decrease.

2. Conglomerate absorption spectrum of $Sm(BrO_3)_3 . 9H_2O$ at four different temperatures. A conglomerate or mass of small crystals has a longer optical path than a single crystal, so that it brings out the weak lines, but it blurs the stronger multiplets.

(Photographs lent by Prof. F. H. Spedding.)

Plate VI

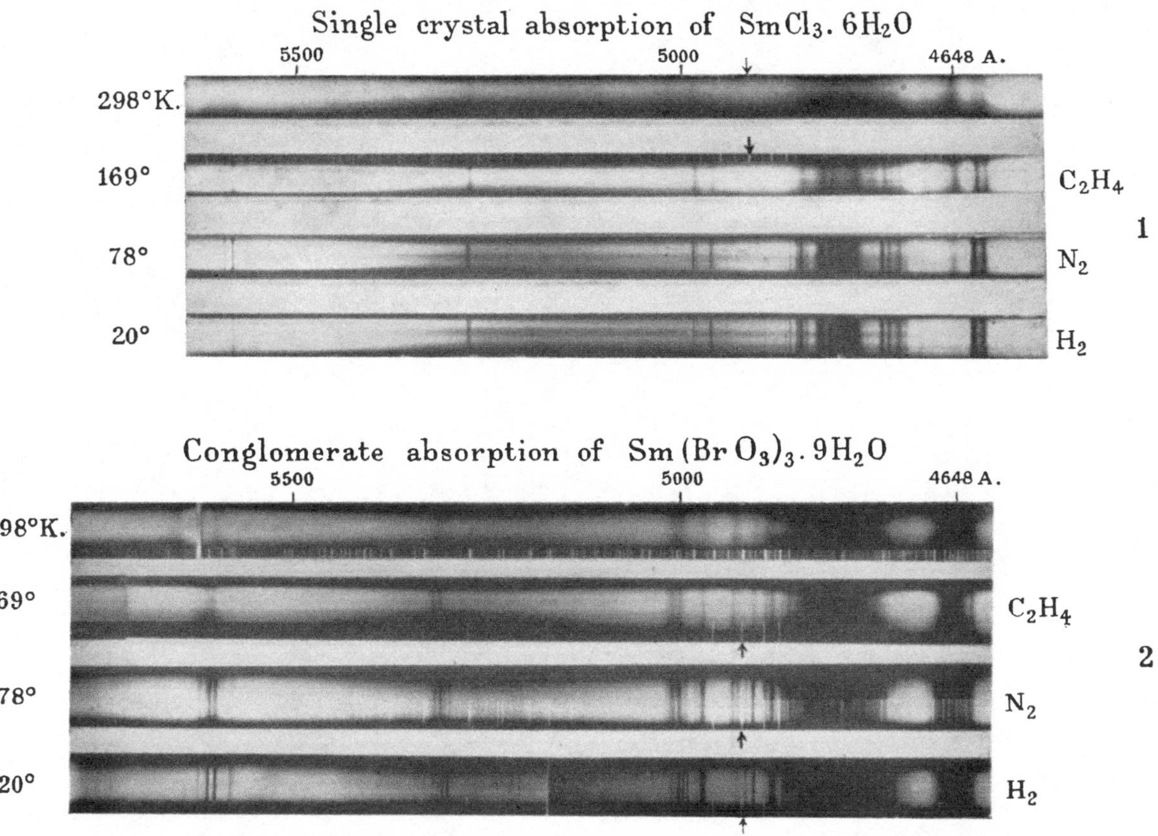

Single crystal absorption of $SmCl_3 . 6H_2O$
5500
5000
4648 A.
298°K.
169°
78°
20°
C_2H_4
N_2
H_2
1
Conglomerate absorption of $Sm(BrO_3)_3 . 9H_2O$
5500
5000
4648 A.
298°K.
169°
78°
20°
C_2H_4
N_2
H_2
2

height of the level from 2 to 30 cm.$^{-1}$, and these involve errors rather greater than might be reasonably expected.

Turning again to the photographs, the lines which originate in the 300 cm.$^{-1}$ level are found entirely absent at $-195°$ C., while those arising in the 204 and 217 cm.$^{-1}$ levels fade rapidly as the temperature is further reduced.

Examined by the same methods hexagonal crystals of the bromate, $Sm(BrO_3)_3 . 9H_2O$, behave very much like the crystals of the chloride until the temperature falls to 78° A.; the shift in position and widening of the multiplets may perhaps be slightly

Temperature	Low temperature line	Satellite A	$\Delta\nu_A$	Satellite B	$\Delta\nu_B$
78° A.	17847	17809	38	17783	64
	18857	18817	40	18790	67
	19987	19949	38	19920	67
	20382	20342	40	20315	67
	28125	28087	38	28055	70

Temperature	Low temperature line	A I	$\Delta\nu_{A_I}$	A II	$\Delta\nu_{A_{II}}$	B I	$\Delta\nu_{B_I}$	B II	$\Delta\nu_{B_{II}}$
20° A.	17847·4 (II)	—	—	17802·5	44·9	—	—	17780	67
	17857·2 (II)	—	—	18812	45	—	—	—	—
	17949 (I)	17912	37	—	—	17867	82	—	—
	19985·6 (I, II)	19949·7	35·9	19941	45	19904	82	19919	67
	20385·9 (I)	20348·9	37·0	—	—	—	—	—	—
	28125·2 (II)	—	—	28080	45	—	—	—	—

Fig. 16·7. Low temperature lines and their high temperature satellites in the absorption spectrum of $Sm(BrO_3)_3.9H_2O$; this shows how the energy levels split as the temperature is reduced from 78° to 20° A.

greater, but only slightly; between 78° and 20° A. however most of the lines split into two components, of which the red one is almost certainly complex (Fig. 16·7). Thus above 78° A. the temperature variation of the lines and a search for constant intervals reveals levels at 39 and 68 cm.$^{-1}$ above the ground level, with perhaps other levels between 100 and 230 cm.$^{-1}$; but at a temperature of 20° A. the levels revealed are at 0, 37, 45, 67 and 82 cm.$^{-1}$ (Fig. 16·8).

Moreover, the transitions observed at 20° A. suggest that these levels form two independent sets; 0, 37, 82 combine with one group of high levels, and 0, 45, 67 with another (Fig. 16·8). That

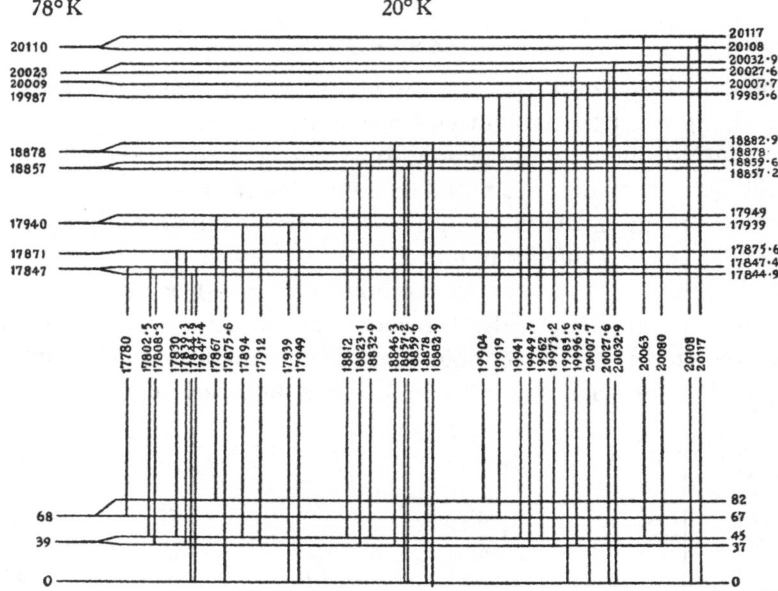

Fig. 16·8. Energy levels of $Sm(BrO_3)_3 . 9H_2O$ at 20° A., with the lines arising in transitions between them. The figures are wave numbers. (After Spedding and Bear, *PR*, 1933, **44** 290.)

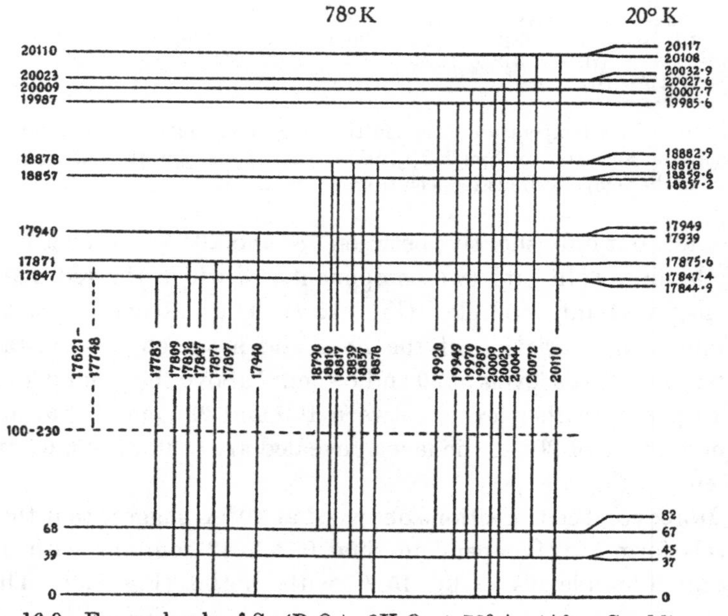

Fig. 16·9. Energy levels of $Sm(BrO_3)_3 . 9H_2O$ at 78° A. (After Spedding and Bear. *PR*, 1933, **44** 290.)

the ground term appears in both groups means nothing, for in fact the method, by which the splitting of the low levels is calculated, automatically reduces any real interval to zero; if all the low levels developed the same interval, one would be free to ascribe this splitting entirely to the high levels.*

The existence of more than three low levels in the samarium salts makes it probable that in Sm^{3+} there is a second electronic level lying close to the $^6H_{2\frac{1}{2}}$ term predicted by Hund, for in an electric field this term splits into only three components; and this fits in well with the magnetic susceptibility which does not agree with the value predicted, if $^6H_{2\frac{1}{2}}$ is the only low term.†

BIBLIOGRAPHY

The last three sections of this chapter should be read with chapter XXII. Spencer, J. F., *The metals of the rare earths*, 1919, is a very thorough study with full references; but it was written before Bohr had outlined the electronic structure of the elements. Hevesy, *Die seltenen Erden*, 1926, makes full use of Bohr's theory.

* Spedding and Bear, *PR*, 1933, **44** 287. $Sm(BrO_3)_3.9H_2O$.

† Spedding, *Am. Chem. Soc. J.* 1932, **54** 2593.

CHAPTER XVII

INTENSITY RELATIONS

1. Experimental

For rough estimates of intensity the spectroscopist has often relied on his eye; but the eye is subjective and far from accurate, so that in recent years much attention has been paid to methods of estimating the density of a photographic plate. In general if the intensity of one component of a multiplet is expressed as a percentage of the brightest line, then these methods ensure that the percentage is correct to the nearest integer; but this statement is subject to a few restrictions, of which the most important is that the lines must not be too far apart, for no one knows quite how the sensitivity of a photographic plate varies with wave-length.

This is not the place to indulge in a description of experimental procedure, especially as it is fully described elsewhere;* but the wedge method may be briefly reviewed as illustrating the chief points of interest. As the density of a photographic image is not proportional to the length of exposure or the intensity of the incident light, but shows an initial lag (Fig. 17·1), one is not justified in comparing two densities and then saying that the intensities must have been in the same ratio. Instead, one may only say that if the density at one point is equal to that at another, then the intensities were also equal. The word 'density' is here a technical term, being defined as

$$\log\left\{\frac{\text{intensity of incident light}}{\text{intensity of transmitted light}}\right\}.$$

This means that a scale ought to appear on each plate, and one way of obtaining this is to photograph a wedge of dull grey glass; the density will then be proportional to the distance from the edge of the wedge, and this may be measured by a micrometer fixed to the microscope. Thus one may say that the density of a

* Dobson, Griffith and Harrison, *Photographic photometry*, 1926.

certain line is equal to that at a certain place on the wedge, and the latter may be measured on a scale which is a linear scale of intensity.

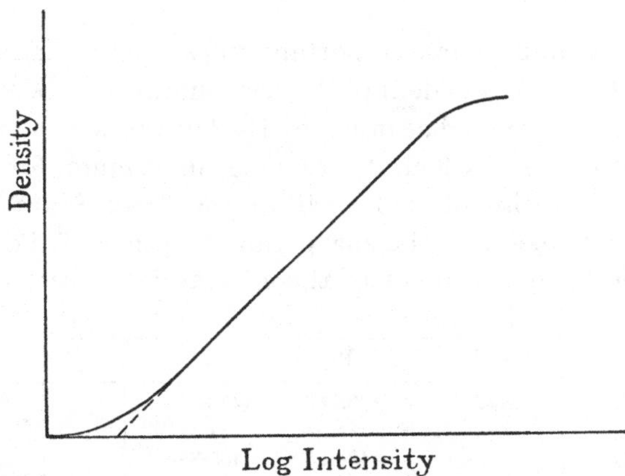

Fig. 17·1. Density-intensity graph of a photographic plate.

2. The normal multiplet

The Sommerfeld intensity rule has already been discussed, but in a form applicable only to multiplets in which $\Delta L = \pm 1$. Multiplets of the $P \rightarrow P^\circ$ type, in which $\Delta L = 0$, give their strongest lines when ΔJ is zero, while the two groups arising from $\Delta J = \pm 1$ are less intense than the chief lines but equal to one another. To include this type of transition the Sommerfeld rule may be restated:*

The chief lines of any multiplet are due to those transitions in which $\Delta J = \Delta L$; a weaker group, technically known as satellites of the first order, arise when $\Delta J = \Delta L \pm 1$, while satellites of the second order occur when $\Delta J = \Delta L \pm 2$.

This rule is qualitative only; attempting to make it quantitative Sommerfeld in 1923 considered first those multiplets which arise by the combination of a single level with a multiplet level; the three SP lines in spectra of all multiplicities are of this type, and experiment† shows that the intensities due to the three transitions

* Sommerfeld and Heisenberg, *ZP*, 1922, **11** 131.

† Dorgelo, *ZP*, 1924, **22** 170.

vary as $(2J+1)$ for the three P terms; thus the intensities of the triplet lines $^3S_1 \rightarrow {}^3P_{2,1,0}$ of Mg are in the ratio of $5:3:1$, while those of the octet lines $^8S_{3\frac{1}{2}} \rightarrow {}^8P_{4\frac{1}{2},3\frac{1}{2},2\frac{1}{2}}$ of Mn are in the ratio of $10:8:6$.

From this result three important rules emerge. First, the intensity ratio is independent of the serial number n. Secondly, the intensity is determined as much by the level to which an electron is going as the level which it is leaving; in calcium, for example, the intensity ratios of $2\,^3S \rightarrow 2\,^3P$ and $3\,^3P \rightarrow 2\,^3S$ are identical, though one triplet is of the sharp and the other of the principal series. The third is a point of theory; the intensities are in the

Term	P_0	P_1	P_2	
D_1	4425·43 25	4435·67 19	4456·61 1	45
D_2	—	4434·95 54	4455·88 18	72
D_3	—	—	4454·77 100	100
	25	73	119	Intensity sum

Fig. 17·2. Intensities in a diffuse triplet of calcium; the upper number is the wave-length, the lower the intensity measured by Burger and Dorgelo.

ratios of the number of Zeeman components possessed by the level, and this fits in well with the practice of taking this number as the statistical weight of a level; a further significance will appear when the intensities of Zeeman multiplets are considered.

The next step was to examine the transitions between two multiplet levels. Working in this direction Burger and Dorgelo* first verified the validity of the above rules for the PD doublet of sodium where the D levels are not resolved, and then turned to the PD multiplet of calcium; in it they found that the sum of the intensities of the lines originating from one P level was to the sum of the intensities from another P level as the statistical weights of the levels. And the same held true for the three D levels.

Fig. 17·2 gives their measurements, the upper figure in each

* Burger and Dorgelo, *ZP*, 1924, **23** 258.

square being the wave-length and the lower the intensity adjusted to a scale in which 100 represents the brightest line. The sums of the intensities of the lines originating from the various P levels are given at the bottom, 25:73:119, and it will be observed that they are roughly in the ratio of 1:3:5. Similarly, the intensities, 45, 72, 100, arising from the three D levels are roughly in the ratio 3:5:7.

The summation rule used alone enables us to calculate the intensities resulting from the combination of two doublet levels (Fig. 17·3), but it will not suffice for more complicated multiplets.

Term	$P_{\frac{1}{2}}$	$P_{1\frac{1}{2}}$	
$D_{1\frac{1}{2}}$	5	1	6
$D_{2\frac{1}{2}}$	—	9	9
	5	10	ΣI

Term	$D_{1\frac{1}{2}}$	$D_{2\frac{1}{2}}$	
$F_{2\frac{1}{2}}$	14	1	15
$F_{3\frac{1}{2}}$	—	20	20
	14	21	ΣI

Fig. 17·3. Theoretical intensity ratios of diffuse and fundamental doublets.

Thus, consider a PD triplet in which the intensities are those shown in Fig. 17·4. Burger and Dorgelo's summation rule shows that

$$\left.\begin{aligned} \frac{a_1}{1} = \frac{b_1+b_2}{3} = \frac{c_1+c_2+c_3}{5} \\ \frac{a_1+b_1+c_1}{3} = \frac{b_2+c_2}{5} = \frac{c_3}{7} \end{aligned}\right\}.$$

But this gives only four equations to determine five ratios. To resolve the problem Russell,* among others, called in the correspondence principle. His argument cannot be given here, but his result may be reviewed.

	$^3P_0$	$^3P_1$	$^3P_2$	
$^3D_1$	a_1	b_1	c_1	
$^3D_2$	—	b_2	c_2	
$^3D_3$	—	—	c_3	

Fig. 17·4. Assumed intensities of a diffuse triplet.

	$^3P_0$	$^3P_1$	$^3P_2$	
$^3D_1$	20	15	1	36
$^3D_2$	—	45	15	60
$^3D_3$	—	—	84	84
	20	60	100	

Fig. 17·5. Theoretical intensities of a diffuse triplet.

* Russell, *Nat. Acad. Sci. Proc.* 1925, **11** 314, 322; Sommerfeld and Hönl, *Preuss. Akad. Wiss. Berlin*, 1925, **9** 141; Hönl, *AP*, 1926, **79** 274; Kronig, *ZP*, 1925, **31** 885, **33** 261; Dirac, *PRS*, 1926, **111** 281.

As the intensities of a multiplet are determined by the levels between which the electron jumps and not by the direction of the jump, it is not necessary to consider $\Delta L = \pm 1$ but only one of these; arbitrarily, then, we elect to consider only $L \to (L-1)$ and $L \to L$. For each of these transitions there are three values of ΔJ, so that six formulae may be expected, and these take their simplest form if J is defined as the larger of the two quantum numbers concerned.

In the transition $L \to (L-1)$ the correspondence principle gives

for $J \to (J-1)$ $$I_- = \frac{S}{4L} \cdot \frac{1}{J} P(J) . P(J-1),$$

for $J \to J$ $$I_0 = \frac{S}{4L} \cdot \frac{(2J+1)}{J(J+1)} . P(J) . Q(J),$$

for $(J-1) \to J$ $$I_+ = \frac{S}{4L} \cdot \frac{1}{J} Q(J) . Q(J-1).$$

In the transition $L \to L$ the summation rule shows that the two groups of satellites arising from $\Delta J = \pm 1$ must be identical, so that only two formulae are needed:

for $J \to J$ $$I_0 = \frac{S(2L+1)}{4L(L+1)} \cdot \frac{(2J+1)}{J(J+1)} . R^2(J),$$

for $J \to (J-1)$ or $(J-1) \to J$ $$I_\pm = \frac{S(2L+1)}{4L(L+1)} \cdot \frac{1}{J} P(J) . Q(J-1).$$

In these equations $P(J)$, $Q(J)$ and $R(J)$ are convenient abbreviations defined as

$$P(J) = (J+L)(J+L+1) - S(S+1),$$
$$Q(J) = S(S+1) - (J-L)(J-L+1),$$
$$R(J) = J(J+1) + L(L+1) - S(S+1).$$

In theory these formulae should compare the intensities of any two lines arising in a transition from one configuration to another, but they are valid only when the coupling is Russell-Saunders and there are no inter-system lines. As very few spectra satisfy this condition the formulae are normally applied only to the intensities of components of a single multiplet, and for this purpose the first factor, which is a function of S and L only, may be dropped.

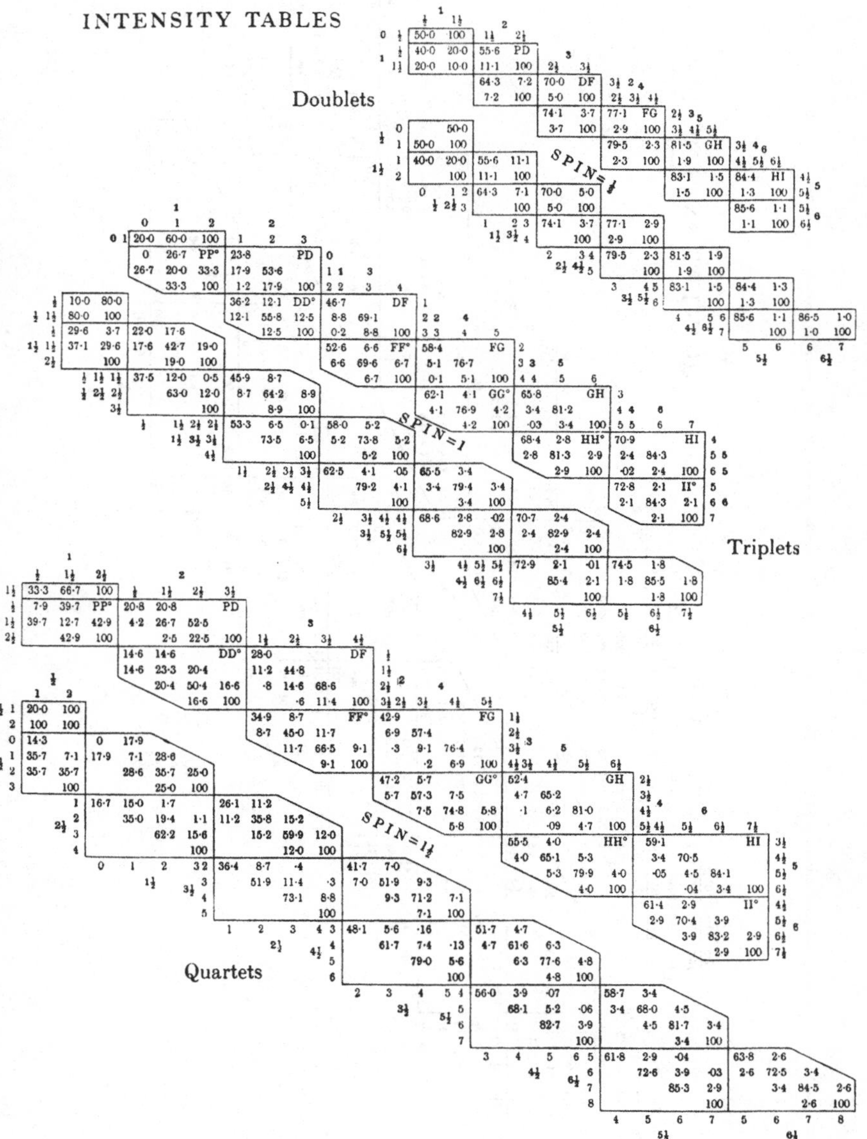

Fig. 17·6. Multiplet intensities. The intensity of each component is given as a percentage of the strongest line of the multiplet. The numbers outside the frames are L and J, the former in heavy type. The tables may be applied to (**jj**) coupling (p. 150), related multiplets (p. 104) and hyperfine structure (p. 183). (After White and Eliason, *PR*, 1933, **44** 753.)

QUINTET INTENSITIES

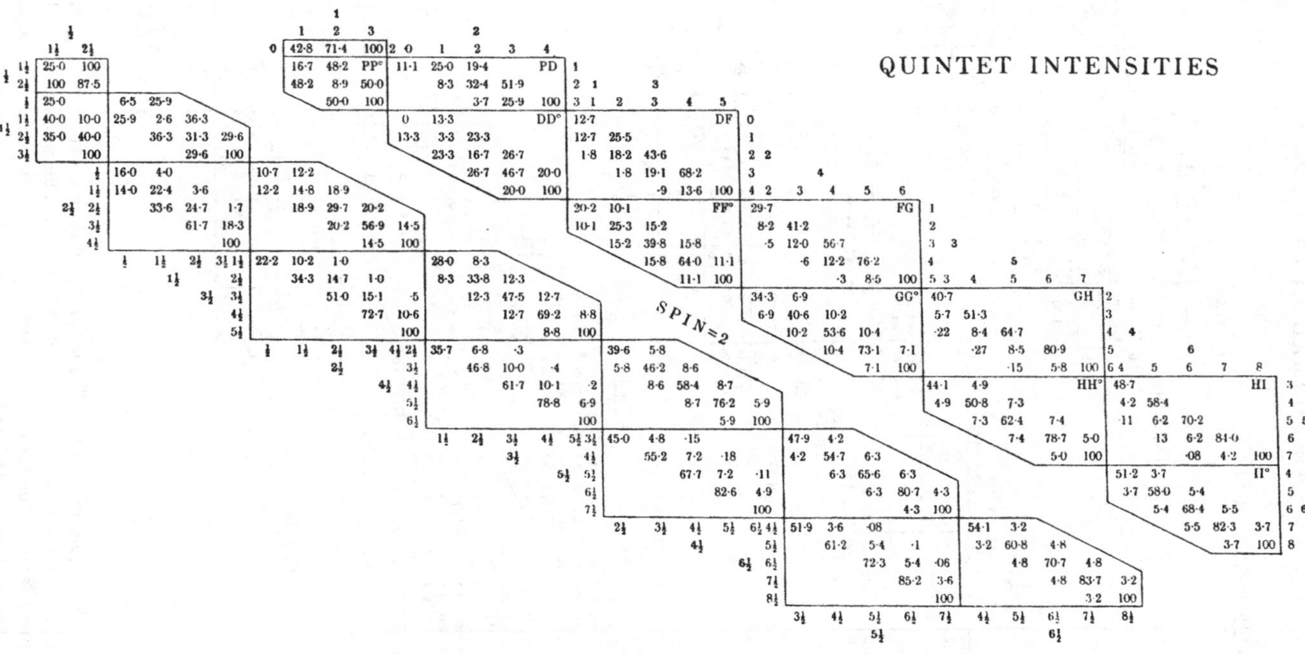

SEXTET INTENSITIES

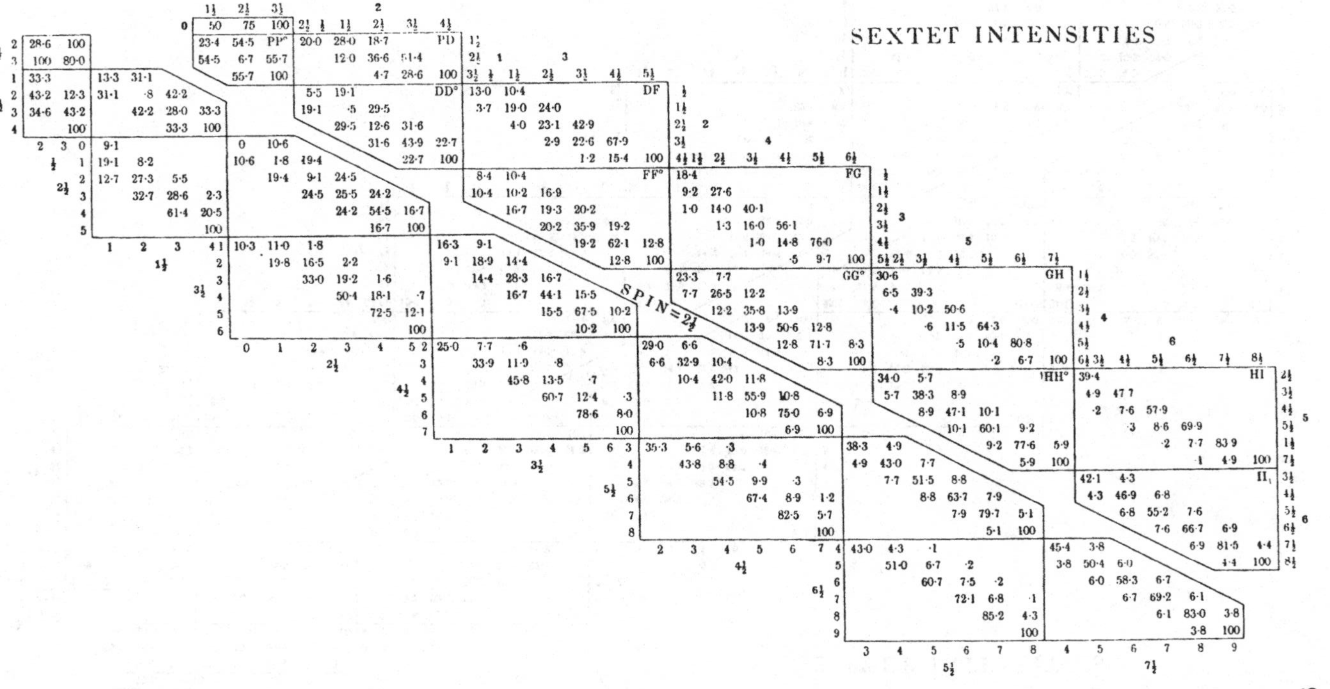

SEPTET INTENSITIES

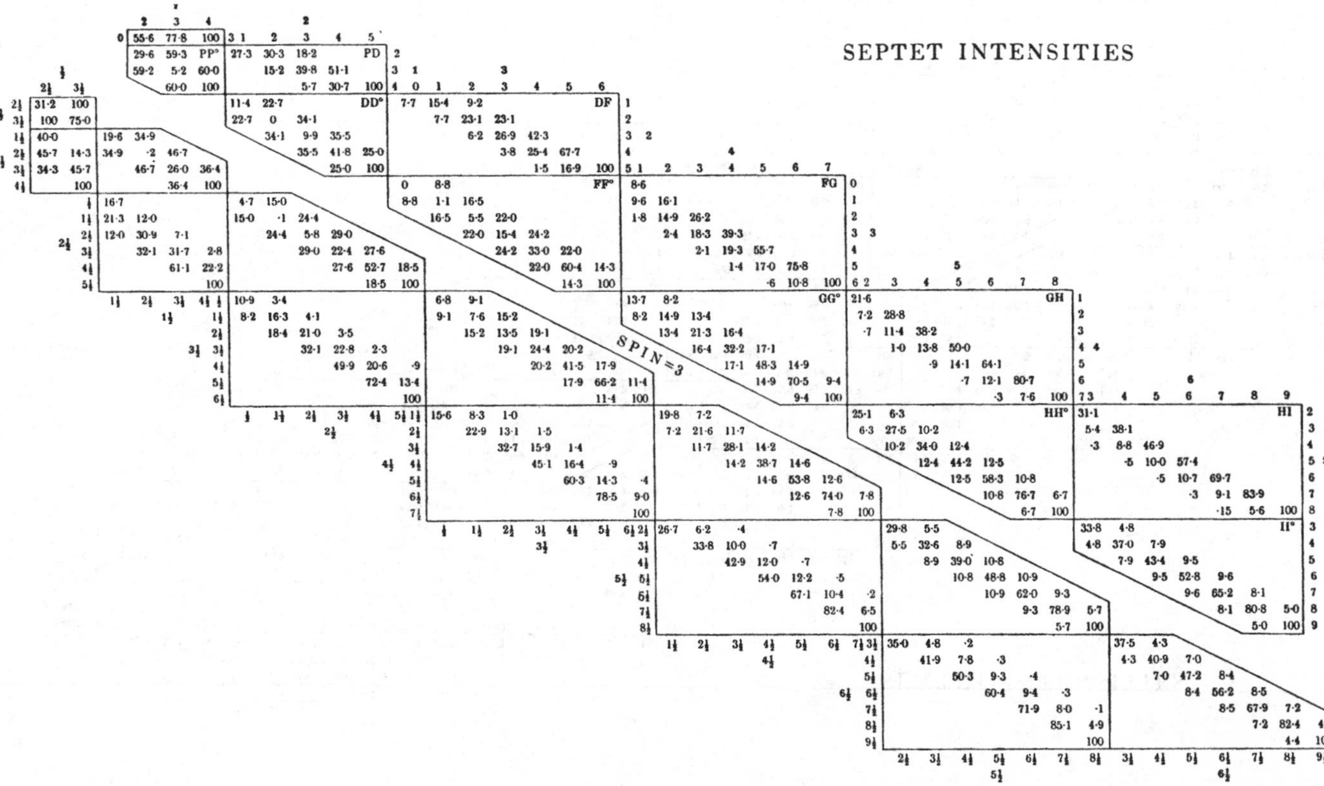

	2	3	4
0	55·6	77·8	100
2	29·6	59·3	PP°
3	59·2	5·2	60·0
4		60·0	100

3	1	2	3	4	5
2	27·3	30·3	18·2		PD
3		15·2	39·8	51·1	
4			5·7	30·7	100

	1	2	3	4	5
1	11·4	22·7			DD°
2	22·7	0	34·1		
3		34·1	9·9	35·5	
4			35·5	41·8	25·0
5				25·0	100

	0	1	2	3	4	5	6
1	7·7	15·4	9·2				DF
2		7·7	23·1	23·1			
3			6·2	26·9	42·3		
4				3·8	25·4	67·7	
5					1·5	16·9	100

	0	1	2	3	4	5	6
0	0	8·8					FF°
1	8·8	1·1	16·5				
2		16·5	5·5	22·0			
3			22·0	15·4	24·2		
4				24·2	33·0	22·0	
5					22·0	60·4	14·3
6						14·3	100

	1	2	3	4	5	6	7
0	8·6						FG
1	9·6	16·1					
2	1·8	14·9	26·2				
3		2·4	18·3	39·3			
4			2·1	19·3	55·7		
5				1·4	17·0	75·8	
6					·6	10·8	100

	1	2	3	4	5	6	7
1	13·7	8·2					GG°
2	8·2	14·9	13·4				
3		13·4	21·3	16·4			
4			16·4	32·2	17·1		
5				17·1	48·3	14·9	
6					14·9	70·5	9·4
7						9·4	100

	2	3	4	5	6	7	8
1	21·6						GH
2	7·2	28·8					
3	·7	11·4	38·2				
4		1·0	13·8	50·0			
5			·9	14·1	64·1		
6				·7	12·1	80·7	
7					·3	7·6	100

	2	3	4	5	6	7	8
2	25·1	6·3					HH°
3	6·3	27·5	10·2				
4		10·2	34·0	12·4			
5			12·4	44·2	12·5		
6				12·5	58·3	10·8	
7					10·8	76·7	6·7
8						6·7	100

	3	4	5	6	7	8	9
2	31·1						HI
3	5·4	38·1					
4	·3	8·8	46·9				
5		·5	10·0	57·4			
6			·5	10·7	69·7		
7				·3	9·1	83·9	
8					·15	5·6	100

	3	4	5	6	7	8	9
3	33·8	4·8					II°
4	4·8	37·0	7·9				
5		7·9	43·4	9·5			
6			9·5	52·8	9·6		
7				9·6	65·2	8·1	
8					8·1	80·8	5·0
9						5·0	100

	2½	3½
2½	31·2	100
3½	100	75·0
1½	40·0	
2½	45·7	14·3
3½	34·3	45·7
4½		100

	1½	2½	3½	4½
1½	19·6	34·9		
2½	34·9	·2	46·7	
3½		46·7	26·0	36·4
4½			36·4	100
½	16·7			
1½	21·3	12·0		
2½	12·0	30·9	7·1	
3½		32·1	31·7	2·8
4½			61·1	22·2
5½				100

	½	1½	2½	3½	4½	5½
½	4·7	15·0				
1½	15·0	·1	24·4			
2½		24·4	5·8	29·0		
3½			29·0	22·4	27·6	
4½				27·6	52·7	18·5
5½					18·5	100
½	10·9	3·4				
1½	8·2	16·3	4·1			
2½		18·4	21·0	3·5		
3½			32·1	22·8	2·3	
4½				49·9	20·6	·9
5½					72·4	13·4
6½						100

	½	1½	2½	3½	4½	5½	6½
½	6·8	9·1					
1½	9·1	7·6	15·2				
2½		15·2	13·5	19·1			
3½			19·1	24·4	20·2		
4½				20·2	41·5	17·9	
5½					17·9	66·2	11·4
6½						11·4	100
1½	15·6	8·3	1·0				
2½		22·9	13·1	1·5			
3½			32·7	15·9	1·4		
4½				45·1	16·4	·9	
5½					60·3	14·3	·4
6½						78·5	9·0
7½							100

	1½	2½	3½	4½	5½	6½	7½
1½	19·8	7·2					
2½	7·2	21·6	11·7				
3½		11·7	28·1	14·2			
4½			14·2	38·7	14·6		
5½				14·6	53·8	12·6	
6½					12·6	74·0	7·8
7½						7·8	100
2½	26·7	6·2	·4				
3½		33·8	10·0	·7			
4½			42·9	12·0	·7		
5½				54·0	12·2	·5	
6½					67·1	10·4	·2
7½						82·4	6·5
8½							100

	2½	3½	4½	5½	6½	7½	8½
2½	29·8	5·5					
3½	5·5	32·6	8·9				
4½		8·9	39·0	10·8			
5½			10·8	48·8	10·9		
6½				10·9	62·0	9·3	
7½					9·3	78·9	5·7
8½						5·7	100
3½	35·0	4·8	·2				
4½		41·9	7·8	·3			
5½			50·3	9·3	·4		
6½				60·4	9·4	·3	
7½					71·9	8·0	·1
8½						85·1	4·9
9½							100

	3½	4½	5½	6½	7½	8½	9½
3½	37·5	4·3					
4½	4·3	40·9	7·0				
5½		7·0	47·2	8·4			
6½			8·4	56·2	8·5		
7½				8·5	67·9	7·2	
8½					7·2	82·4	4·4
9½						4·4	100

OCTET INTENSITIES

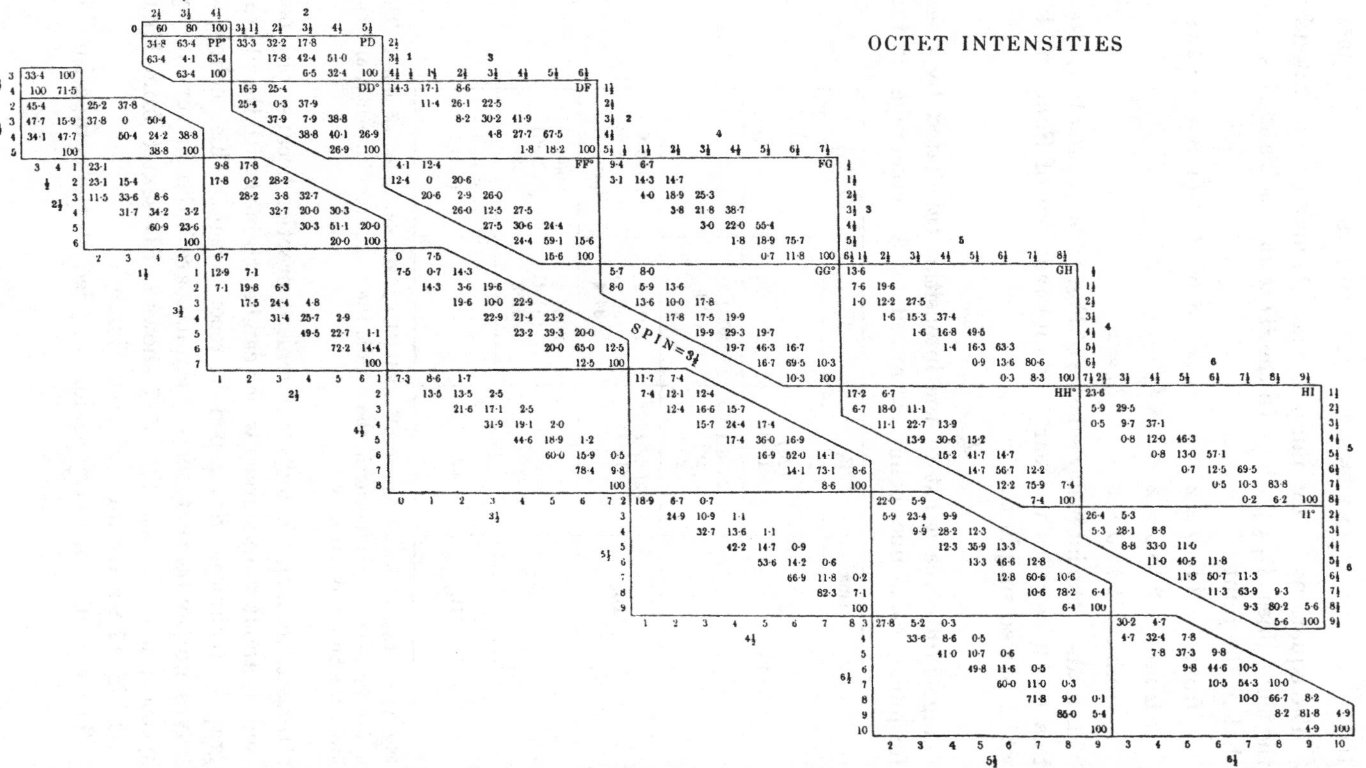

The equations for the jump $L \rightarrow (L-1)$ may be illustrated by the three lines $^3D_{1,2,3} \rightarrow {}^3P_2$. In the ^3D term L is 2 and S is 1, so that in the jump

from $J=3$ to $J=2$ $\qquad I_- = \frac{1}{3}(5.6-1.2)(4.5-2)=168,$

from $J=2$ to $J=2$ $\qquad I_0 = 30,$

from $J=1$ to $J=2$ $\qquad I_+ = 2.$

This leads to the intensity scheme of Fig. 17·5. Incidentally, too, this result satisfies the early qualitative rule of Sommerfeld, which stated simply that

$$I_- > I_0 > I_+.$$

Fig. 17·6 gives a list of the intensities calculated by these formulae, each line being expressed as a percentage of the

Terms	$^6P_{1\frac{1}{2}}$	$^6P_{2\frac{1}{2}}$	$^6P_{3\frac{1}{2}}$
$^6D_{\frac{1}{2}}$	3079·6 (20) 18·3	—	—
$^6D_{1\frac{1}{2}}$	3073·1 (28) 25·8	3081·3 (12) 13·1	—
$^6D_{2\frac{1}{2}}$	3062·1 (18·7) 18·7	3070·3 (36·6) 34·4	3082·1 (4·8) Lapped
$^6D_{3\frac{1}{2}}$	—	3054·4 (51·4) 51·3	3066·0 (28·6) 26·1
$^6D_{4\frac{1}{2}}$	—	—	3044·6 (100) 100

Fig. 17·7. Intensities of the w $^6P^\circ \rightarrow 3d^6\,(^5D)\,4s\,^6D$ multiplet of Mn I. The upper figure is the wave-length, the middle the theoretical intensity, and the lower the observed intensity.

strongest line, this being a convenient practice because one per cent. is about the accuracy which can be attained with the photographic technique developed in recent years. This technique leaves no doubt that these formulae are valid in very many spectra; as examples, Fig. 17·7 shows a PD sextet from Mn I, and Fig. 17·8 a DF quartet from Ti II.

The formulae are very seldom applied to the comparison of

different multiplets, but Wijk* has tried them out on the quartets and doublets of O II, a light atom in which the inter-system lines are very weak. This work roughly confirms the theoretical

$3d^3$ / $3d^2.4p$	$^4F_{1\frac{1}{2}}$	$^4F_{2\frac{1}{2}}$	$^4F_{3\frac{1}{2}}$	$^4F_{4\frac{1}{2}}$
$^4D_{\frac{1}{2}}$	3161·19 (70) 72	—	—	—
$^4D_{1\frac{1}{2}}$	3154·18 (28) 28	3161·76 (112) 112	—	—
$^4D_{2\frac{1}{2}}$	? (2·0) ?	3152·24 (37) 35	3162·56 (171) 168	—
$^4D_{3\frac{1}{2}}$	—	? (1·4) ?	3155·65 (29) 30	3168·52 (250) 249

Fig. 17·8. Intensities of the $3d^2 4p\ ^4D \rightarrow 3d^3\ ^4F$ multiplet of Ti II. The theoretical intensity is in brackets.

quartet-doublet intensity ratio of 2 : 1; thus the intensities of the $^4P \rightarrow {}^4S$ and $^2P \rightarrow {}^2S$ lines were found to be as 1·6 : 1, and two other empirical ratios were 2·6 and 2·1.

3. The super-multiplet

In the mercury spectrum the division into singlets and triplets has little experimental justification, and many inter-system lines are strong; moreover, the sum rule in its simple form is but poorly obeyed. Accordingly, Ornstein and Burger† suggested that when the coupling is no longer Russell-Saunders, the sum rule ought to be extended. For this there is a precedent in the laws of the Zeeman effect, which define certain sums for a single multiplet when the coupling is Russell-Saunders, but only for all terms of a configuration when the coupling is abnormal.

In order to make the experimental work as significant as possible, Ornstein and Burger examined first a configuration of

* Wijk, *ZP*, 1928, **47** 622.
† Ornstein and Burger, *ZP*, 1926, **40** 403.

two electrons, which gives rise to a triplet and a singlet term. Moreover, they were careful that one of the two combining triplet terms should be unresolved. A super-multiplet satisfying these requirements is found in some DF combinations of calcium and strontium; if the intensities of the singlet and triplet lines are

Terms	$^1D_2$	$^3D_1$	$^3D_2$	$^3D_3$
^1F	s	—	—	—
^3F	—	t_1	t_2	t_3

Fig. 17·9. Assumed intensities in a fundamental super-multiplet.

written s and t_1, t_2 and t_3 respectively, as in Fig. 17·9, the sum rule applied to the vertical columns states that

$$\frac{s}{5}=\frac{t_1}{3}=\frac{t_2}{5}=\frac{t_3}{7}.$$

Thus theory suggests that the intensity of the singlet should be equal to the mean intensity of the three triplet lines, whereas the intensities actually observed* and quoted in Fig. 17·10

	Ca		Ca		Sr	
Transition	λ	Int.	λ	Int.	λ	Int.
$^1D_2$–^1F	4878	4·3	4355	4·7	5156	4·4
$^3D_3$–^3F	4586	7·0	4099	6·9	4892	6·8
$^3D_2$–^3F	4581	5·0	4095	5·0	4869	5·0
$^3D_1$–^3F	4579	2·5	4093	2·8	4855	3·4
Mean triplet intensity		4·8		4·9		5·1

Fig. 17·10. Observed intensities in three fundamental super-multiplets.

show that it is rather weaker. But the singlet is of considerably longer wave-length than the triplet, and it has been shown* that when the intervals are large the sum rule should be applied not to the intensity itself, but to the intensity divided by ν^4; and, in fact, if the intensity 4·3 of the singlet line is multiplied by $(4878/4582)^4$ the corrected intensity is 5·5, a figure which is as much larger than the mean intensity of the triplet as the first figure was too small.

* Ornstein, Coelingh and Eymers, *ZP*, 1927, **44** 653.

With this satisfactory result Ornstein and Burger* were ready to tackle a group of lines in the mercury spectrum. The lines

Terms	$^1D_2$	$^3D_3$	$^3D_2$	$^3D_1$
$^1P_1$	s	—	i_1	i_2
$^3P_2$	i_4	t_1	t_3	t_6
$^3P_1$	i_3	—	t_2	t_5
$^3P_0$	—	—	—	t_4

Fig. 17·11. Assumed intensities in a diffuse super-multiplet.

which should appear are shown in Fig. 17·11; and the sum rule applied to this figure leads at once to the equations

$$\left.\begin{aligned} \frac{s+i_1+i_2}{3}=\frac{i_4+t_1+t_3+t_6}{5}=\frac{i_3+t_2+t_5}{3}=\frac{t_4}{1} \\ \frac{s+i_4+i_3}{5}=\frac{t_1}{7}=\frac{i_1+t_3+t_2}{5}=\frac{i_2+t_6+t_5+t_4}{3} \end{aligned}\right\}.$$

The eleven lines concerned in the relation differ so in wave-length that if all the intensities were measured close agreement with experiment could not be expected unless the ν^4 correction were applied, and at the time this work was done the ν^4 correction had not been tested out. But the intervals of the D terms are small, and as these equations require that

$$\frac{t_3+t_6+i_4}{t_1}=\frac{4}{21}=19 \text{ per cent.},$$

a check may be applied with measurements on only those four lines which arise from the combination of $^3P_2$ with the D terms.

Wave-length	Combination	Intensity
3650	t_1	100
3655	t_3	10·8 } 18·7
3663	t_6	} 7·9 }
3663	i_4	}

Fig. 17·12. Observed intensities in a diffuse super-multiplet of Hg I.

The empirical results are shown in Fig. 17·12. The line t_6 is so weak that if measured alone its intensity would certainly be less

* Ornstein and Burger, *ZP*, 1926, **40** 403.

than 1 per cent. of t_1, but it lies so close to i_4 that the two may be conveniently measured together after widening the slit of the spectroscope. The figures show that when the inter-system line is included the sum rule is satisfied, whereas if it is omitted (t_3+t_6) is only 12 per cent. of t_1 instead of the 19 per cent. predicted.

These experiments clearly show that in certain spectra the sum rule is valid only if the singlet and triplet lines are treated as parts of a single whole; if this is a general phenomenon, then the intensity of successive lines in singlet and triplet series should decrease according to the same law, so that the relative intensity may be independent of the serial number; and in fact Ornstein and Burger* have confirmed this prediction.

4. The iron-frame elements

In the analysis of the iron frame elements at the Bureau of Standards, Russell and his co-workers have relied more on intensities than on multiplet intervals or magnetic splitting factors. Indeed, if the general intensity laws were not obeyed these spectra would probably still await analysis. The strongest lines arise from transitions in which only a single electron orbit changes; and this is in agreement with the correspondence principle, which indicates that those terms between which strong combinations appear must be built up from the same state of the ion. In contrast, transitions involving a change in the ion are much less probable, and the lines resulting are either absent or fainter even than the weak inter-system lines. Moreover, when only one electron jumps, the intensities of related multiplets can be obtained from the formulae designed to give the intensities of related lines; Russell† suggested a law of this kind, while Kronig used theory to show that if the coupling is Russell-Saunders the figures are identical and Fig. 17·6 can be used. For this purpose S is replaced by ${}_iL$, L by l and J by L, where ${}_iL$ is the orbital moment of the ionic term, l the orbital moment of the jumping electron and L their resultant.

In the iron frame elements, though the relative intensities of

* Ornstein and Burger, *ZP*, 1926, **40** 403.

† Russell and Meggers, *Bur. of Standards*, *Sci. P.* 1927, **22** 332.

the multiplets are admirably regular, the relative intensities within the multiplets are often abnormal. Russell's* visual estimates showed that in Ni I components of small J give fainter lines than they should, and later measurements amply confirm him. Not only are individual intensities irregular, but also the sums taken over all the terms of a multiplet with the same J.

^{5}G

	Statistical weight	5	7	9	11	13	Sum	Quotient
	3	(30) 9·5	—	—	—	Theory Expt.	(29·7) 9·5	(9·9) 3·2
	5	(8) 11	(41) 16	—	—	—	(49·5) 27	(9·9) 5·4
^{5}F	7	(0·5) ~1	(12) 24·5	(57) 25	—	—	(69·2) 50·5	(9·9) 7·2
	9	—	(0·6) ~1	(12) 29	(76) 44	—	(89·0) 74	(9·9) 8·2
	11	Theory Expt.	—	(0·3) ~1	(8) 41	(100) 100	(108) 142	(9·9) 12·9
	Sum	(38·5) 21·5	(53·8) 41·5	(69·2) 55	(84·6) 85	(100) 100	—	—
	Quotient	(7·7) 4·3	(7·7) 5·9	(7·7) 6·1	(7·7) 7·7	(7·7) 7·7	—	—

Fig. 17·13. Comparison of theoretical and experimental intensities in a F→G quintet of Ni I; this quintet arises as d^{8}s (^{4}F) 5s ^{5}F→d^{8}s (^{4}F) 4p ^{5}G°.

Fig. 17·13 shows this; for when the sum rule is valid, the intensity sum divided by the statistical weight yields a constant, but in Ni I and Co I the quotient varies from one J to another.

In a general way spectroscopists have long realised that the Russell-Saunders coupling, which predominates on the left-hand side of the periodic table, gives way to less regular coupling as one passes to the right; so that no one was surprised when titanium

* Russell, *PR*, 1929, **34** 825.

was shown to obey the intensity laws more closely than nickel. Nickel and cobalt indeed form the ultimate members of a series, which grows progressively less regular. In Ti II 62 per cent. of the lines obeyed the formulae to within 5 per cent., but in Ti I, where Harrison* measured twenty-six strong multiplets, the proportion was down to 58 per cent.; in chromium† and manganese‡ still more violations were observed, while in cobalt and nickel hardly a single multiplet is regular.§

Various efforts have been made to trace the cause of the irregularity. In Zr I many intensities are abnormal because two terms of the same configuration and the same J have also nearly the same energy; these terms share their intensities, just as they share their magnetic splitting factors. In particular, the transition $4d^2.5s.5p\ {}^1F_3{}^\circ \rightarrow 4d^3.5s\ {}^5F$ is observed because $d^2.sp\ {}^1F_3{}^\circ$ lies near $d^2.sp\ {}^5D_3{}^\circ$, their energies being 24,387 and 23,889 cm.$^{-1}$ respectively.|| And as the multiplet separations become greater as one passes from left to right across the table, the multiplets overlap and perturb one another more.¶ But adequate as this explanation may be in its place, it is necessarily unable to show why lines involving small values of J are weak compared with those in which J is larger—unless indeed many lines of lower multiplicity and therefore in general smaller J remain unidentified. Moreover, this explanation would suggest strong correlation with departures from the interval and Zeeman rules; but in fact Frerichs,** having examined selected multiplets from some elements of the iron frame, found that the correlation with the interval rule is poor, while Harrison in his detailed study of titanium found no correlation with either rule. Often those multiplets which split irregularly in the magnetic field and have irregular intervals obey the intensity laws well, while those which have regular values of g obey the intensity laws badly. And an attempt to

* Harrison, *JOSA*, 1928, **17** 389.

† Allen and Hesthal, *PR*, 1935, **47** 926.

‡ Seward, *PR*, 1931, **37** 344.

§ Ornstein and Buoma, *PR*, 1930, **36** 679.

|| Kiess, C. C. and Kiess, H. K., *BSJ*, 1931, **6** 621.

¶ Harrison and Johnson, *PR*, 1931, **38** 773.

** Frerichs, *AP*, 1926, **81** 842.

attribute the abnormal weakness of certain lines to the abstraction of energy by inter-system lines was no more successful.

5. Alkali doublets

In general the intensity ratio in a series of multiplets is independent of the serial number, but this is not true of the principal doublets of the alkalis nor of the similar doublets of Tl I.

The controversy* about the alkali doublets has lingered on for many years because of the great experimental difficulties, the most serious being self-absorption, which can be avoided only by working at low temperatures and low current densities; but the general features are now clear (Fig. 17·14). The diffuse and funda-

Element	Doublet		Intensity ratio	
	Transition	Wave-length, A.	Calculated, Fermi	Experimental
Na	$2\,^2P \rightarrow 1\,^2S$	5890–96	2·0	Ro.[1] 1·98
K	$2\,^2P \rightarrow 1\,^2S$	7665–99	2·0	Ra. 1·91
	$3\,^2P \rightarrow 1\,^2S$	4044–47	2·16	Ra. 2·10
Rb	$2\,^2P \rightarrow 1\,^2S$	7800–947	?	Ra. 1·85
	$3\,^2P \rightarrow 1\,^2S$	4202–16	2·60	H. 2·55, K.H. 2·71, Ro.[2] 2·58
	$4\,^2P \rightarrow 1\,^2S$	3587–92	2·97	H. 3·25, K.H. 3·32, Ro.[2] 2·90
Cs	$3\,^2P \rightarrow 1\,^2S$	4555–93	4·3	F.W. 3·3, Ra. 3·85
	$4\,^2P \rightarrow 1\,^2S$	3877–89	7·15	K.H. 8·50, Ro.[2] 7·40, F.W. 4·6, H. 8·0

Fig. 17·14. Intensity ratios of some principal doublets of the alkalis.

Observers

F.W. Füchtbauer and Wolff, *AP*. 1929, **3** 359. Extrapolated to allow for self-absorption.
H. Hübner, *AP*, 1933, **17** 781. Photographic comparison of lines emitted by burner at 2800°.
K.H. Kohn and Hübner, *PZ*, 1933, **34** 278. Emission spectrum.
Ra. Rasetti, *N. Cim*. 1924, **1** 115. Anomalous dispersion.
Ro.[1] Roschdostwenski, *AP*, 1912, **39** 307. Anomalous dispersion.
Ro.[2] Roschdostwenski, *T. Opt. I., Petrograd*, 1921, **13** 1. Anomalous dispersion.

mental doublets give the normal ratios of 9:5:1 and 3:2 in caesium,† and presumably in all other spectra; in the principal series, on the other hand, the normal ratio of 2 : 1 is found only in sodium and potassium‡; the deviation increases rapidly with atomic number

* Joos, *Hb. d. Expt. Phys.* 1929, **22** 313.
† Filippov, *ZP*, 1927, **42** 495.
‡ Füchtbauer and Wolff, *AP*, 1929, **3** 359.

and is quite unmistakable in caesium. In the latter Rasetti* and Roschdostwenski,† both of whom used the accurate method of anomalous dispersion, found ratios of 3·85 and 7·40 in the second and third principal doublets respectively; while Sambursky‡ states that after rising to a maximum value of 25 : 1 in the fifth doublet, the intensity ratio decreases to 5 : 1 in the eighth. In the first doublet the ratio deviates very little from the normal value of 2 even in caesium.

This much was known when Fermi§ applied the quantum mechanics to the problem, and showed that if certain terms, ordinarily neglected, are taken into account, deviations very similar to those observed should arise. Thus theory shows that the deviation will increase rapidly with atomic number, but should not affect the first doublet of the principal series; while the introduction of numerical values leads to intensity ratios of 4·3 and 7·15 for the second and third doublets of caesium; for the first doublets theory gives a ratio somewhat less than 2, but the difference is too small to measure. The agreement here obtained with experiment is probably as close as can be expected.

In Tl I, which like the alkalis has a single electron outside closed shells, similar deviations occur. In the $m\,{}^2\mathrm{P}_{1\frac{1}{2},\frac{1}{2}} \rightarrow 2\,{}^2\mathrm{S}_{\frac{1}{2}}$ series, experiment shows that the intensity ratios when $m = 4, 5, 6, 7$ are 4·4, 6·6, 6·0 and 5·2.||

6. The Zeeman multiplet

As with the normal so with the Zeeman multiplet, certain simple rules have been established, but these suffice to determine the intensity ratios only in the simpler transition; in the more complex, reliance must be placed on formulae deduced with the aid of the quantum mechanics.

Three rules are usually cited,¶ but of these the first states only the well-known fact that the Zeeman multiplet is symmetrical about the undisplaced line. The second adds that the intensity

* Rasetti, *N. Cim.* 1924, **1** 115.

† Roschdostwenski, *T. Opt. I., Petrograd*, 1921, **13** 1.

‡ Sambursky, *ZP*, 1928, **49** 731. § Fermi, *ZP*, 1930, **59** 680.

|| Williams and Herlihy, *PR*, 1932, **39** 802.

¶ Ornstein and Burger, *ZP*, 1924, **28** 135.

sum of all lines originating in one Zeeman level is equal to the intensity sum of all lines originating in any other Zeeman level; and this holds true if the word 'ending' is substituted for 'originating'. This law appears at first analogous to the Burger and Dorgelo sum rule, the statistical weight of each Zeeman level being unity; but further examination shows that the relation is closer than analogy, for a term splits to $(2J+1)$ Zeeman components, so that the normal multiplet rule is a necessary consequence of the Zeeman rule.

M		Total intensities	
$^3S_1$	$^3P_2$	σ	π
1	2	a_3	—
1	1	—	b_1
0	1	a_2	—
1	0	a_1	—
0	0	—	b_0
-1	0	a_1	—
0	-1	a_2	—
-1	-1	—	b_1
-1	-2	a_3	—

Fig. 17·15. Assumed intensities of the Zeeman components of a $^3S_1 \rightarrow {}^3P_2$ line.

The third rule concerns the polarisation, and states that if the various components emitted in any direction are combined the resulting beam must be unpolarised. Thus in the normal Zeeman triplet observed transverse to the magnetic field, the sum of the intensities of the two σ components must be equal to the intensity of the π component. These rules appear simple enough, but the simplicity is in part only apparent, for the intensities mentioned in the sum and polarisation rules are not the same; the sum rule applies to the total radiated intensity, while the polarisation rule applies to the intensity observed in a particular direction. Thus when a pattern is observed transverse to the field, those oscillations which produce σ components vibrate in a circle whose plane is perpendicular to the magnetic axis; one component of this vibration is along the line of sight and so invisible; thus only half the radiated intensity of a σ component is observed. On the other hand the oscillators which are producing π components vibrate along the magnetic axis, so that the whole of the radiated intensity reaches an eye looking transverse to the field.

As an example of the way in which these three rules are applied consider the $^3S_1 \rightarrow {}^3P_2$ transition.* Fig. 17·15 gives on the left the possible values of the magnetic quantum number M, and on the right the total intensity radiated in each transition, the π and σ components being separated for convenience. The consequences of the symmetry rule are embodied in the notation; the sum rule, applied to the components of the $^3P_2$ term, states that

$$b_0 + 2a_1 = b_1 + a_2 + a_3,$$

while applied to the $^3S_1$ term it shows that

$$b_0 + 2a_2 = b_1 + a_1 + a_3.$$

In order to apply the polarisation rule, elect to observe the pattern transverse to the magnetic field; then the argument

Element	Line	Intensity	σ			π			σ			$\Sigma\pi$	$\Sigma\sigma$
		Theoretical	5	15	30	30	40	30	30	15	5	100	100
Mg	5183	Observed	5	16	30	32	39	32	30	14	3	103	98
Ca	6162	Observed	4	13	29	30	41	28	28	14	6	99	94
Zn	4810	Observed	5	15	29	32	41	31	29	15	3·5	104	97
Cd	5085	Observed	5	15	29	32	41	31	27	14	5	104	95

Fig. 17·16. Observed intensities of the Zeeman components of a $^3S_1 \rightarrow {}^3P_2$ line in various spectra.

given above shows that the total radiated intensity of the π components must be equal to half the total radiated intensity of the σ components; that is,

$$\begin{aligned} b_0 + 2b_1 &= \tfrac{1}{2}\{2a_1 + 2a_2 + 2a_3\} \\ &= a_1 + a_2 + a_3. \end{aligned}$$

The three simple rules thus determine the four unknown ratios, being satisfied by the values

$$a_1 = 1, \quad a_2 = 3, \quad a_3 = 6,$$
$$b_0 = 4, \quad b_1 = 3.$$

These predictions for the $^3S_1 \rightarrow {}^3P_2$ line have been amply confirmed by Van Geel,† as Fig. 17·16 shows; moreover, similar

* Ornstein and Burger, *ZP*, 1924, **29** 241.

† Van Geel, Diss. Utrecht, 1928, 60.

predictions for the two other lines of the triplet are equally satisfactory (Fig. 17·17).

Line	Transition	Intensity	σ		π		σ		$\Sigma\pi$	$\Sigma\sigma$
6122	$^3S_1 \rightarrow {}^3P_1$	Theoretical	15	15	30	30	15	15	60	60
		Observed	17	16	30	28	15	15	58	63
6102	$^3S_1 \rightarrow {}^3P_0$	Theoretical	10		20		10		20	20
		Observed	9		20		10		20	19

Fig. 17·17. Intensities of the Zeeman components of two calcium lines.

Though these three rules suffice when the J values of the two terms concerned are small, in more complex transitions resort must be had to the correspondence principle* or to the quantum mechanics.† Calculations based on these principles show that for the transitions $J \rightarrow (J-1)$, the intensities are given by

$$\begin{array}{lll} \sigma \text{ Jump} & M \rightarrow (M-1) & I_- = \tfrac{1}{2} p(M)\, p(M-1), \\ \pi \text{ Jump} & M \rightarrow M & I_0 = p(M)\, q(M), \\ \sigma \text{ Jump} & (M-1) \rightarrow M & I_+ = \tfrac{1}{2} q(M)\, q(M-1); \end{array}$$

while for the transitions $J \rightarrow J$

$$\begin{array}{lll} \pi \text{ Jump} & M \rightarrow M & I_0 = r^2(M), \\ \sigma \text{ Jump} & M \rightarrow (M-1) & I\pm = p(M)\, q(M-1). \\ & \text{or } (M-1) \rightarrow M & \end{array}$$

In these equations the transitions considered make J and M the larger of the two quantum numbers concerned; while p, q and r are abbreviations, defined by

$$p(M) = J + M,$$
$$q(M) = J - M,$$
$$r(M) = M.$$

The intensities given by these formulae are the total radiated intensities. The formulae for the transition from $(J-1)$ to J are not quoted, because the intensities are independent of the direction in which the electrons jump, so that one may consider always $J \rightarrow (J-1)$.

Intensities calculated from these formulae are given in Fig. 17·18; as the formulae do not contain L or S the Zeeman inten-

* Hönl, *ZP*, 1925, **31** 340; *AP*, 1926, **79** 288; Kronig, *ZP*, 1925, **31** 885.

† Heisenberg and Jordan, *ZP*, 1926, **37** 263.

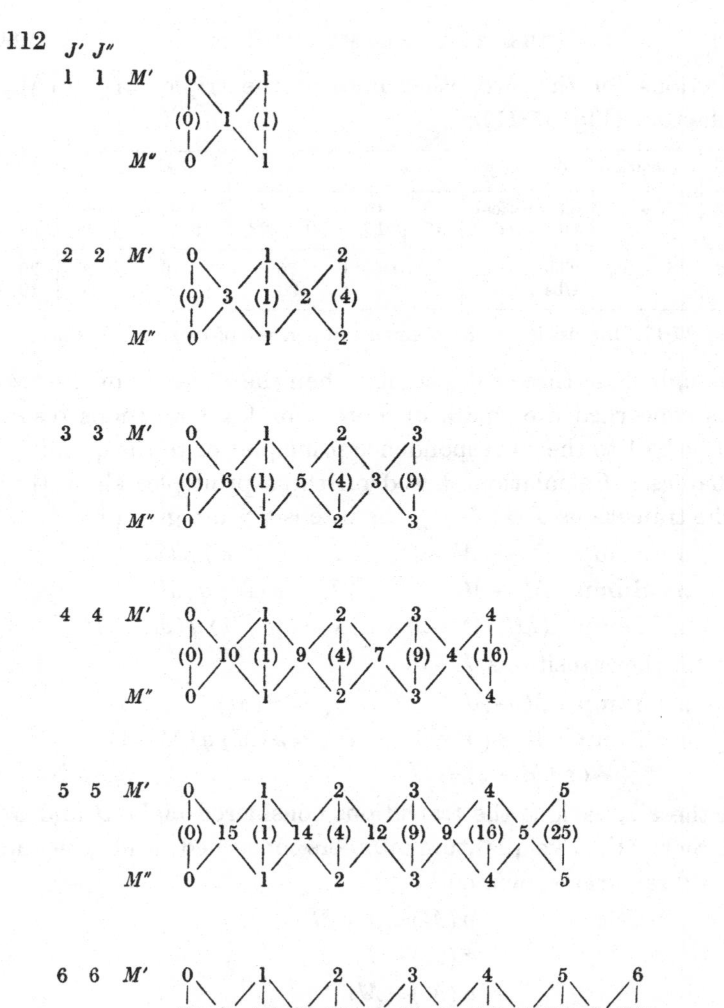

Fig. 17·18 *a*. Integral; $\Delta J = 0$.

Table of Zeeman intensities. The table is divided into four sections according as J is integral or half integral, and according as ΔJ is 0 or ± 1.

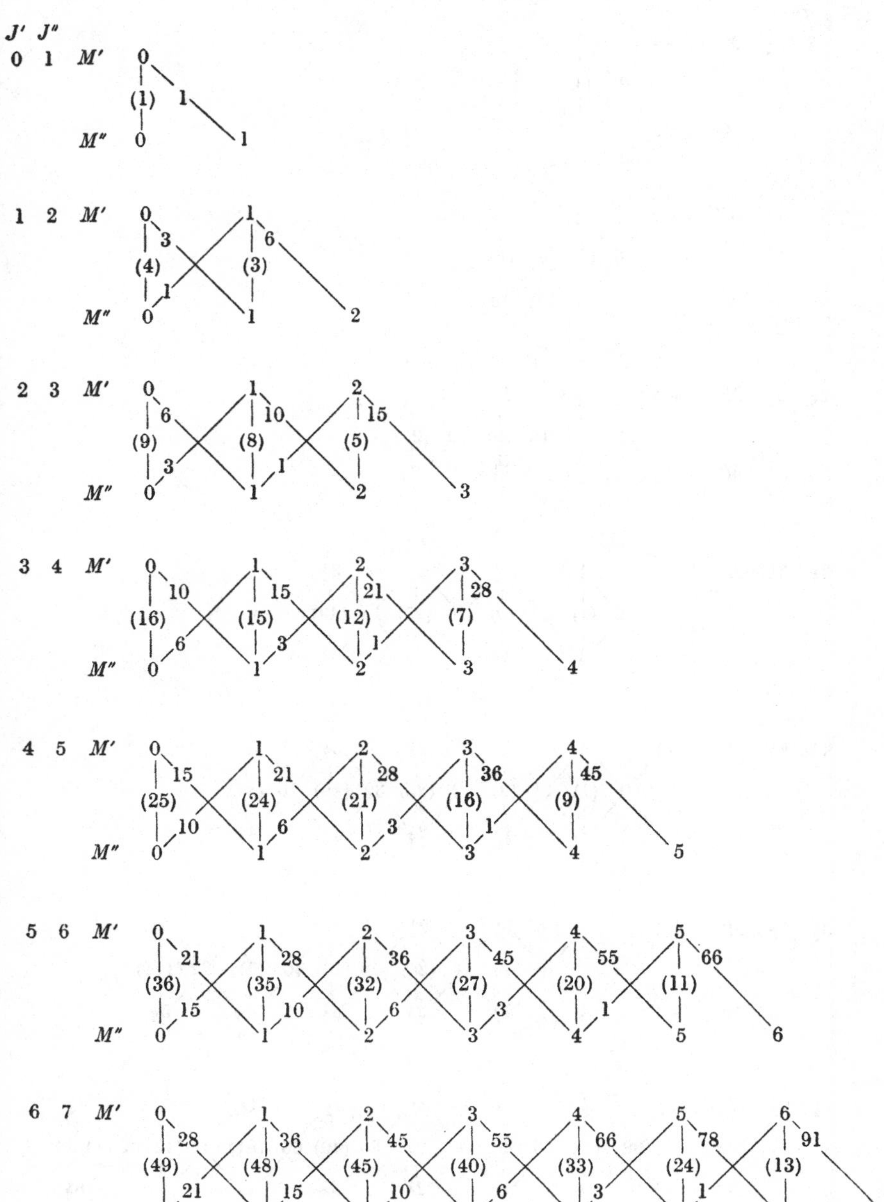

Fig. 17·18 *b*. J integral; $\Delta J = \pm 1$.

J' J''

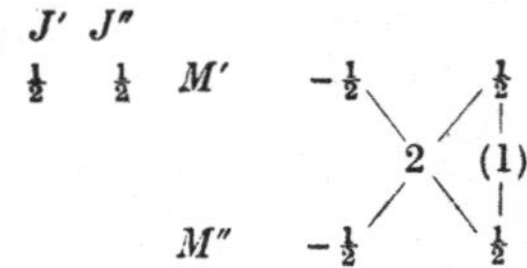

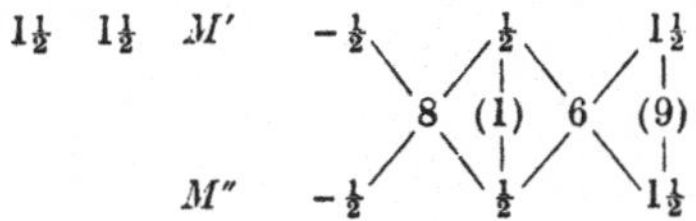

2½ 2½ M' −½ ½ 1½ 2½
18 (1) 16 (9) 10 (25)
M'' −½ ½ 1½ 2½

3½ 3½ M' −½ ½ 1½ 2½ 3½
32 (1) 30 (9) 24 (25) 14 (49)
M'' −½ ½ 1½ 2½ 3½

4½ 4½ M' −½ ½ 1½ 2½ 3½ 4½
50 (1) 48 (9) 42 (25) 32 (49) 18 (81)
M'' −½ ½ 1½ 2½ 3½ 4½

5½ 5½ M' −½ ½ 1½ 2½ 3½ 4½ 5½
72 (1) 70 (9) 64 (25) 54 (49) 40 (81) 22 (121)
M'' −½ ½ 1½ 2½ 3½ 4½ 5½

6½ 6½ M' −½ ½ 1½ 2½ 3½ 4½ 5½ 6½
98 (1) 96 (9) 90 (25) 80 (49) 66 (81) 48 (121) 26 (169)
M'' −½ ½ 1½ 2½ 3½ 4½ 5½ 6½

Fig. 17·18 *c*. J half-integral; $\Delta J = 0$.

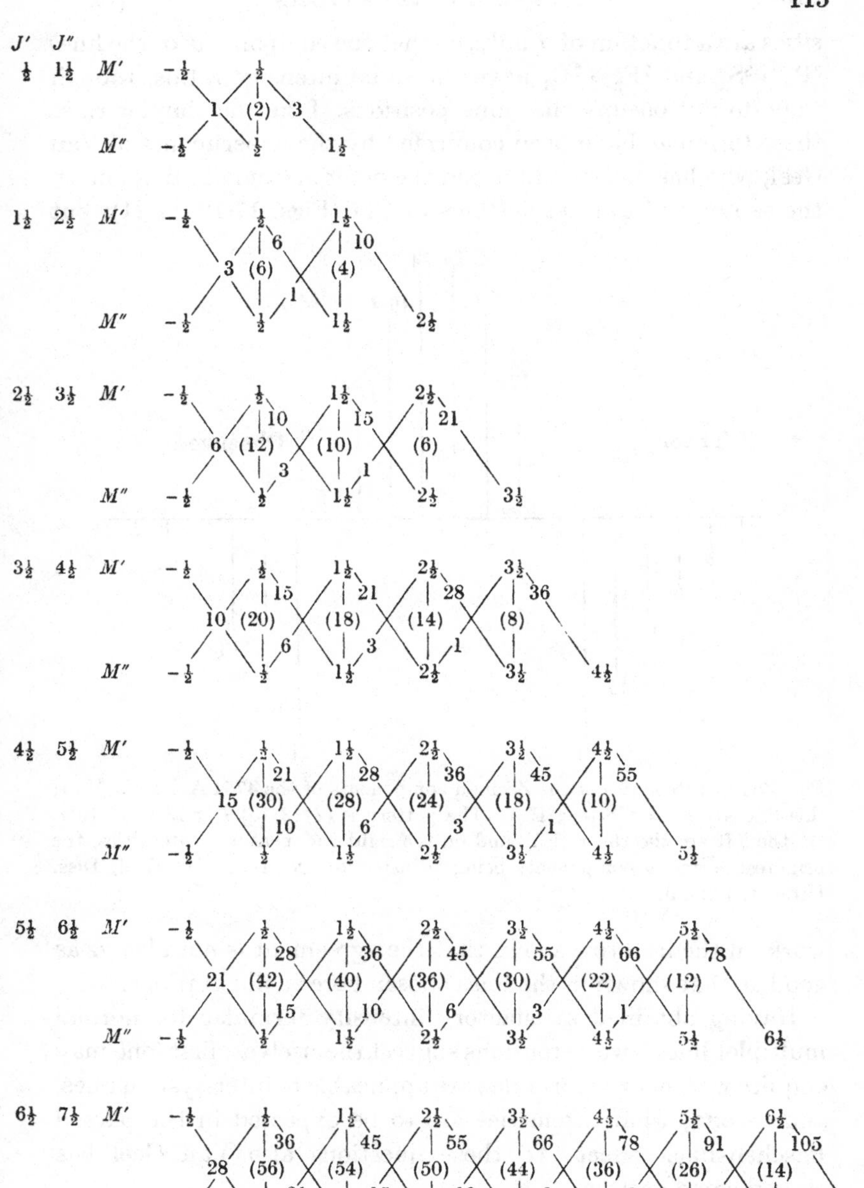

Fig. 17·18 *d*. *J* half-integral; $\Delta J = \pm 1$.

sities are a function of J only, so that the components of the lines $^3P_2 \to {^3S_1}$ and $^7F_2 \to {^7D_1}$ have the same intensity ratios, though they do not occupy the same positions. Like the simpler rules these formulae have been confirmed by the experiments of Van Geel, who has measured lines in the octet system of Mn I, and in the septet and quintet systems of Cr I (Figs. 17·19–17·21); but

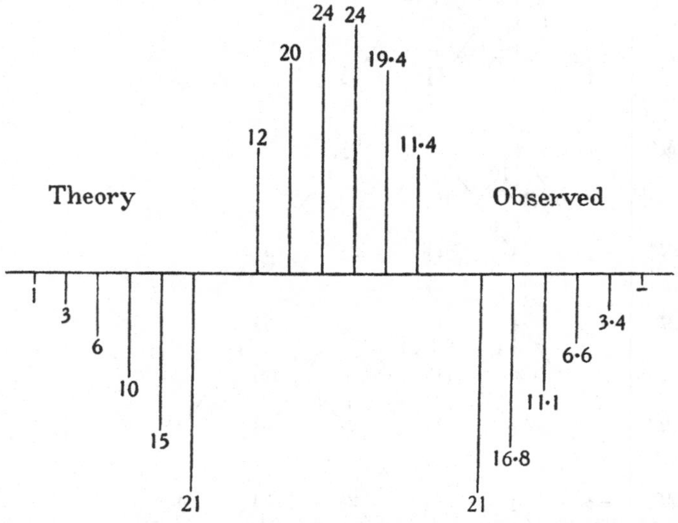

Fig. 17·19. Intensities of the Zeeman components of the 4754 A. line of Mn I; this line arises as $a\,^8S_{3\frac{1}{2}} \to z\,^8P^{\circ}_{2\frac{1}{2}}$. The pattern is *(1)* (3) (5) *9* 11 13 15 17 19/7. On the left are the theoretical, and on the right the measured intensities, the brightest π and σ components being adjusted to fit. After Van Geel, Diss. Utrecht, 1928, 65.

work on the iron row shows that the agreement is not always as good as that shown in the lines chosen here as illustrations.

Having obtained satisfactory intensity formulae for normal multiplet lines, two extensions suggest themselves; first, one may enquire whether the formulae are applicable to inter-system lines, and second, what intensities are to be expected in the partial Paschen-Back effect. To these questions also Van Geel has offered some answer.*

As an inter-system line he chose $^1D_2 \to {^3P_2}$, 3663·3 A., of Hg I,

* Van Geel, Diss. Utrecht, 1928 and *ZP*, 1928, **47** 615.

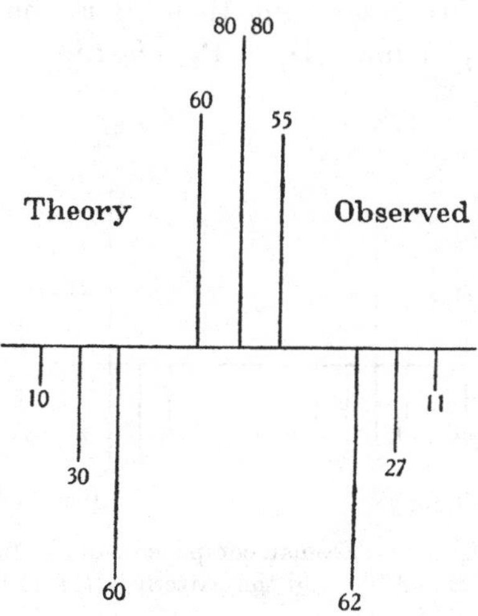

Fig. 17·20. Intensities of the Zeeman components of the 5205 A. line of Cr I; this line arises as $z\,{}^5P^\circ_1 \rightarrow a\,{}^5S_2$; the pattern is (*0*) (1) *3* 4 5/2. After Van Geel, Diss. Utrecht, 1928, 68.

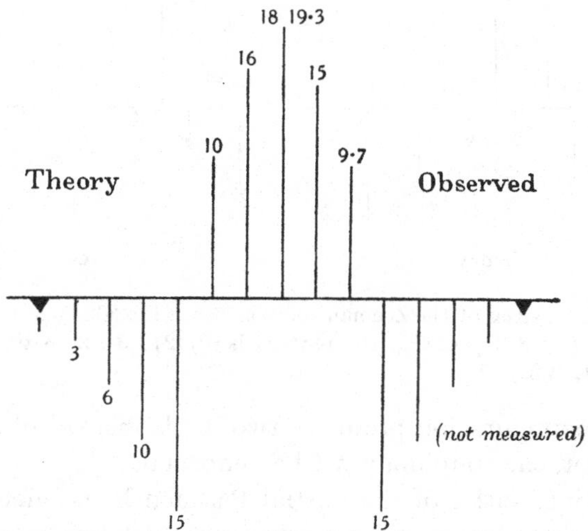

Fig. 17·21. Intensities of the Zeeman components of the 5208 A. line of Cr I; this line arises as $z\,{}^5P^\circ_3 \rightarrow a\,{}^5S_2$; the pattern is (*0*) (1) (2) *3* 4 5 6 7/3. After Van Geel, Diss. Utrecht, 1928, 67.

and obtained the intensity pattern shown in Fig. 17·22. This is certainly rather irregular, but then so is the pattern of the neighbouring triplet line, $^3D_1 \rightarrow {}^3P_2$, 3662·9 A. (Fig. 17·23), and

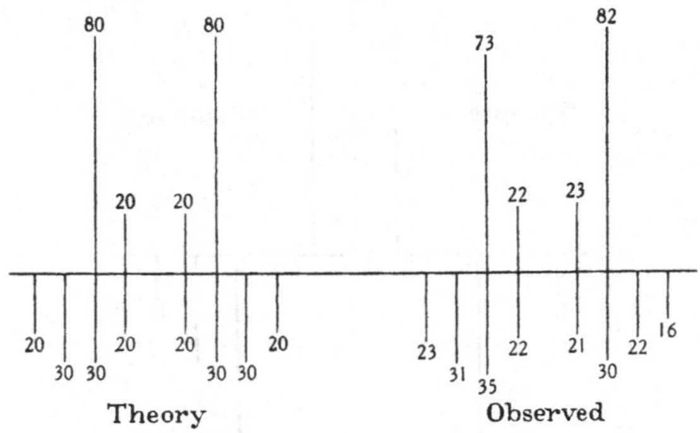

Fig. 17·22. Intensities of the Zeeman components of the 3663·3 A. line of Hg I; this line arises as 3 $^1D_2 \rightarrow 2\ ^3P_2$, and the pattern is (1) (*2*) 1 *2 3* 4/2. After Van Geel, Diss. Utrecht, 1928, 77.

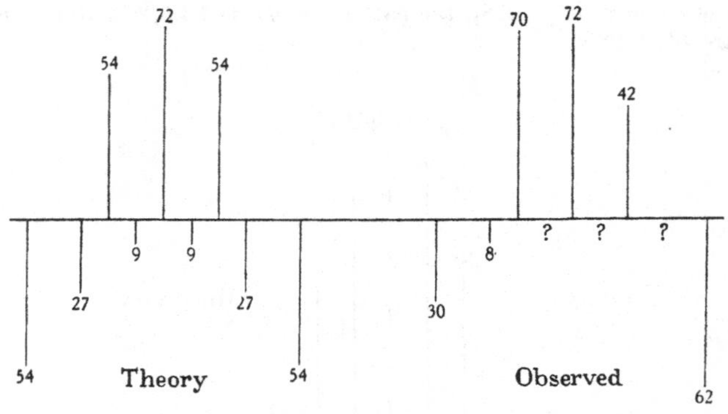

Fig. 17·23. Intensities of the Zeeman components of the 3662·9 A. line of Hg I; this line arises as 3 $^3D_1 \rightarrow 2\ ^3P_2$; the pattern is (*0*) (2) 1 3 *5*/2. After Van Geel, Diss. Utrecht, 1928, 77.

as this sd configuration produces two diads instead of a singlet and a triplet, the coupling must be abnormal.

For the intensities of the partial Paschen-Back effect appeal must be had to the quantum mechanics, which has been applied by Darwin* and others to predict both the intensity and displace-

* Darwin, C. G., *PRS*, 1927, **115** 1.

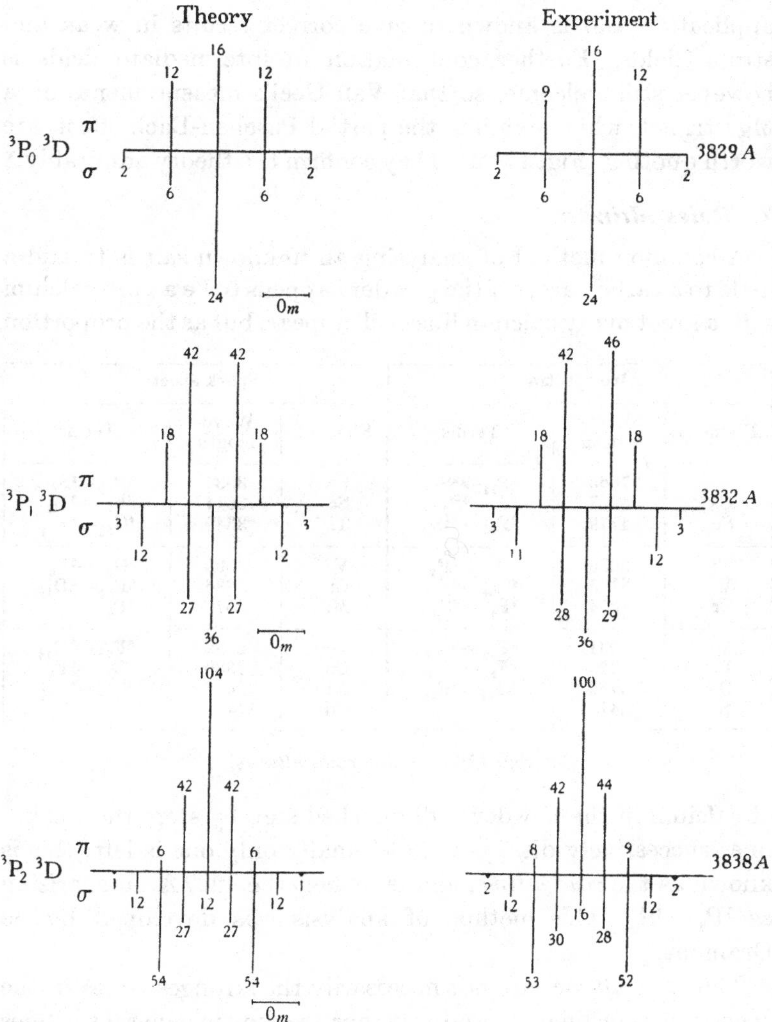

Fig. 17·24. Intensities of the Zeeman components of a diffuse triplet of Mg I; the lines arise as $3\,^3D \rightarrow 2\,^3P_{0,1,2}$. This is an example of the partial Paschen Back effect. After Van Geel, Diss. Utrecht, 1928, 84.

ment of the magnetic components of any normal multiplet in any field. That this theory is valid few doubt, for it is of very general application and is known to give correct results in weak and strong fields. Further confirmation in intermediate fields is however still welcome, so that Van Geel's measurements on a Mg I triplet, which exhibits the partial Paschen-Back effect, are worth quoting (Fig. 17·24). They confirm the theory admirably.*

7. *Raies ultimes*

A common method of analysing an unknown salt is to add a little to a carbon arc;† if the powder happens to be a pure calcium salt, a great many calcium lines will appear, but as the proportion

Arc spectra			Spark spectra		
Element	Wave-length	Terms	Element	Wave-length	Terms
K	7665	$^2P_{1\frac{1}{2}} \rightarrow {^2S_{\frac{1}{2}}}$	Ca	3934	$^2P_{1\frac{1}{2}} \rightarrow {^2S_{\frac{1}{2}}}$
Ca	4227	$^1P_1 \rightarrow {^1S_0}$	Sc	3614	$^3F_4 \rightarrow {^3D_3}$
Sc	4779	$^2F_{3\frac{1}{2}} \rightarrow {^2D_{2\frac{1}{2}}}$	Ti	3349	$^4G_{5\frac{1}{2}} \rightarrow {^4F_{4\frac{1}{2}}}$
Ti	3635	$^3G_5 \rightarrow {^3F_4}$	V	3093	$^5G_6 \rightarrow {^5F_5}$
V	3185	$^4G_{5\frac{1}{2}} \rightarrow {^4F_{4\frac{1}{2}}}$	Cr	2836	$^6F_{5\frac{1}{2}} \rightarrow {^6D_{4\frac{1}{2}}}$
Cr	4254	$^7P_4 \rightarrow {^7S_3}$	Mn	2576	$^7P_4 \rightarrow {^7S_3}$
Mn	4031	$^6P_{3\frac{1}{2}} \rightarrow {^6S_{2\frac{1}{2}}}$	Fe	2382	$^6F_{5\frac{1}{2}} \rightarrow {^6D_{4\frac{1}{2}}}$
Fe	3720	$^5F_5 \rightarrow {^5D_4}$	Co	2389	$^5G_6 \rightarrow {^5F_5}$
Co	3452	$^4G_{5\frac{1}{2}} \rightarrow {^4F_{4\frac{1}{2}}}$	Ni	2416	$^4G_{5\frac{1}{2}} \rightarrow {^4F_{4\frac{1}{2}}}$
Ni	3415	—	Cu	—	—

Fig. 17·25. Some *raies ultimes.*

of calcium in the powder is diminished step by step, the weaker lines successively disappear until finally only one is left; this is known as the *raie ultime*, and is in fact the 4227 A. line, arising as $^1P_1 \rightarrow {^1S_0}$. This method of analysis was developed by de Gramont.‡

The *raies ultimes* are not necessarily the strongest lines in the spectrum as ordinarily produced, nor are they in general the lines

* See also Back's measurements on a diffuse doublet of copper, Vol. I, 117.

† Twyman and Smith, *Wavelength tables of spectrum analysis*, 1931. Ryde and Jenkins, *Sensitive arc lines of 50 elements*, 1930. Both book and pamphlet are published by Adam Hilger.

‡ De Gramont, *Comptes Rendus*, 1920, **171** 1106.

which require the least energy to excite them, the resonance lines. Instead they are determined by four conditions, of which two are energy conditions.* The first states that the lower term is always the ground term, and the second that the higher term is of the same system. As inter-system lines are in general weaker than lines arising within a system, this condition is surprising only as excluding lines, such as the 2536 A. $^3P_1 \rightarrow {}^1S_0$ line of Hg I, which are exceptionally strong for other reasons. Subject to these two conditions, and in part also to the fourth, the energy required to excite the line must be as small as possible. The fourth condition states that when the energies are nearly equal, the *raie ultime* will usually arise from a transition in which ΔL is -1 in preference to one in which it is $+1$; but in aluminium and its homologues the ^2D term lies so much higher than the ^2S that the *raie ultime* is $^2S_{\frac{1}{2}} \rightarrow {}^2P_{\frac{1}{2}}$, the fourth condition notwithstanding.

BIBLIOGRAPHY

Frerichs in the *Handbuch der Physik*, 1929, **21**, deals with theory and its empirical justification. Van Geel, *Intensiteitsverhoudingen van Magnetisch Gesplitste Spectraallijnen*, 1928, contains a thorough study of Zeeman intensities; this is a dissertation presented to the University of Utrecht.

For experimental methods:

Dobson, Griffith and Harrison, *Photographic photometry*, 1926;
Ornstein, Moll and Burger, *Objektive Spektralphotometrie*, 1932.

* Laporte and Meggers, *JOSA*, 1925, **11** 459. Meggers and Scribner, however, have proposed a new rule: 'A *raie ultime* originates in a simple interchange of a single electron between an s and a p state, usually preferring configurations in which only one electron occurs in these states.' The *raie ultime* of Hf II is $5d^2 . 6p\,{}^4G_{5\frac{1}{2}} \ldots 5d^2 . 6s\,{}^4F_{4\frac{1}{2}}$, though ^4F is not the ground term. *BSJ*, 1934, **13** 657; 1935, **14** 629.

CHAPTER XVIII

THE SUM RULES AND (jj) COUPLING

1. Deviations from the Russell-Saunders coupling

The low terms of the light elements are easily divided into multiplets, and the multiplets into configurations. In general the components of a multiplet differ in energy by an amount which is small compared with the height which separates one multiplet from another; and similarly the interval separating two terms of a configuration is usually small compared with the height which separates one configuration from another. Each level is characterised by a certain value of J, and each multiplet by values of L and S. J and L are determined primarily by selection rules and S by the number of components in a multiplet; but in the allotment of quantum numbers the interval ratios, magnetic splitting factors and intensities all have to be considered. Moreover, J is the vector sum of L and S. All these regularities are regarded as arising in the Russell-Saunders coupling of the electrons, and conversely any irregularity is attributed to some distortion of the coupling.

Deviations from these rules occur frequently in elements of high atomic weight, and in the high terms of a spectrum; they are rather more common on the right-hand side of the periodic table than on the left. But the various rules are not equally sensitive; the interval ratios are often abnormal when the g factors and intensities are normal; while the g factors are rather more easily disturbed than the intensities. This must not be taken to mean that terms in which the interval ratio is normal and the intensities abnormal do not sometimes occur; but in the analysis of an irregular spectrum like Ni I intensities usually give the surest clue to the names of the terms. There are spectra, however, such as argon, in which none of these clues are worth much, for the terms are no longer divided into multiplets but only into configurations; values of L and S cannot then be assigned. Finally, there exist a few spectra in which different configurations lie at the same height and so perturb one another.

Yet however abnormal a spectrum, the J values of each configuration remain unchanged, and there is strong evidence that if the displacements, g factors and intensities are summed for terms having the same values of J, then the result is the same in all spectra. The question remains over how many terms the sum is to be taken, and in this the energy seems to be crucial; in simple spectra in which each multiplet is well separated from other multiplets, the rule is valid for each multiplet separately; when the multiplets are intermingled, but the configurations distinct, then the sum must be taken over all terms of the configuration; this is the form in which the sum rules are usually met. But when two configurations overlap, then the sum must extend over both.

2. Invariance of the *g* sum

In the Zeeman and Paschen-Back effects the sum of the magnetic energies of all terms, which have the same M and which arise from a single multiplet, is proportional to the field; or if the increments of energy are measured in terms of the normal increment cho_m, then their sum is independent of the field, and

$$\sum_J Mg = \sum_{L.S} (M_L + 2M_S).$$

This sum rule makes possible the calculation of the weak field splitting factor without the aid of the quantum mechanics; applying, as it does, however, only to a single multiplet, it is valid only when the coupling is Russell-Saunders.

The rule may be generalised, however, for an electron configuration by summing over all terms of a configuration. For if the sum is independent of the field intensity, it should not change in a field so strong that each electron vector precesses independently about the magnetic axis; the coupling is then

$$\{(\mathbf{l}_1\mathbf{H})(\mathbf{l}_2\mathbf{H})\dots(\mathbf{s}_1\mathbf{H})(\mathbf{s}_2\mathbf{H})\dots\}$$

and the sum rule reads

$$\Sigma Mg = \Sigma (m_l + 2m_s).$$

M is of course $\Sigma (m_l + m_s)$. As an example, consider a pd configuration giving rise to the six terms ^{3}PDF, ^{1}PDF (Fig. 18·1).

Among these there are four with components in which M is 3, and in these $\Sigma(m_l + 2m_s)$ is 14. In a weak field these components appear as parts of the $^3F_4$, $^3F_3$, $^3D_3$ and $^1F_3$ terms, whose magnetic splitting factors are $\frac{5}{4}$, $\frac{13}{12}$, $\frac{4}{3}$ and 1 respectively, so that Σg is $4\frac{2}{3}$ and ΣMg is 14 as theory requires.

l_1 l_2	m_{l_1}	m_{l_2}	m_{s_1} m_{s_2}	$\Sigma(m_l+m_s)$				$\Sigma(m_l+2m_s)$				$\Sigma m_l m_s$			
1 2	1	2	$\pm\frac{1}{2}$ $\pm\frac{1}{2}$	4	3	3	2	5	3	3	1	$1\frac{1}{2}$	$\frac{1}{2}$	$-\frac{1}{2}$	$-1\frac{1}{2}$
		1		3	2	2	1	4	2	2	0	1	0	0	−1
		0		2	1	1	0	3	1	1	−1	$\frac{1}{2}$	$\frac{1}{2}$	$-\frac{1}{2}$	$-\frac{1}{2}$
		−1		1	0	0	−1	2	0	0	−2	0	1	−1	0
		−2		0	−1	−1	−2	1	−1	−1	−3	$-\frac{1}{2}$	$1\frac{1}{2}$	$-1\frac{1}{2}$	$\frac{1}{2}$
	0	2	$\pm\frac{1}{2}$ $\pm\frac{1}{2}$	3	2	2	1	4	2	2	0	1	−1	1	−1
		1		2	1	1	0	3	1	1	−1	$\frac{1}{2}$	$-\frac{1}{2}$	$\frac{1}{2}$	$-\frac{1}{2}$
		0		1	0	0	−1	2	0	0	−2	0	0	0	0
		−1		0	−1	−1	−2	1	−1	−1	−3	$-\frac{1}{2}$	$\frac{1}{2}$	$-\frac{1}{2}$	$\frac{1}{2}$
		−2		−1	−2	−2	−3	0	−2	−2	−4	−1	1	−1	1
	−1	2	$\pm\frac{1}{2}$ $\pm\frac{1}{2}$	2	1	1	0	3	1	1	−1	$\frac{1}{2}$	$-1\frac{1}{2}$	$1\frac{1}{2}$	$-\frac{1}{2}$
		1		1	0	0	−1	2	0	0	−2	0	−1	1	0
		0		0	−1	−1	−2	1	−1	−1	−3	$-\frac{1}{2}$	$-\frac{1}{2}$	$\frac{1}{2}$	$\frac{1}{2}$
		−1		−1	−2	−2	−3	0	−2	−2	−4	−1	0	0	1
		−2		−2	−3	−3	−4	−1	−3	−3	−5	$-1\frac{1}{2}$	$-\frac{1}{2}$	$\frac{1}{2}$	$1\frac{1}{2}$

Fig. 18·1. Magnetic quantum numbers of a pd configuration in a field so strong that the electronic vectors precess independently.

Thus far the sum has been assumed independent only of the field, but if it remains unchanged in a field so strong that all electronic vectors are uncoupled, it must surely be independent of the coupling. And a logical consequence of this hypothesis is that the sum of the splitting factors is independent of the coupling when summed over all those terms of a configuration which have the same value of J. For let the sum of the g values of terms having $J=a$ be Σg_a, and when $M=b$ let $\Sigma Mg=\sigma_b$; then if the highest value of J is J', when $M=J'$,

$$J'\,\Sigma g_{J'}=\sigma_{J'}$$

or

$$\Sigma g_{J'}=\sigma_{J'}/J',$$

and as the right-hand side is independent of the coupling, so also is $\Sigma g_{J'}$. When $M=(J'-1)$,

$$\Sigma g_{J'}+\Sigma g_{(J'-1)}=\sigma_{(J'-1)}/(J'-1).$$

Again, the right-hand side is independent of the coupling, and so therefore must $\Sigma g_{(J'-1)}$ be. Clearly this argument can be extended to all values of J.

The constancy of the g sum was first deduced by Pauli* and Landé† from theory, but in the last decade experiment has amply confirmed it. In the spark spectrum of chromium‡ the $^4P_{\frac{1}{2}}$ and

Term	g factor	
	Landé	Empirical
$^6F_{\frac{1}{2}}$	−0·667	−0·671
$^6D_{\frac{1}{2}}$	3·333	2·841
$^4D_{\frac{1}{2}}$	0	0
$^4P_{\frac{1}{2}}$	2·667	3·145
Sum Σg	5·333	5·315

Fig. 18·2. g factors of those terms of the 3d⁴ (^{5}D) 4p configuration of Cr II, which have $J=\frac{1}{2}$. An example of the sum rule.

Term	g values		g sums	
	Landé	Empirical	Landé	Empirical
$^4P_{2\frac{1}{2}}$	1·60	1·60	1·60	1·60
$^4P_{1\frac{1}{2}}$	1·73	1·63	3·06	3·06
$^2P_{1\frac{1}{2}}$	1·33	1·43		
$^4P_{\frac{1}{2}}$	2·67	2·53	3·34	3·34
$^2P_{\frac{1}{2}}$	0·67	0·81		

Fig. 18·3. g factors of the p^4 (^{3}P) 5s terms of A II, showing that the sum rule is obeyed.

$^6D_{\frac{1}{2}}$ terms of the 3d^4 (^{5}D) 4p configuration show very irregular g values, but the sum for the four terms which have a J value of $\frac{1}{2}$ agrees closely with that predicted by Landé (Fig. 18·2). Other examples of the sum rule are to be found in the spark spectra of neon,§ argon|| and krypton,¶ in La I, La II** and Ag II.†† Fig. 18·3, which cites the p^4 (^{3}P) 5s configuration of A II,

* Pauli, *ZP*, 1923, **16** 155.

† Landé, *ZP*, 1923, **19** 112.

‡ Krömer, *ZP*, 1928, **52** 542. Cr II.

§ Bakker and de Bruin, *ZP*, 1931, **69** 19. Ne II.

|| Bakker, *K. Akad. Amsterdam, Proc.* 1928, **31** 1041. A II.

¶ Bakker and de Bruin, *ZP*, 1931, **69** 36. Kr II.

** Russell and Meggers, *BSJ*, 1932, **9** 665. La I, La II.

†† Shenstone and Blair, *PM*, 1929, **8** 765. Ag II.

brings out also one important corollary of the sum rule; $^4P_{2\frac{1}{2}}$ is the only term arising from this configuration which has a J value of $2\frac{1}{2}$, and so its g value must be that predicted by Landé whatever the coupling.

Configurations in which the g sum is abnormal are rare, but not unknown; usually the one which deviates is known to overlap another in the energy scale. In K II* the g sum of terms having a J value of 1 in the configuration $3p^5\,(^2P)\,4s$ is $2{\cdot}57 \pm 0{\cdot}02$ instead of 1·50; this configuration overlaps $3p^5.3d$, but not all the 3d levels are yet known, so that the combined g sum cannot be calculated. In Rb II a similar deviation occurs, where two configurations overlap.†

That the g sum rule is broken when two configurations overlap is of interest, because the quantum mechanics shows that the sum of the magnetic energies of all states having the same projection of angular momentum on the axis of the magnetic field is independent of the coupling between the vectors. As the sum has to be taken over all states of the atom, the theory in this form does not give much information; if, however, only states which have nearly the same energy influence one another,‡ all empirical results are seen to hang together. The rule is valid for a single multiplet in spectra whose coupling is Russell-Saunders, since a single multiplet is there isolated; it is valid for a configuration when a configuration is isolated; and when two configurations overlap it is still probably true if the sum is extended over both, though this point has not yet been tested.

3. Invariance of the Γ sum

In the Paschen-Back effect evidence is found that the sum of the displacements of those components of a multiplet, which have the same M, remains constant when the strength of the magnetic field varies. That the similar law for the g sum remains valid for any coupling provided the sum is extended to all terms of a configuration, suggests a similar extrapolation for the Γ sum.

Compared with the g sum, the Γ sum is unsatisfactory; whereas

* Whitford, *PR*, 1932, **39** 898.

† Laporte, Miller and Sawyer, *PR*, 1931, **38** 843.

‡ Goudsmit, *PR*, 1931, **37** 664.

theory can dictate absolute g values, it can dictate displacements only relative to an undetermined centroid, and only as a multiple of an undetermined constant A. Accordingly, the Γ sum rule is not susceptible of direct verification as the g sum rule is; and indeed had there not been some regularities which cried aloud for explanation, it seems hardly probable that Goudsmit could have pushed ahead on so flimsy a scaffolding.

Parent spectrum		Derived spectrum	
Spectrum	Interval	Spectrum	Interval
	$p.\Delta^2P$		$(^2P)\ ns\ \Delta^3P$
C II	64	C I	3s 60 4s 45 5s 54
Si II	287	Si I	4s 275 5s 283
P III	560	P II	4s 527 5s 547
S IV	950	S III	4s 450 5s 924
Cl V	1495	Cl IV	4s 1446
Ge II	1768	Ge I	5s 1661 6s 1740 7s 1765
Sn II	4253	Sn I	6s 3988 7s 4201
	$p^3.\Delta^2D$		$(^2D)\ ns.\Delta^3D$
O II	30	O I	3s 20
	$p^3.\Delta^2P$		$(^2P)\ ns.\Delta^3P$
	5		3s 17
	$p^5.\Delta^2P$		$(^2P)\ ns.\Delta^3P$
Ne II	782	Ne I	3s 777 4s 780 5s 778 6s 781
Na III	1371	Na II	3s 1357
A II	1431	A I	4s 1410 5s 1397 6s 1414 7s 1433
K III	2164	K II	4s 2642
Kr II	5371	Kr I	5s 5220
Xe II	10117	Xe I	6s 9129
	$d^9.\Delta^2D$		$(^2D)\ ns.\Delta^3D$
Ni II	1507	Ni I	4s 1508 5s 1506 6s 1506
Pd II	3539	Pd I	5s 3530 6s 3532 7s 3539
Pt II	—	Pt I	6s 10132

Fig. 18·4. Comparison of doublet intervals with those of the triplets derived from them, when an s electron is added.

The regularities mentioned connect the arc and spark spectra of an element. Thus in a number of spectra the triplet resulting from the addition of an s electron to a doublet ground term has the same extreme interval as the doublet. The $^3P_{2,0}$ interval from the $s^2.p\,(^2P)\,ns$ configuration of Si I is equal to the ^2P interval of Si II, and this is true also of isoelectronic spectra. Again, the $(^2P)\,ns\,^3P_{2,0}$ interval of the inert gas spectra is independent of the chief quantum number n and equal to the ^{2}P interval of the ionic ground term; and the same is true of the very similar $d^9\,(^2D)\,ns$ configuration in nickel and palladium (Fig. 18·4). Standing in contrast to these three groups of spectra in which the arc and spark intervals are identical stand the alkaline earth triplets whose extreme intervals obey the doublet laws, just as their spark doublets do.

To account for these regularities, assume that the Γ sum rule, like the g sum rule, is valid in a field so strong that the electronic spin and orbital vectors precess independently round the field axis, provided the sum be then taken over all those terms of a configuration which have the same value of M. Then construct a table of all the values which ${}_iM_L$ and ${}_iM_S$ can assume in the ^{2}P term of an ion, and of m_l and m_s for the s electron. Then if a sufficiently strong field be postulated,

$$\Gamma = A\,{}_iM_L\,{}_iM_S$$

and

$$\gamma = am_l m_s,$$

where A and a are the interval quotients of the ground term of the ion and the electron respectively; A is thus determined by the ionic term, and a presumably by the Landé doublet formula; however, when Fig. 18·5 is made out from these equations, it shows that a vanishes provided that m_l is always zero, as it is when the electron moves in an s orbit.

${}_iM_L$	m_l	${}_iM_S$	m_s	${}_i\Gamma/A$	γ/a	M_L	M_S	M	Γ/A
1	0	$\frac{1}{2}$	$\frac{1}{2}$	$\frac{1}{2}$	0	1	1	2	$\frac{1}{2}$
0				0		0		1	0
−1				$-\frac{1}{2}$		−1		0	$-\frac{1}{2}$
1	0	$\frac{1}{2}$	$-\frac{1}{2}$	$\frac{1}{2}$	0	1	0	1	$\frac{1}{2}$
0				0		0		0	0
−1				$-\frac{1}{2}$		−1		−1	$-\frac{1}{2}$
1	0	$-\frac{1}{2}$	$\frac{1}{2}$	$-\frac{1}{2}$	0	1	0	1	$-\frac{1}{2}$
0				0		0		0	0
−1				$\frac{1}{2}$		−1		−1	$\frac{1}{2}$
1	0	$-\frac{1}{2}$	$-\frac{1}{2}$	$-\frac{1}{2}$	0	1	−1	0	$-\frac{1}{2}$
0				0		0		−1	0
−1				$\frac{1}{2}$		−1		−2	$\frac{1}{2}$

Fig. 18·5. Electronic and atomic displacements in a system consisting of an ion in a ^{2}P state and an s electron.

The Γ sums are to be taken over terms having the same value of M, so that this table must be re-arranged in the form of Fig. 18·6.

M_S \ M	−2	−1	0	1	2
1			$-\frac{1}{2}$	0	$\frac{1}{2}$
0		$-\frac{1}{2}$	0	$\frac{1}{2}$	
0		$\frac{1}{2}$	0	$-\frac{1}{2}$	
−1	$\frac{1}{2}$	0	$-\frac{1}{2}$		
$\Sigma\Gamma/A$	$\frac{1}{2}$	0	−1	0	$\frac{1}{2}$

Fig. 18·6. Atomic displacements of the system of Fig. 18·5 in a strong field.

But the Γ sum is independent of the coupling, so that these sums must be those found empirically in a weak field. If Γ_2, Γ_1, Γ_1' and Γ_0 are the displacements of the $^3P_2$, $^3P_1$, $^1P_1$ and $^3P_0$ terms respectively, then arranging these by their M values we obtain Fig. 18·7.

J \ M	−2	−1	0	1	2
2	Γ_2	Γ_2	Γ_2	Γ_2	Γ_2
1		Γ_1	Γ_1	Γ_1	
1		Γ_1'	Γ_1'	Γ_1'	
0			Γ_0		

Fig. 18·7. Atomic displacements of the system of Fig. 18·5 in a weak field.

Accordingly

$$\left.\begin{aligned}\Gamma_2 &= \tfrac{1}{2}A\\ \Gamma_1+\Gamma_1' &= -\tfrac{1}{2}A\\ \Gamma_0 &= -\ A\end{aligned}\right\}.$$

These equations show that in the atom

$$^3P_2 - {}^3P_0 = \Gamma_2 - \Gamma_0 = \tfrac{3}{2}A,$$

whatever the coupling; and moreover that this is equal to the ^2P interval of the ion.

When the addition of an s electron to a ^2D term is considered, a similar argument shows that

$$^3D_3 - {}^3D_1 = \tfrac{5}{2}A = {}^2D_{2\frac{1}{2}} - {}^2D_{1\frac{1}{2}};$$

A being now the interval quotient of the ionic ^2D term. While when the method is applied to the alkaline earth triplets, in which a p or d electron is added to a ^{2}S ground term, it shows that the extreme triplet interval should depend only on the electronic quotient a and not at all on the ionic quotient A. Thus the invariance of the displacement sum satisfactorily explains the intervals observed in three different columns of the periodic table.

But if these were the points chiefly needing explanation, they do not limit the usefulness of the theory; with its aid Slater* has explained why the alkali doublets fit a formula developed for X-ray spin doublets, while Goudsmit† has related together the interval quotients of the numerous terms which may arise from a single configuration.

As an example of an X-ray spin doublet, consider the $L_{\text{II}}\,L_{\text{III}}$ doublet which arises from a p^5 group of electrons. The orbital and spin vectors of this group, restricted as they are by the exclusion principle, appear in Fig. 18·8.

m_l					m_s					M_L	M_S	γ/a					M	Γ/a
1	1	0	0	-1	$\frac12$	$-\frac12$	$\frac12$	$-\frac12$	$\frac12$	1	$\frac12$	$\frac12$	$-\frac12$	0	0	$-\frac12$	$1\frac12$	$-\frac12$
									$-\frac12$		$-\frac12$					$\frac12$	$\frac12$	$\frac12$
1	1	0	-1	-1	$\frac12$	$-\frac12$	$\frac12$	$\frac12$	$-\frac12$	0	$\frac12$	$\frac12$	$-\frac12$	0	$-\frac12$	$\frac12$	$\frac12$	0
							$-\frac12$				$-\frac12$			0			$-\frac12$	0
1	0	0	-1	-1	$\frac12$	$\frac12$	$-\frac12$	$\frac12$	$-\frac12$	-1	$\frac12$	$\frac12$	0	0	$-\frac12$	$\frac12$	$-\frac12$	$\frac12$
					$-\frac12$						$-\frac12$	$-\frac12$					$-1\frac12$	$-\frac12$

Fig. 18·8. Electronic and atomic displacements of a p^5 configuration.

These permitted combinations are rearranged according to their M values in Fig. 18·9, and the sum of the displacements is obtained. But the only two permitted values of J are $\frac12$ and $1\frac12$, so

M_S \ M	$-1\frac12$	$-\frac12$	$\frac12$	$1\frac12$
$\frac12$	—	$\frac12$	0	$-\frac12$
$-\frac12$	$-\frac12$	0	$\frac12$	—
$\Sigma\Gamma_{st/a}$	$-\frac12$	$\frac12$	$\frac12$	$-\frac12$

Fig. 18·9. Atomic displacements of a p^5 configuration in a strong field arranged by the corresponding values of M and M_S.

* Slater, *PR*, 1926, **28** 291.
† Goudsmit, *PR*, 1929, **31** 946.

that if the displacements in zero field are $\Gamma_{\frac{1}{2}}$ and $\Gamma_{1\frac{1}{2}}$, the displacements in a weak field must be those shown in Fig. 18·10.

J \ M	$-1\frac{1}{2}$	$-\frac{1}{2}$	$\frac{1}{2}$	$1\frac{1}{2}$
$1\frac{1}{2}$	$\Gamma_{1\frac{1}{2}}$	$\Gamma_{1\frac{1}{2}}$	$\Gamma_{1\frac{1}{2}}$	$\Gamma_{1\frac{1}{2}}$
$\frac{1}{2}$	—	$\Gamma_{\frac{1}{2}}$	$\Gamma_{\frac{1}{2}}$	—
$\Sigma\Gamma_w$	$\Gamma_{1\frac{1}{2}}$	$(\Gamma_{\frac{1}{2}}+\Gamma_{1\frac{1}{2}})$	$(\Gamma_{\frac{1}{2}}+\Gamma_{1\frac{1}{2}})$	$\Gamma_{1\frac{1}{2}}$

Fig. 18·10. Atomic displacements of a p^5 configuration in a weak field.

Comparison of these two tables shows that, if the Γ sum is independent of the coupling, then

$$\left.\begin{aligned}\Gamma_{1\frac{1}{2}} &= -\tfrac{1}{2}a\\ \Gamma_{\frac{1}{2}}+\Gamma_{1\frac{1}{2}} &= \tfrac{1}{2}a\end{aligned}\right\},$$

so that

$$\Gamma_{1\frac{1}{2}}-\Gamma_{\frac{1}{2}} = -\tfrac{3}{2}a,$$

showing that the interval is equal to that of an alkali P doublet.

To extend this method to other groups of equivalent electrons Goudsmit had to postulate Russell-Saunders coupling of the atomic vectors. Thus the d^2 configuration gives rise to the terms ^3F, ^3P, ^1G, ^1D and ^1S, or six unknowns, since Γ is zero in all singlet terms. Calculations based on an assumed strong field however give only five equations, one for each value of M from 0 to 4; so that the Γ values are indeterminate. Should we assume however that the ^3F and ^3P terms obey the Landé interval rule, then the

M_S \ M	-4	-3	-2	-1	0	1	2	3	4
1					$-\frac{1}{2}$	0	$\frac{1}{2}$	1	$1\frac{1}{2}$
			$-1\frac{1}{2}$	-1	$-\frac{1}{2}$	0	$\frac{1}{2}$		
0					2	$1\frac{1}{2}$	1	$\frac{1}{2}$	0
				$1\frac{1}{2}$	1	$\frac{1}{2}$	0	$-\frac{1}{2}$	
			1	$\frac{1}{2}$	0	$-\frac{1}{2}$	-1		
		$\frac{1}{2}$	0	$-\frac{1}{2}$	-1	$-1\frac{1}{2}$			
	0	$-\frac{1}{2}$	-1	$-1\frac{1}{2}$	-2				
-1			$\frac{1}{2}$	0	$-\frac{1}{2}$	-1	$-1\frac{1}{2}$		
	$1\frac{1}{2}$	1	$\frac{1}{2}$	0	$-\frac{1}{2}$				
$\Sigma\Gamma/a$	$1\frac{1}{2}$	1	$-\frac{1}{2}$	-1	-2	-1	$-\frac{1}{2}$	1	$1\frac{1}{2}$

Fig. 18·11. Atomic displacements of a d^2 configuration in a strong field, arranged by the values of M and M_S.

displacements of their components may be stated in terms of only two unknowns A_F and A_P. This is shown in Figs. 18·11, 18·12.

$L.J$ \ M	−4	−3	−2	−1	0	1	2	3	4
$^3F_4$	$3A_F$	$3A_F$	$3A_F$	$3A_F$	$3A_F$	$3A_F$	$3A_F$	$3A_F$	$3A_F$
$^3F_3$		$-A_F$	$-A_F$	$-A_F$	$-A_F$	$-A_F$	$-A_F$	$-A_F$	
$^3F_2$			$-4A_F$	$-4A_F$	$-4A_F$	$-4A_F$	$-4A_F$		
$^3P_2$			A_P	A_P	A_P	A_P	A_P		
$^3P_1$				$-A_P$	$-A_P$	$-A_P$			
$^3P_0$					$-2A_P$				
$^1G_4$	0	0	0	0	0	0	0	0	0
$^1D_2$			0	0	0	0	0		
$^1S_0$					0				

Fig. 18·12. Atomic displacements of a d^2 configuration in a weak field, arranged by the values of M and L_J.

The Γ sum rule shows that

$$\begin{aligned} \text{when } M &= 4 & 3A_F &= 1\tfrac{1}{2}a, \\ &= 3 & 2A_F &= a, \\ &= 2 & -2A_F + A_P &= -\tfrac{1}{2}a, \\ &= 1 & -2A_F &= -a, \\ &= 0 & -2A_F - 2A_P &= -2a. \end{aligned}$$

And these equations are satisfied if

$$A_P = A_F = \tfrac{1}{2}a.$$

Proceeding in this way Goudsmit* was able to calculate the interval quotient A of different components of many super-multiplets in terms of the interval quotients a of p or d electrons; the results are summarised in Fig. 18·13.

This theory has been strikingly successful in predicting experimental facts. In the d^7 configuration of Ru II all the terms should be inverted except the ^2F term, and so in fact experiment shows that they are.† While in the d^3 configuration of Ti II‡ the relative separations are in very fair agreement with theory; Fig. 18·14 shows this, the separation of the ^2H term having been fitted to the experimental value.§

* Goudsmit, *PR*, 1928, **31** 948.

† Meggers and Shenstone, *PR*, 1930, **35** 868*a*.

‡ Russell, *Ast. P.J.* 1927, **66** 283.

§ Pauling and Goudsmit, *Structure of line spectra*, 1930, 163.

Configuration	Multiplet	Interval quotient	Extreme interval	Configuration
p	^{2}P	a	$1\frac{1}{2}a$	$-$p^5
p^2	^{3}P	$\frac{1}{2}a$	$1\frac{1}{2}a$	$-$p^4
p^3	^{2}D	0	0	$-$p^3
	^{2}P	0	0	
d	^{2}D	a	$2\frac{1}{2}a$	$-$d^9
d^2	^{3}F	$\frac{1}{2}a$	$3\frac{1}{2}a$	$-$d^8
	^{3}P	$\frac{1}{2}a$	$1\frac{1}{2}a$	
d^3	^{4}F	$\frac{1}{3}a$	$3\frac{1}{2}a$	$-$d^7
	^{4}P	$\frac{1}{3}a$	$\frac{4}{3}a$	
	^{2}H	$\frac{1}{5}a$	$\frac{11}{10}a$	
	^{2}G	$\frac{3}{10}a$	$\frac{27}{20}a$	
	^{2}F	$-\frac{1}{6}a$	$-\frac{7}{12}a$	
	^{2}D	$\frac{1}{3}a$	$\frac{5}{6}a$	
	^{2}P	$\frac{2}{3}a$	a	
d^4	^{5}D	$\frac{1}{4}a$	$\frac{5}{2}a$	$-$d^6
	^{3}H	$\frac{1}{10}a$	$\frac{11}{10}a$	
	^{3}G	$\frac{3}{20}a$	$\frac{27}{20}a$	
	^{3}F	$\frac{1}{12}a$	$\frac{7}{12}a$	
	^{3}D	$-\frac{1}{12}a$	$-\frac{5}{12}a$	
	^{3}P	$\frac{1}{2}a$	$1\frac{1}{2}a$	
d^5	All	0	0	$-$d^5

Fig. 18·13. Atomic interval quotients of terms arising from shells of p and d electrons in terms of the electronic interval quotients.

Term \ J	$\frac{1}{2}$ – $1\frac{1}{2}$	$1\frac{1}{2}$ – $2\frac{1}{2}$	$2\frac{1}{2}$ – $3\frac{1}{2}$	$3\frac{1}{2}$ – $4\frac{1}{2}$	$4\frac{1}{2}$ – $5\frac{1}{2}$
b ^{4}F	Observed	75·8	103·4	128·4	—
	Calculated	*74·1*	*103·6*	*133·2*	—
a ^{4}P	32·0	122·3	—	—	—
	44·4	*74·1*	—	—	—
a ^{2}H	—	—	—	—	97·8
					(*97·8*)
a ^{2}G	—	—	—	120·5	—
				120·0	
b ^{2}F	—	—	$-$59·9	—	—
			$-$*51·8*		
b ^{2}D	—	129·4	—	—	—
— ^{2}D	—	—[a]	—	—	—
		Σ = *74·1*	—	—	—
a ^{2}P	125·0	—	—	—	—
	88·9				

[a] This state has not been identified.

Fig. 18·14. Term intervals in the d^3 configuration of Ti II; to obtain the calculated values the ^{2}H term has been fitted to the experimental value.

Nevertheless, in spite of its successes the theory has its difficulties; of these one example will suffice. In the $p^5.ns$ terms of neon, theory shows that Γ_0, $(\Gamma_1+\Gamma_1')$ and Γ_2 are independent of the coupling and therefore of the chief quantum number. Now Γ has always been interpreted as the displacement of a term from the centroid of the multiplet, so that for the ^{1}P term Γ_1' should be zero; indeed Γ has been assumed zero for all singlet terms in some of the above calculations. But if this assumption is made $(\Gamma_1-\Gamma_0)$

Term	$^1P_1$ ms_2		$^3P_0$ ms_3		$^3P_1$ ms_4		$^3P_2$ ms_5		
Interval / m		$\Gamma_1'-\Gamma_0$		$\Gamma_0-\Gamma_1$		$\Gamma_1-\Gamma_2$		$\Gamma_0-\Gamma_2$	$\Gamma_1'+\Gamma_1-2\Gamma_0$
1	38040·7	*1070·1*	39110·8	*359·4*	39470·2	*417·4*	39887·6	*776·8*	*710·7*
2	(14506·5)	*(145·4)*	(14651·9)	*(489·6)*	(15141·5)	*(290·7)*	(15432·2)	*(780·3)*	*(−344·2)*
3	7272·9	*50·2*	7323·1	*693·6*	8016·7	*84·6*	8101·3	*778·2*	*−6*
4	4201·8	*21·7*	4223·5	*738·6*	4962·1	*42·7*	5004·8	*781·3*	*−716·9*
5	2605·4	*11·2*	2616·6	*755·8*	3372·4	*24·3*	3396·7	*780·1*	*−744·6*
6	1667·7	*7·4*	1675·1	*764·8*	2440·0	*16·2*	2456·1	*781·0*	*−757·4*
7	1072·4	*4·9*	1077·3	*771·2*	1848·5	*9·6*	1858·1	*780·8*	*−766·3*

Figures in brackets have been intrapolated from the series formula.

Fig. 18·15. Energies and intervals of the s terms of neon; the term values are measured down from the series limit.

should be independent of n just as $\Gamma_2-\Gamma_0$ is; a prediction which experiment does not support. Moreover, the most obvious way of dodging the difficulty is blocked; for if one suggests plausibly enough that when the coupling is no longer Russell-Saunders one may not rightly speak of a singlet term, still though Γ_1' need not be zero, yet $(\Gamma_1+\Gamma_1')$ must be constant. And if this be admitted, theory predicts that

$$(\Gamma_1+\Gamma_1')-2\Gamma_0=(\Gamma_1-\Gamma_0)+(\Gamma_1'-\Gamma_0)$$

should be constant, a prediction quite at variance with experiment (Fig. 18·15).

4. The intensity sum

That the intensity sum is proportional to the statistical weight $(2J+1)$ is a thesis, which has been developed by successive stages from the SP combination, in which it is true of a single line, through the general multiplet, PD, to the super-multiplet or configuration in which the sum must be taken over all lines which originate in terms having the same value of J.*

This development is so closely analogous to the development of the g and Γ sums, that one expects similar deviations to occur when configurations overlap. And in fact abnormal intensities in Ba I have been ascribed to this cause.†

5. General coupling of two electrons

The coupling of electron vectors postulated by Russell and Saunders explains that division of levels into multiplets which is characteristic of light atoms. It explains, for example, why in the $s^2 p.s$ configuration of C I the four levels are divided into a triplet below and a singlet above; it even explains the 2 : 1 interval ratio; but when in Pb I these four levels divide into two diads, the Russell-Saunders coupling fails. What then is to be put in its place?

In the model the interaction of ion and electron is represented as the coupling of four vectors. This means that there are six interactions, but these belong to only four different types, since $(\mathbf{l}_1\mathbf{s}_1)$ and $(\mathbf{l}_2\mathbf{s}_2)$ are identical and so are $(\mathbf{l}_1\mathbf{s}_2)$ and $(\mathbf{l}_2\mathbf{s}_1)$; of these four types, $(\mathbf{s}_1\mathbf{s}_2)$, $(\mathbf{l}_1\mathbf{l}_2)$, $(\mathbf{l}_1\mathbf{s}_1)$ and $(\mathbf{l}_1\mathbf{s}_2)$, the quantum mechanics states that the fourth may be neglected, unless the atom is extremely light, but the other three are all important. In spectra which conform to the Russell-Saunders type, the spin coupling $(\mathbf{s}_1\mathbf{s}_2)$ determines the separation of terms of different multiplicity, the orbital coupling $(\mathbf{l}_1\mathbf{l}_2)$ that of terms of different name, and the orbit-spin coupling $(\mathbf{l}_1\mathbf{s}_1)$ the multiplet intervals. As the separation of the terms is a measure of the coupling, and in Russell-Saunders spectra terms of different multiplicity are

* Harrison and Johnson, *PR*, 1931, **38** 758, give the full theory and compare it with experiment.

† Langstroth, *PRS*, 1933, **142** 286.

widely separated, the $(\mathbf{s}_1\mathbf{s}_2)$ coupling must be strong. The $(\mathbf{l}_1\mathbf{l}_2)$ coupling is weaker than the $(\mathbf{s}_1\mathbf{s}_2)$ but stronger than the $(\mathbf{l}_1\mathbf{s}_1)$, for terms with different names but the same multiplicity are less widely separated than terms of different multiplicity but more widely than different components of the same multiplet.*

If the evidence of the quantum mechanics is to be accepted, and the $(\mathbf{l}_1\mathbf{s}_2)$ coupling ignored, there are still three alternatives to the coupling postulated by Russell and Saunders. If the latter is written

(A) $$\{(\mathbf{l}_1\mathbf{l}_2)(\mathbf{s}_1\mathbf{s}_2)\} = \{\mathbf{LS}\} = \mathbf{J},$$

the other three are

(B) $$\{(\mathbf{l}_1\mathbf{s}_1)(\mathbf{l}_2\mathbf{s}_2)\} = \{\mathbf{j}_1\mathbf{j}_2\} = \mathbf{J},$$

(C) $$\{[(\mathbf{l}_1\mathbf{s}_1)\mathbf{s}_2]\mathbf{l}_2\} = \{[\mathbf{j}_1\mathbf{s}_2]\mathbf{l}_2\} = \{\mathbf{j}'\mathbf{l}_2\} = \mathbf{J},$$

(D) $$\{[(\mathbf{l}_1\mathbf{s}_1)\mathbf{l}_2]\mathbf{s}_2\} = \{[\mathbf{j}_1\mathbf{l}_2]\mathbf{s}_2\} = \{\mathbf{j}''\mathbf{s}_2\} = \mathbf{J}.$$

These three all show an ion, whose orbital and spin vectors are not at once unlinked when a second electron is added, but differ in the influence the ion has on the coupling of the series electron. Thus in (B) the coupling $(\mathbf{l}_2\mathbf{s}_2)$ is preserved and the atomic resultant $\mathbf{J}$ appears as the sum of two electronic vectors $\mathbf{j}_1\mathbf{j}_2$. In (C) the ion shows a special attraction for the spin vector $\mathbf{s}_2$ of the second electron, and so breaks the $(\mathbf{l}_2\mathbf{s}_2)$ coupling; while in (D) the ion attracts particularly the $\mathbf{l}_2$ vector.

All three coupling types have been considered by those who have tried to interpret abnormal spectra; in particular, the magnetic g factors have been calculated and compared with experiment in more than one spectrum;† but the result of this work has so far been to show that types (C) and (D) have no advantage over the simpler (jj) coupling of type (B). Incidentally, too, the three are identical when the electron added occupies an **s** orbit.

The further discussion of abnormal spectra will therefore be simplified by treating the (jj) coupling as though it were the only alternative to that postulated by Russell and Saunders.

* Hund, *Linienspektren*, 1927, 91 f.

† For g factors of all four coupling types, see p. 145.

6. (jj) coupling

In the change from weak to strong magnetic field, while the energies change, the number of states and their magnetic quantum numbers, M, do not; similarly, in the change from (**LS**) to (**jj**) coupling, while the relative energies change, the J values do not. Accordingly, in the complex spectrum of a heavy metal, such as platinum, values of J can be assigned to the empirical terms, but the Russell-Saunders notation, depending as it does on L and S, is of little use.

l_1	l_2	j_1	j_2	J	Names if coupling is R.S.
0	0	$\frac{1}{2}$	$\frac{1}{2}$	0	$^1S_0$
				1	$^3S_1$
0	1	$\frac{1}{2}$	$\frac{1}{2}$	0	$^3P_0$
				1	$^3P_1$ $^1P_1$
		$\frac{1}{2}$	$1\frac{1}{2}$	1	
				2	$^3P_2$
1	1	$\frac{1}{2}$	$\frac{1}{2}$	0	
				1	$^1S_0$ $^3P_0$
		$\frac{1}{2}$	$1\frac{1}{2}$	1	
				2	$^1P_1$ $^3S_1$ $^3P_1$ $^3D_1$
		$1\frac{1}{2}$	$\frac{1}{2}$	1	
				2	$^1D_2$ $^3P_2$ $^3D_2$
		$1\frac{1}{2}$	$1\frac{1}{2}$	0	
				1	
				2	
				3	$^3D_3$

Fig. 18·16. Terms arising from the jj coupling of two unlike electrons in s.s, sp and p.p states; the J values are the same as those of the corresponding Russell-Saunders terms.

To illustrate the invariance of J, consider the terms arising from the (jj) coupling of two electrons. The values of j permitted a single electron are $(l \pm \frac{1}{2})$, and the terms arising from the combination of two electrons can be deduced by combining the two **j** vectors. When the electrons are not equivalent, **J** is simply the vectorial sum of $\mathbf{j}_1$ and $\mathbf{j}_2$, and there are no restrictions; thus two electrons having j values of $1\frac{1}{2}$ and $2\frac{1}{2}$ produce four states having J values of 1, 2, 3, 4. Other examples are given in Fig. 18·16. On the other hand, when the two electrons are equivalent, the Pauli exclusion principle does not allow two electrons with the same values of n, l, j and m; but when this condition is introduced, as it is in the derivation of Figs. 18·17, 18·18, the resulting J values are still identical to those developed by the (**LS**) coupling. The J

l_1	l_2	j_1	j_2	m_1	m_2	M	J	Names with R.S. coupling
1	1	$\frac{1}{2}$	$\frac{1}{2}$	$\frac{1}{2}$	$-\frac{1}{2}$	0	0	
		$\frac{1}{2}$	$1\frac{1}{2}$				1 2	$^1S_0$ $^3P_0$
		$1\frac{1}{2}$	$1\frac{1}{2}$	$1\frac{1}{2}$	$\frac{1}{2}$	2		
					$-\frac{1}{2}$	1		$^3P_1$
					$-1\frac{1}{2}$	0	0 2	
				$\frac{1}{2}$	$-\frac{1}{2}$	0		$^1D_2$ $^3P_2$
					$-1\frac{1}{2}$	-1		
				$-\frac{1}{2}$	$-1\frac{1}{2}$	-2		

Fig. 18·17. Terms arising from the jj coupling of two equivalent p electrons.

l_1	l_2	l_3	j_1	j_2	j_3	m_1	m_2	m_3	M	J	Names with R.S. coupling
1	1	1	$\frac{1}{2}$	$\frac{1}{2}$	$1\frac{1}{2}$	$\frac{1}{2}$	$-\frac{1}{2}$	$1\frac{1}{2}$	$1\frac{1}{2}$		
								$\frac{1}{2}$	$\frac{1}{2}$	$1\frac{1}{2}$	
								$-\frac{1}{2}$	$-\frac{1}{2}$		
								$-1\frac{1}{2}$	$-1\frac{1}{2}$		
			$\frac{1}{2}$	$1\frac{1}{2}$	$1\frac{1}{2}$	$\frac{1}{2}$	$1\frac{1}{2}$	$\frac{1}{2}$	$2\frac{1}{2}$		
								$-\frac{1}{2}$	$1\frac{1}{2}$		
								$-1\frac{1}{2}$	$\frac{1}{2}$		
							$\frac{1}{2}$	$-\frac{1}{2}$	$\frac{1}{2}$		
								$-1\frac{1}{2}$	$-\frac{1}{2}$	$2\frac{1}{2}$	$^2P_{\frac{1}{2}}$
							$-\frac{1}{2}$	$-1\frac{1}{2}$	$-1\frac{1}{2}$		
										$1\frac{1}{2}$	$^2P_{1\frac{1}{2}}$ $^2D_{1\frac{1}{2}}$ $^4S_{1\frac{1}{2}}$
						$-\frac{1}{2}$	$1\frac{1}{2}$	$\frac{1}{2}$	$1\frac{1}{2}$		
								$-\frac{1}{2}$	$\frac{1}{2}$		
								$-1\frac{1}{2}$	$-\frac{1}{2}$	$\frac{1}{2}$	$^2D_{2\frac{1}{2}}$
							$\frac{1}{2}$	$-\frac{1}{2}$	$-\frac{1}{2}$		
								$-1\frac{1}{2}$	$-1\frac{1}{2}$		
							$-\frac{1}{2}$	$-1\frac{1}{2}$	$-2\frac{1}{2}$		
			$1\frac{1}{2}$	$1\frac{1}{2}$	$1\frac{1}{2}$	$1\frac{1}{2}$	$\frac{1}{2}$	$-\frac{1}{2}$	$1\frac{1}{2}$		
								$-1\frac{1}{2}$	$\frac{1}{2}$		
							$-\frac{1}{2}$	$-1\frac{1}{2}$	$-\frac{1}{2}$	$1\frac{1}{2}$	
						$\frac{1}{2}$	$-\frac{1}{2}$	$-1\frac{1}{2}$	$-1\frac{1}{2}$		

Fig. 18·18. Terms arising from the jj coupling of three equivalent p electrons.

values to be expected from more complex configurations are shown in Fig. 18·19.

Turning from the states themselves to their energies, consider first the addition of an s electron to an ion in the ^2P state; this actually occurs in columns IV and VIII, and much is known of the resulting terms. The two ionic levels are defined by j_1, which here assumes the values $1\frac{1}{2}$ and $\frac{1}{2}$. The addition of the second electron splits each of the levels in two, for $\mathbf{j}_2$ can orient itself parallel or anti-parallel to $\mathbf{j}_1$. $\mathbf{j}_2$ is here identical with $\mathbf{s}_2$, since $\mathbf{l}_2$ is zero. Thus the atomic level scheme consists of two 'diads'; and the

Configuration	Number of electrons in which j is			Number of levels in which J is						
	$\frac{1}{2}$	$1\frac{1}{2}$	$2\frac{1}{2}$	0	1	2	3	4	5	6
p^2	2			1						
	1	1			1	1				
		2		1		1				
d^2		2		1		1				
		1	1		1	1	1	1		
			2	1		1		1		
d^4		4		1						
		3	1		1	1	1	1		
		2	2	2	1	4	2	3	1	1
		1	3	1	2	2	3	2	1	1
			4	1		1		1		
Configuration	j			J						
	$\frac{1}{2}$	$1\frac{1}{2}$	$2\frac{1}{2}$	$\frac{1}{2}$	$1\frac{1}{2}$	$2\frac{1}{2}$	$3\frac{1}{2}$	$4\frac{1}{2}$	$5\frac{1}{2}$	$6\frac{1}{2}$
p^3	2	1			1					
	1	2		1	1	1				
		3			1					
d^3		3			1					
		2	1	1	1	2	1	1		
		1	2	1	2	2	2	1	1	
			3		1	1		1		
d^5		4	1			1				
		3	2	1	2	2	2	1	1	
		2	3	2	3	4	3	3	1	1
		1	4	1	2	2	2	1	1	
			5			1				

Fig. 18·19. Terms arising from various numbers of equivalent p and d electrons.

distance between the diads, being due to the coupling $(\mathbf{l}_1\mathbf{s}_1)$, must be greater than the interval of either diad, for the latter is due to $(\mathbf{j}_1\mathbf{j}_2)$. Further, the relative positions of the two terms can be foretold, for the energy is due primarily to the interaction of $\mathbf{l}_1$ and $\mathbf{s}_1$ and secondarily to the interaction of $\mathbf{s}_1$ and $\mathbf{s}_2$. Hund's rule states that the energy of interaction of $\mathbf{s}_1$ and $\mathbf{s}_2$ is small when the angle between them is small, so that in the lower level of each diad $\mathbf{s}_1$ and $\mathbf{s}_2$ will be parallel. The levels constructed on this principle are shown in Figs. 18·20 and 18·22, the former

Ionic levels	Vector coupling	Atomic levels	J
$\mathbf{j}_1=\mathbf{l}_1+\mathbf{s}_1$	$\mathbf{l}_1 \longrightarrow$, $\mathbf{s}_1 \rightarrow$, $\mathbf{s}_2 \leftarrow$	$\mathbf{J}=\mathbf{j}_1-\mathbf{s}_2$	l_1
	$\mathbf{l}_1 \longrightarrow$, $\mathbf{s}_1 \rightarrow$, $\mathbf{s}_2 \rightarrow$	$\mathbf{J}=\mathbf{j}_1+\mathbf{s}_2$	l_1+1
$\mathbf{j}_1=\mathbf{l}_1-\mathbf{s}_1$	$\mathbf{l}_1 \longrightarrow$, $\mathbf{s}_1 \leftarrow$, $\mathbf{s}_2 \rightarrow$	$\mathbf{J}=\mathbf{j}_1+\mathbf{s}_2$	l_1
	$\mathbf{l}_1 \longrightarrow$, $\mathbf{s}_2 \leftarrow$, $\mathbf{s}_1 \leftarrow$	$\mathbf{J}=\mathbf{j}_1-\mathbf{s}_2$	l_1-1

Fig. 18·20. Energy levels resulting when an s electron is added to an ion in an erect doublet state; jj coupling is assumed.

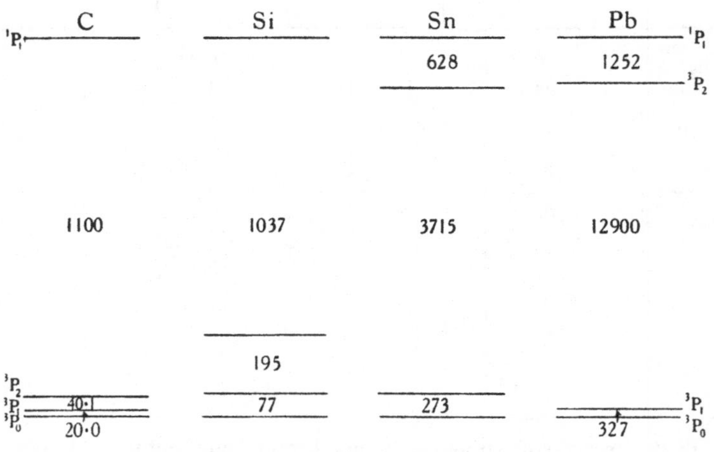

Fig. 18·21. Intervals of low $s^2p.s$ terms in four elements of column IV; the scale is adjusted so as to make the $^1P_1-{}^3P_0$ interval the same in all elements. Note the change from **LS** coupling in C I to jj in Pb I.

representing an erect and the latter an inverted parent term; the first agrees well with the empirical $s^2\,p.ns$ terms of tin and lead (Fig. 18·21), and the second with the $p^5\,(^2P)\,ns$ terms of the inert gases (Fig. 18·23).

Ionic levels	Vector coupling	Atomic levels	J
	$\mathbf{l}_1 \longrightarrow$, $\mathbf{s}_1 \leftarrow$, $\mathbf{s}_2 \rightarrow$	$\mathbf{J}=\mathbf{j}_1+\mathbf{s}_2$ ——	l_1
$\mathbf{j}_1=\mathbf{l}_1-\mathbf{s}_1$ ——			
	$\mathbf{l}_1 \longrightarrow$, $\leftarrow \leftarrow$ $\mathbf{s}_2$ $\mathbf{s}_1$	$\mathbf{J}=\mathbf{j}_1-\mathbf{s}_2$ ——	l_1-1
	$\mathbf{l}_1 \longrightarrow$ $\mathbf{s}_1 \rightarrow$, $\mathbf{s}_2 \leftarrow$	$\mathbf{J}=\mathbf{j}_1-\mathbf{s}_2$ ——	l_1
$\mathbf{j}_1=\mathbf{l}_1+\mathbf{s}_1$ ——			
	$\mathbf{l}_1 \longrightarrow$ $\mathbf{s}_1 \rightarrow$ $\mathbf{s}_2 \rightarrow$	$\mathbf{J}=\mathbf{j}_1+\mathbf{s}_2$ ——	l_1+1

Fig. 18·22. Energy levels resulting when an s electron is added to an ion in an inverted doublet state; jj coupling is assumed.

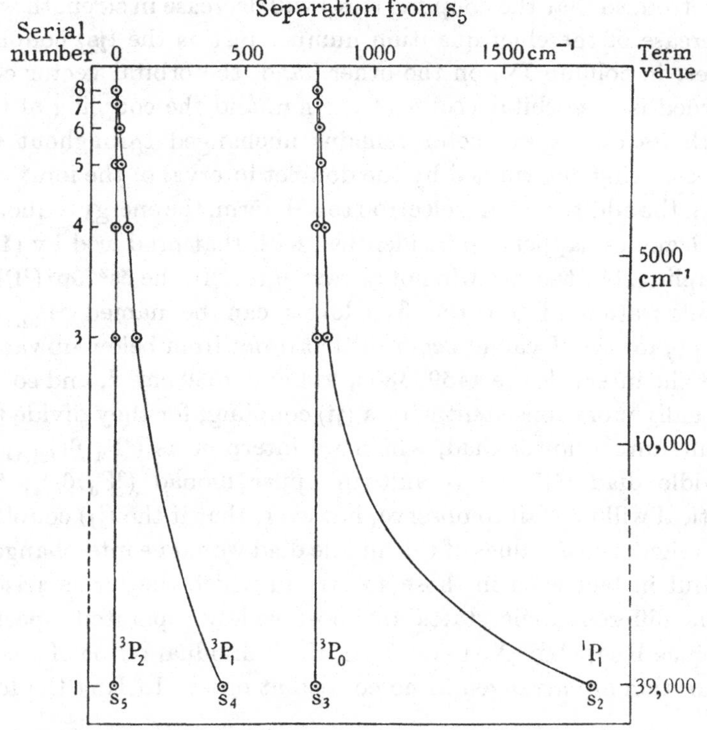

Fig. 18·23. Intervals of the s terms of neon; in each configuration the s_5 or $^3P_2$ term lies lowest and the s_2 or $^1P_1$ term highest. Note the change from **LS** coupling in the lowest term to jj coupling in the high terms.

But one may ask, why does the addition of an s electron to a ^{2}P ground term produce two diads in columns IV and VIII, when the addition of a p electron to a ^{2}S ground term produces triplets and singlets in column II. The couplings which have to be contrasted are

$$\{\mathbf{s}_1(\mathbf{l}_2\mathbf{s}_2)\} = \{\mathbf{s}_1\mathbf{j}_2\} \quad \text{in the alkaline earths,}$$

and

$$\{(\mathbf{l}_1\mathbf{s}_1)\mathbf{s}_2\} = \{\mathbf{j}_1\mathbf{s}_2\} \quad \text{in column IV.}$$

In both the coupling of the electronic vector **j** with the isolated spin vector will decrease rapidly in strength with increase of the chief quantum number, the coupling energy being roughly proportional to $n^{\star -3}$. But the interaction of the orbital vector with its own spin vector differs in the two columns; in the alkaline earths the orbital vector concerned is the orbital vector of the series electron, so that the coupling $(\mathbf{l}_2\mathbf{s}_2)$ will decrease in strength with increase of the chief quantum number just as the $\{\mathbf{js}\}$ coupling does; in column IV, on the other hand, the orbital vector concerned is the orbital vector of the ion, and the coupling of this with its own spin vector remains unchanged throughout the series, being determined by the doublet interval of the ion.*

In the addition of an s electron to a ^{2}P term, the energy sequence of J values happens to be identical with that produced by (**LS**) coupling, but the two will not always agree. In the 5s^2.5p^4 (^{3}P) 6s configuration of I I† the five levels can be named $^4P_{2\frac{1}{2}, 1\frac{1}{2}, \frac{1}{2}}$, $^2P_{1\frac{1}{2}, \frac{1}{2}}$, for the J values occur in that order from below upwards; but the intervals are 1459, 4803, 924 and 4530 cm.$^{-1}$, and so are actually more consistent with a (**jj**) coupling; for they divide the terms into a lower diad, which we interpret as $(^3P_2 . 6s)_{2\frac{1}{2}, 1\frac{1}{2}}$, a middle diad $(^3P_1 . 6s)_{\frac{1}{2}, 1\frac{1}{2}}$ and an upper monad $(^3P_0 . 6s)_{\frac{1}{2}}$; the critical will not fail to observe, however, that if the (**jj**) coupling was rigid, the J values of the middle diad would be interchanged.

But in fact even in those spectra in which the terms arising from different ionic states are most widely separated, spectra such as Rn I, Cs II, Xe I and Xe II, the individual terms of a configuration are arranged in no consistent order. In Rn I the four

* Pauling and Goudsmit, *Structure of line spectra*, 1930, 104.

† Evans, S. F., *PRS*, 1931, **133** 417.

terms ($^2P_{1\frac{1}{2}}.p_{1\frac{1}{2}}$) might lie consistently below the two terms ($^2P_{1\frac{1}{2}}.p_{\frac{1}{2}}$), but in fact the six empirical terms cannot be separated into two groups.

7. Calculation of *g* for any coupling

Less important than the sum rule at the moment because data is scanty, but destined perhaps to be equally useful, is the calculation of g for any coupling. Following Landé, Goudsmit and Uhlenbeck* have shown that if any vector **Z** is the resultant of two other vectors **X** and **Y**, so that the permissible values of Z are determined by

$$|X+Y| \geqslant Z \geqslant |X-Y|,$$

then

$$g(Z)=\frac{Z(Z+1)+X(X+1)-Y(Y+1)}{2Z(Z+1)}g(X)+\frac{Z(Z+1)+Y(Y+1)-X(X+1)}{2Z(Z+1)}g(Y).$$

With this assumption consider the four coupling types of an earlier section.

(A) $$\{(\mathbf{l}_1\mathbf{l}_2)(\mathbf{s}_1\mathbf{s}_2)\}=\{\mathbf{LS}\}=J.$$

J is the sum of the **L** and **S**, so

$$g(J)=\frac{J(J+1)+L(L+1)-S(S+1)}{2J(J+1)}g(L)+\frac{J(J+1)+S(S+1)-L(L+1)}{2J(J+1)}g(S).$$

Now **L** is the sum of the vectors $\mathbf{l}_1$ and $\mathbf{l}_2$, so

$$g(L)=\frac{L(L+1)+l_1(l_1+1)-l_2(l_2+1)}{2L(L+1)}g(l_1)+\frac{L(L+1)+l_2(l_2+1)-l_1(l_1+1)}{2L(L+1)}g(l_2)$$

$$=1, \text{ if } g(l)=1 \text{ for all values of } l.$$

And similarly $g(S)=2$, if $g(s)=2$ for all values of s. Substituting these values in the expression for $g(J)$ gives at once the usual Landé formula

$$g=1+\frac{J(J+1)+S(S+1)-L(L+1)}{2J(J+1)}.$$

* Goudsmit and Uhlenbeck, *ZP*, 1926, **35** 618.

This analysis brings out, perhaps more clearly than any other, the cause of the whole anomalous Zeeman effect, which is nothing else than the double magnetism of the electron.

If there are more than two electrons active one may assume that the binding of $\mathbf{s}_1$ and $\mathbf{s}_2$ is tighter than that of their resultant $\mathbf{s}_{1,2}$ with $\mathbf{s}_3$; then one may first work out $g(s_{1,2})$ and then combine $\mathbf{s}_{1,2}$ with $\mathbf{s}_3$. But for the Russell-Saunders coupling $g(L)$ is always 1 and $g(S)$ always 2.

(B) $$\{(\mathbf{l}_1\mathbf{s}_1)(\mathbf{l}_2\mathbf{s}_2)\}=(\mathbf{j}_1\mathbf{j}_2\}=\mathbf{J}.$$

$\mathbf{J}$ is compounded of $\mathbf{j}_1$ and $\mathbf{j}_2$, so

$$g(J)=\frac{J(J+1)+j_1(j_1+1)-j_2(j_2+1)}{2J(J+1)}g(j_1)+\frac{J(J+1)+j_2(j_2+1)-j_1(j_1+1)}{2J(J+1)}g(j_2).$$

The $g(j_1)$ of this expression appears in Landé's table simply as the g factor of some doublet term, for $\mathbf{j}_1$ is the sum of $\mathbf{l}_1$ and $\mathbf{s}_1$, only one electron being active.

Very similar to this (jj) coupling for two electrons is the binding of the series electron near the limit of a term sequence; indeed the equation is unaltered, save that the ground term of the ion is not necessarily a doublet.

(C) In the coupling

$$\{[(\mathbf{l}_1\mathbf{s}_1)\mathbf{s}_2]\mathbf{l}_2\}=\{[\mathbf{j}_1\mathbf{s}_2]\mathbf{l}_2\}=\{\mathbf{j}'\mathbf{l}_2\}=\mathbf{J},$$

$\mathbf{J}$ is the resultant of $\mathbf{j}'$ and $\mathbf{l}_2$, so

$$g(J)=\frac{J(J+1)+j'(j'+1)-l_2(l_2+1)}{2J(J+1)}g(j')+\frac{J(J+1)+l_2(l_2+1)-j'(j'+1)}{2J(J+1)}g(l_2).$$

In this equation $g(l_2)=1$, and $\mathbf{j}'$ is the resultant of $\mathbf{j}_1$ and $\mathbf{s}_2$, so that

$$g(j')=\frac{j'(j'+1)+j_1(j_1+1)-s_2(s_2+1)}{2j'(j'+1)}g(j_1)+\frac{j'(j'+1)+s_2(s_2+1)-j_1(j_1+1)}{2j'(j'+1)}g(s_2).$$

Here $g(j_1)$ may be taken from Landé's doublet table, while $g(s_2)$ is 2, as always for a single electron.

(D) The coupling $\{[(\mathbf{l}_1\mathbf{s}_1)\mathbf{l}_2]\mathbf{s}_2\}$ is so similar to (C) that it need not be worked out in detail.

The values to be predicted by each of these four coupling types have been worked out for certain spectra whose Zeeman splitting is irregular, but no close agreement with experiment has ever been found. Coupling types (B), (C) and (D) often give better

J	Coupling scheme							g values					J
	A	*B* j_1	*B* j_2	*C* j'	*C* l_2	*D* j''	*D* s_2	*A*	*B*	*C*	*D*	Empirical	
3½	$^4\mathrm{D}_{3\frac{1}{2}}$	2	1½	2½	1	3	½	1·43	1·43	1·43	1·43	1·43	3½
2½	$^4\mathrm{P}_{2\frac{1}{2}}$	2	1½	2½	1	2	½	1·60	1·44	1·53	1·53	1·60	2½
	$^4\mathrm{D}_{2\frac{1}{2}}$	1	1½	1½	1	2	½	1·37	1·40	1·40	1·40	1·33	
	$^2\mathrm{D}_{2\frac{1}{2}}$	2	½	1½	1	3	½	1·20	1·33	1·24	1·24	1·24	
								Σg 4·17	4·17	4·17	4·17	4·17	
1½	$^4\mathrm{P}_{1\frac{1}{2}}$	2	1½	1½	1	1	½	1·73	1·47	1·49	1·50	1·73	1½
	$^4\mathrm{D}_{1\frac{1}{2}}$	0	1½	1½	1	2	½	1·20	1·33	1·29	1·30	1·20	
	$^4\mathrm{S}_{1\frac{1}{2}}$	2	½	2½	1	1	½	2·00	1·67	1·84	1·83	2·00	
	$^2\mathrm{D}_{1\frac{1}{2}}$	1	½	½	1	2	½	0·80	1·22	1·11	1·10	0·90	
	$^2\mathrm{P}_{1\frac{1}{2}}$	1	1½	½	1	1	½	1·33	1·38	1·33	1·33	1·23	
								Σg 7·06	7·06	7·06	7·06	7·06	
½	$^4\mathrm{P}_{\frac{1}{2}}$	1	½	1½	1	0	½	2·67	1·78	2·11	2·00	2·67	½
	$^4\mathrm{D}_{\frac{1}{2}}$	0	½	½	1	1	½	0·00	0·67	0·67	0·67	0·00	
	$^2\mathrm{P}_{\frac{1}{2}}$	1	1½	½	1	1	½	0·67	1·22	0·89	1·00	0·99	
	$^2\mathrm{S}_{\frac{1}{2}}$	2	1½	1½	1	1	½	2·00	1·67	1·67	1·67	1·68	
								Σg 5·33	5·33	5·33	5·33	5·34	

Fig. 18·24. g factors of the p^4 (^{3}P) 4p terms of A II compared with values calculated for four different couplings. The sum rule is valid although the g factors are abnormal. (Bakker, *K. Akad. Amsterdam*, 1928, **31** 1041.)

agreement for certain terms, but none of them has been found to give close agreement even for one configuration in one element. A good example of results obtained is afforded by the (^{3}P) 4p configuration of A II (Fig. 18·24).

As there is evidence that departures from the interval rule and Landé's g formula are both signs of a break from the Russell-Saunders coupling, a study of the variation of g in a series or in a sequence of homologous spectra has been a long felt want. Recently Pogány* has supplied the need by studying the change in g in the p^5. (^{2}P) ns terms of the inert gases. This configuration produces two terms having a J value of 1, and in neon both in-

* Pogány, *ZP*, 1935, **93** 376.

tensity laws and magnetic splitting factors show that the lower is $^3P_1$ and the higher $^1P_1$.* Of these $^3P_1$ approaches the lower limit of the ion, and must therefore flow to the (jj) coupling term of $(^2P_{1\frac{1}{2}}.s)$, with a g value of 7/6, while the higher term approaches the higher limit and must flow to $(^2P_{\frac{1}{2}}.s)_1$ with a g value of 4/3. Thus as the coupling changes the magnetic splitting factor of

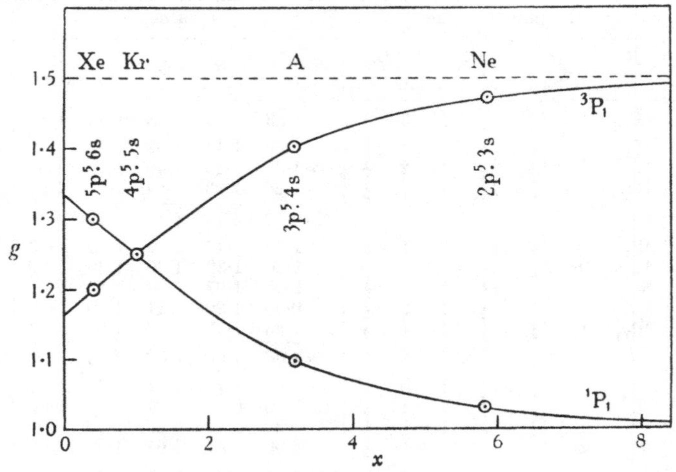

Fig. 18·25. g factors of the s terms of the inert gases, showing the transition from **LS** to jj coupling. (After Pogány, *ZP*, 1935, **93** 376.)

the $^3P_1$ term flows from 1·5 to 1·17, while the g factor of the $^1P_1$ term flows from 1·0 to 1·33; this shows that the g values cross and that for some particular coupling the two g values must be equal. Experiment has not yet revealed the steps in this transition in the series of neon, but the low terms of all inert gases are readily accessible and Pogány has shown five steps in the analogous transition from the (**LS**) coupling characteristic of light to the (**jj**) characteristic of heavy atoms; in neon the g values approach the ideal values of 1·5 and 1·0, being 1·47 and 1·03, whereas in krypton they are equal, and in xenon become 1·20 and 1·30 (Fig. 18·25). A similar transition can be detected in the p terms (Fig. 18·26).

* See p. 42.

(LS) coupling		Neon (^{2}P) 3p		Argon (^{2}P) 4p		Krypton (^{2}P) 5p		6p	Xenon (^{2}P) 6p		(jj) coupling		
Term	g	Term	g	Term	g	Term	g	g	Term	g	j_1	j_2	g
$^3P_0$	0	p_1	0	p_1	0	p_1	0	0	p_1	0	$\frac{1}{2}$	$\frac{1}{2}$	0
$^1S_0$	0	p_3	0	p_5	0	p_5	0	0	p_5	0	$1\frac{1}{2}$	$1\frac{1}{2}$	0
$^3P_1$	1·5	p_2	1·340	p_2	1·379	p_3	1·425	1·384	p_2		$\frac{1}{2}$	$1\frac{1}{2}$	1·5
$^1P_1$	1·0	p_5	0·999	p_4	0·819	p_4	0·619	0·635	p_4		$\frac{1}{2}$	$\frac{1}{2}$	0·67
$^3D_1$	0·5	p_7	0·699	p_7	0·840	p_7	1·028	1·046	p_7	1·02	$1\frac{1}{2}$	$1\frac{1}{2}$	1·33
$^3S_1$	2·0	p_{10}	1·984	p_{10}	1·962	p_{10}	1·891	1·820	p_{10}		$1\frac{1}{2}$	$\frac{1}{2}$	1·5
$^3P_2$	1·5	p_4	1·301	p_3	1·248	p_2	1·163		p_3	1·183	$\frac{1}{2}$	$1\frac{1}{2}$	1·17
$^3D_2$	1·17	p_6	1·229	p_6	1·302	p_6	1·400	1·406	p_6	1·402	$1\frac{1}{2}$	$1\frac{1}{2}$	1·33
$^1D_2$	1·0	p_8	1·137	p_8	1·121	p_8	1·116	1·110	p_9	1·113	$1\frac{1}{2}$	$\frac{1}{2}$	1·17
$^3D_3$	1·33	p_9	1·329	p_9	1·333	p_9	1·333	1·333	p_8		$1\frac{1}{2}$	$1\frac{1}{2}$	1·33

Fig. 18·26. g factors of the p terms of the inert gases, showing the transition from **LS** to **jj** coupling. These terms arise as p^5 (^{2}P) np.

Sources. Ne: Back, *AP*, 1925, **76** 330, which is in good agreement with Murakawa and Iwama, *Inst. Phys. and Chem.* Tokyo, 1930, **13** 289. A: Pogány, *ZP*, 1935, **93** 364, which agrees well with Terrien and Dijkstra, *J. de Phys.* 1934, **5** 443. Kr and Xe: Pogány, *ZP*, 1935, **93** 376.

8. Electronic displacements

In an earlier chapter, while trying to explain why terms are inverted if they arise from a shell more than half full, resort was had to the theory that the atomic displacement Γ may be regarded as the sum of electronic displacements γ_1 and γ_2; and it was shown that provided the coupling is Russell-Saunders, the relation between the two is

$$\Gamma = \gamma_1 + \gamma_2 = \cos(\mathbf{LS})\, \Sigma a_1 l_1 \cos(\mathbf{l}_1\mathbf{L})\, s_1 (\mathbf{s}_1\mathbf{S})$$

$$= LS\cos(\mathbf{LS})\, \Sigma a_1 \frac{l_1}{L}\cos(\mathbf{l}_1\mathbf{L})\frac{s_1}{S}\cos(\mathbf{s}_1\mathbf{S}). \qquad \ldots\ldots(13\cdot2)$$

The quantity summed in this expression is commonly abbreviated as the interval quotient A, so that

$$A = \Sigma a_1 \frac{l_1}{L}\cos(\mathbf{l}_1\mathbf{L})\frac{s_1}{S}\cos(\mathbf{s}_1\mathbf{S}). \qquad \ldots\ldots(18\cdot1)$$

If this equation is applied to the addition of an electron to an ion, the expression for the interval quotient A will in general contain two interval quotients a_1 and a_2, and no method is known separating these. But under certain conditions the expression simplifies; thus if either the ion or the electron is in an s state, or if the ion is in a singlet state, one of the interval quotients vanishes; while if the two electrons are equivalent a_1 and a_2 are equal. The method cannot be extended to the combination of three electrons, unless two are assumed so tightly linked that they are quite undisturbed by the third electron.

Thus consider the addition of an electron $(\mathbf{l}_2\mathbf{s}_2)$ to an ionic S term defined by $(\mathbf{l}_1\mathbf{s}_1)$; then since l_1 is zero, a_1 vanishes from the expression for A, and there is left simply

$$A = a_2 \frac{l_2}{L} \cos(\mathbf{l}_2 \mathbf{L}) \frac{s_2}{S} \cos(\mathbf{s}_2 \mathbf{S}). \qquad \ldots\ldots(18\cdot2)$$

In this equation

$$\frac{l_2}{L}\cos(\mathbf{l}_2\mathbf{L}) = \frac{l_2(l_2+1)+L(L+1)-l_1(l_1+1)}{2L(L+1)}$$
$$= 1, \text{ when } l_1 = 0 \text{ and } L = l_2,$$

so that the expression for A simplifies to

$$A = a_2 \frac{s_2(s_2+1)+S(S+1)-s_1(s_1+1)}{2S(S+1)}. \qquad \ldots\ldots(18\cdot3)$$

Now when an electron is added to a term with spin vector $\mathbf{s}_1$, two terms will be produced with spin moments $(s_1+\frac{1}{2})$ and $(s_1-\frac{1}{2})$; if these have interval quotients A_+ and A_-,

$$A_+ = a_2 \frac{\frac{3}{4}+(s_1+\frac{1}{2})(s_1+\frac{3}{2})-s_1(s_1+1)}{2(s_1+\frac{1}{2})(s_1+\frac{3}{2})}$$
$$= a_2/(2s_1+1).$$

And similarly it may be shown that

$$A_- = \frac{-a_2}{(2s_1+1)}.$$

Applied to the alkaline earth spectra the expression for A_+ leads directly to the results already obtained by the sum rule; but applied to the $d^5(^6S)np$ configuration of Cr I and Mo I, the formulae lead to results which are new; thus the ^7P term should be erect and the ^5P term inverted, while both should have the same interval quotient. In Cr I the two terms obey the interval rule well, and the ^5P term is inverted, but the interval quotients instead of being equal bear a ratio of 9 : 1 (Fig. 18·27). In Mo I the terms do not obey the interval rule and both are erect; while if the theory is applied to the p^3 (^{4}S) np configuration of column VI the results are even more unsatisfactory.

Agreement, however, is only to be expected if the p^3 and d^5 groups are so firmly bound in an S state that the addition of the

Terms		Cr I		Mo I	
		$\Delta\nu$	A	$\Delta\nu$	A
$3d^5$ (^{6}S) 4p for Cr I and	$^7P_2$ $^7P_3$ $^7P_4$	81·4 112·5	27·1 28·1	257 449	86 112
$4d^5$ (^{6}S) 5p for Mo I	$^5P_1$ $^5P_2$ $^5P_3$	−5·7 −8·8	−2·9 −2·9	121 87	61 29

Fig. 18·27. Intervals of the d^5 (^{6}S) p configuration in Cr I and Mo I.

p electron does not disturb them, and if the coupling of the ion and electron is Russell-Saunders. That experiment does not here support theory shows only that these assumptions are invalid, for a very valuable application of the same theory has been made to the displaced terms of beryllium, magnesium and isoelectronic spectra. The 3s.3p $^3P^\circ$ and $3p^2\,^3P$ of the magnesium-like spectra have many of them identical intervals as Fig. 18·28 shows. Now according to the above theory the intervals should be determined by

$$\Gamma = ALS\cos(\mathbf{LS}),$$

where

$$A = a_1 \frac{l_1}{L}\cos(\mathbf{l}_1\mathbf{L})\frac{s_1}{S}\cos(\mathbf{s}_1\mathbf{S}) + a_2\frac{l_2}{L}\cos(\mathbf{l}_2\mathbf{L})\frac{s_2}{S}\cos(\mathbf{s}_2\mathbf{S}). \quad \ldots(18\cdot4)$$

For the 3s.3p configuration the first term of this equation is zero, and there remains simply $A = \frac{1}{2}a_2$. For the $3p^2$ configuration, on the other hand, the two terms of this equation are equal, and since $L = S = l_1 = l_2 = 1$ and $s_1 = s_2 = \frac{1}{2}$, there results again $A = \frac{1}{2}a_2$. Thus the interval quotients are equal, and since both are ^{3}P terms their displacements and intervals are also equal.

Bechert* has advanced even further than this and has been able to apply a theory of Goudsmit's† to account for the discrepancies from the 2 : 1 interval ratio, but the theory involves the use of the quantum mechanics and is beyond the scope of this book.

	3s.3p $^3P^\circ$			3p^2 ^3P		
	$\Delta^3P^\circ_{1,2}$	$\Delta^3P^\circ_{0,1}$	$\Delta^3P^\circ_{1,2}/\Delta^3P^\circ_{0,1}$	$\Delta^3P_{1,2}$	$\Delta^3P_{0,1}$	$\Delta^3P_{1,2}/\Delta^3P_{0,1}$
Mg I	40·6	20·1	2·02	40·6	20·5	1·98
Al II	122·6	64·1	1·91	122·6	59·7	2·05
Si III	261·3	130·7	2·00	261·3	130·1	2·01
P IV	469·1	227·1	2·07	469·1	234·7	2·00
S V	767·1	360·4	2·13	767·1	412·4	1·86
Cl VI	1183·7	521·3	2·27	1183·7	668·0	1·77

Fig. 18·28. sp $^3P^\circ$ and p^2 ^3P intervals in spectra isoelectronic with Mg I.

9. Abnormal Intensities

Kronig has shown that the intensity tables given in the previous chapter may be applied to configurations in which the coupling is (jj), if S is replaced by j_1 and L by j_2, where j_1 is taken to be the quantum number which does not change in the transition.‡

Abnormal intensities due to (jj) coupling must be distinguished from those due to the overlapping of multiplets and configurations; to obtain clear evidence of the validity of this formula, one would wish to measure the intensities of lines arising in a configuration well separated from all others and exhibiting the intervals characteristic of (jj) coupling.

10. Perturbed terms

There is one factor producing abnormal series, interval ratios, g factors and intensities, which has only recently received the attention it deserves. It is known as perturbation, and is best introduced by a study of a perturbed series.

Of perturbed series, the diffuse series of calcium affords a clear

* Bechert, *ZP*, 1931, **69** 735.

† Goudsmit, *PR*, 1930, **35** 1325.

‡ White and Eliason, *PR*, 1933, **44** 753. Bartlett, *PR*, 1929, **34** 1247, derived formulae for (jj) coupling.

example. The first four members of the ^{3}D term sequence contract in the usual way, but the next three expand anomalously; the expansion reaches a maximum at the seventh term, the succeeding triplets contracting again as they approach the series limit. This may be shown in two ways, by plotting the displacement of

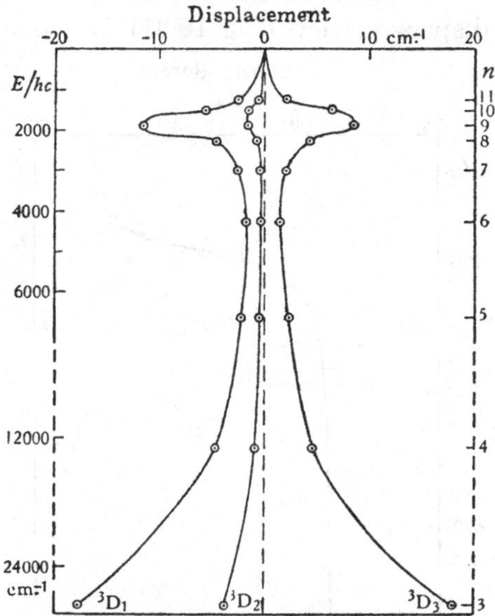

Fig. 18·29. The 4s nd ^{3}D term series of calcium showing perturbation. The abscissae are the displacements from the centroid, which is shown dotted.

each term from the centroid against the energy (Fig. 18·29) and by plotting the effective quantum number against the energy (Fig. 18·30).

Similar abnormalities have been noticed in the ^{3}F series of Al II* and the ^{2}P series of Cu I,† and their cause has been elucidated by Russell and Shenstone.‡ They show that these anomalous series only occur when a term of another series intrudes; thus between the sixth and seventh terms of the 4s nd ^{3}D sequence of

* Sawyer and Paschen, *AP*, 1927, **84** 1.

† Shenstone, *PR*, 1929, **34** 1623.

‡ Russell and Shenstone, *PR*, 1932, **39** 415. White, *Introduction to atomic spectra*, 1934, gives a clear and full account of this work.

calcium there lies the 3d 5s ^{3}D term, a term be it noted with the same values of L, S and J, but with an interval much greater than that to be expected at this height in the 4s nd series (Fig. 18·32). When this intruding term is excluded from the sequence, all succeeding values of the quantum defect are reduced by one, and the graph of the defect against the energy comes to resemble an anomalous dispersion curve (Fig. 18·31). Moreover, if a formula

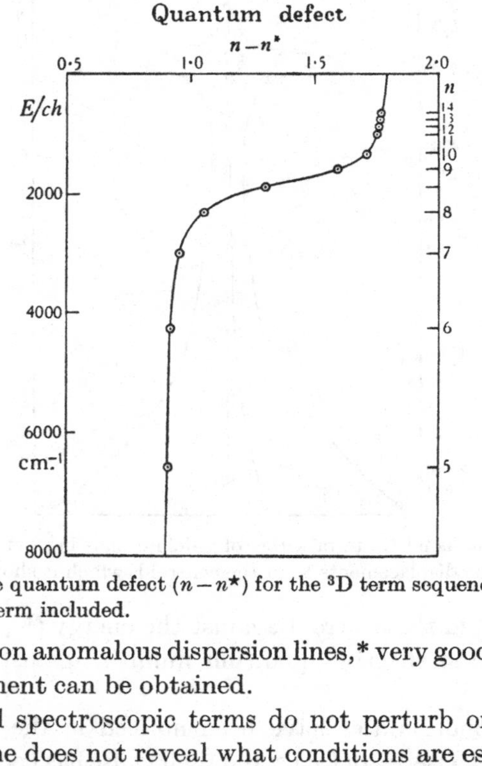

Fig. 18·30. The quantum defect ($n-n^\star$) for the ^{3}D term sequence of Ca I with the intruding term included.

is developed on anomalous dispersion lines,* very good agreement with experiment can be obtained.

Clearly all spectroscopic terms do not perturb one another; analysis alone does not reveal what conditions are essential, but the quantum mechanics provides an answer to this important question. If two levels have the same characteristic quantum numbers and lie close together, the eigenfunction of each level contains some components of the eigenfunction of the other, so that both levels belong in part to both configurations. Stated in other words this means that when two states perturb one another,

* Langer, *PR*, 1930, **35** 649*a*.

there is a certain probability that the atom will jump back and forth between the two without energy being radiated. The assignment of a given level to a definite electron coupling is therefore indefinite; it becomes more definite the further the levels are apart. More precisely two levels perturb one another when both have the same J and both are odd or both even; the two terms

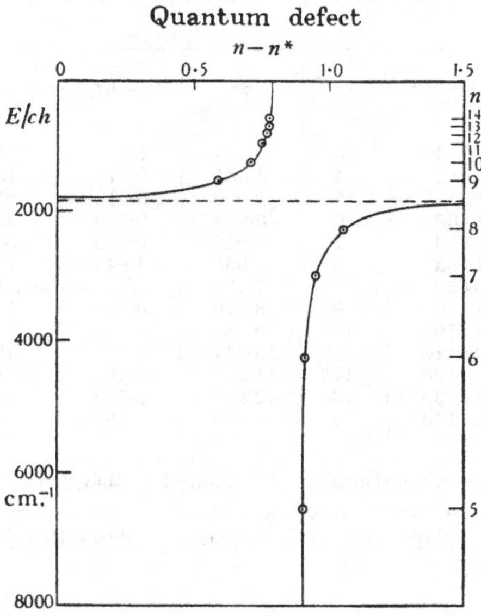

Fig. 18·31. The quantum defect $(n-n^*)$ for the ^3D term sequence of Ca I with the intruding term removed. Note how similar this graph is to that characteristic of anomalous dispersion.

need not belong to the same configuration, or be derived from the same state of the ion, but experiment shows that the perturbation is greater when the two terms have the same values of L and S if the coupling is (**LS**) or the same values of j_1 and j_2 if the coupling is (**jj**).

Since the assignment of electron coupling is indefinite, each perturbed level is likely to have some of the properties of the other, and in fact much evidence has accumulated to show that the perturbing levels share their interval factors, their magnetic splitting factors and their intensities.

The quantum mechanics shows that in terms of energy perturbation manifests itself essentially as a repulsion, but it is not difficult to show that repulsion leads logically to a sharing of intervals. Consider for example a narrow triplet lying close below a wide one; if one of these is odd and the other even, they will not perturb one another, and if the coupling is (**LS**) they will appear as shown on the left of Fig. 18·33. If however both terms are

Observed term value cm.$^{-1}$	Configuration	n	n^*	$n - n^*$	Calculated term value cm.$^{-1}$	T (obs.)– T (calc.)
28969·1	4s 3d	3	1·946	1·054	26465	2504
11556·4	4s 4d	4	3·081	0·919	11556	0
6561·4	4s 5d	5	4·090	0·910	6562	− 1
4255·5	4s 6d	6	5·078	0·922	4252	+ 3
3002·4	4s 7d	7	6·045	0·955	3002	0
2268·3	4s 8d	8	6·955	1·045	2287	− 19
1848·9	3s 5d		7·702		Foreign term	
1551·3	4s 9d	9	8·410	0·590	1551	0
1273·1	4s 10d	10	9·284	0·716	1276	− 3
1045·6	4s 11d	11	10·244	0·756	1048	− 2
869·8	4s 12d	12	11·232	0·768	873	− 3
734·0	4s 13d	13	12·227	0·773	737	− 3
628·0	4s 14d	14	13·219	0·781	630	− 2

Fig. 18·32. Series calculations of the anomalous 4s nd $^3D_1$ series of calcium. The calculated term values are obtained from a formula designed on the lines of those used to explain anomalous dispersion. (After Russell and Shenstone, *PR*, 1932, **39** 415.)

even, the levels with the same J will repel one another, and as the repulsion is greater the nearer the levels are to one another, the $^3P_0$ terms which are nearest undergo the greatest displacement; thus the narrow triplet is widened and the wide triplet is narrowed; this is shown on the right side of the figure. Moreover, if the narrow term lies above the wide, the same sharing of the intervals results.

Of the sharing of magnetic splitting factors much evidence has accumulated. Thus is Zr I the $^3P_2$ and $^1D_2$ terms of the $4d^2.5s^2$ configuration should have g values of 1·5 and 1·0, whereas in fact both have the same g value, 1·25; these terms are separated by 915 cm.$^{-1}$* Further examples from Zr I are shown in Fig. 18·34,

* Kiess, C. C. and Kiess, H. K., *BSJ*, 1931, **6** 660.

while the phenomenon has been observed in Ti I, Cu I, Ni II* and La II. In all these spectra the two perturbing terms belong to the same configuration, but the deviating g sums of K II, Rb II and Pn III make it probable that this is not an essential condition. In Pb III† the g value of the 6s.6p $^3P_2$ term, which

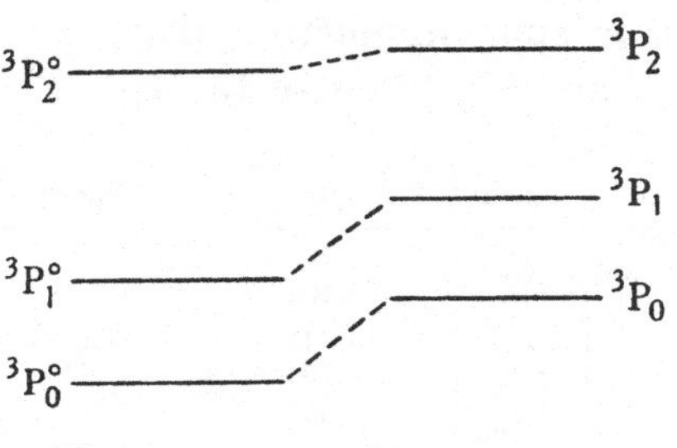

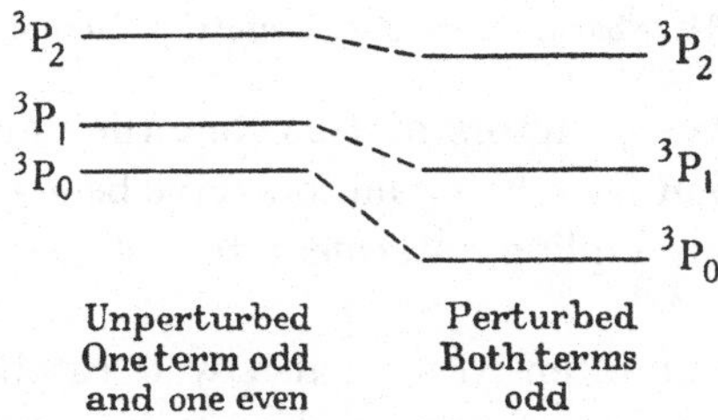

Fig. 18·33. Changes produced by perturbation in wide and narrow triplets when these lie close to one another.

should be 1·50 whatever the coupling, is in fact 1·35. This can be most reasonably explained if the term is perturbed, but what unidentified configuration is likely to lie near enough is not clear.

Short of complete sharing, the g factors may be more or less distorted. A method of calculating the precise change in g caused by neighbouring states was first developed by Houston‡ for the

* Shenstone, *PR*, 1927, **30** 264.

† Green and Loring, *PR*, 1932, **41** 389*a*.

‡ Houston, *PR*, 1929, **33** 297.

ps configuration; Goudsmit* later extended the method to the ds, and Laporte and Inglis† to the $p^5.s$ and $d^9.s$ configurations. The only empirical constant which these formulae contain is one easily determined from the energies of the terms.

Though perturbed terms share their magnetic splitting factors, one must not too readily assume the converse, that when two terms share their splitting factors they are perturbing one another. The $^3P_1$ and $^1P_1$ terms of the $4p^5(^2P)5s$ configuration

Configuration	Term	E/hc	g theory	g observed
$4d^2.5s^2$	$^3P_2$ $^1D_2$	4,186 5,101	1·500 1·000	1·25
$4d^3.5s$	$^3F_4$ $^3G_4$	12,342 12,761	1·250 1·050	1·15
$4d^2.5s$ 5p	$^1G_4$ $^3F_4$	26,931 26,938	1·000 1·250	1·13
$4d^3.5p$	$^5P_2$ $^1D_2$	34,761 34,850	1·833 1·000	1·42
$4d^3.5p$	$^1F_3$ $^3G_3$	36,760 36,942	1·000 0·750	0·87

Fig. 18·34. Sharing of magnetic splitting factors in Zr I.

of Kr I both have g factors of 1·25, but they are separated by 4930 cm.$^{-1}$, and in fact their g factors have been explained as due to a particular coupling intermediate between (**LS**) and (**jj**) (Fig. 18·25).

The sharing of intensities is also well established. In Zr I singlet terms have been found combining with quintets, but only when the singlet has a quintet neighbour or the quintet a singlet and the two perturbing terms have the same J.‡ Moreover, in barium Langstroth § has identified the terms which perturb the first three multiplets of the diffuse and fundamental series, and has shown that the experimental intensity sums are those required for (**LS**) coupling. In this comparison an uncalculable parameter appears connecting the intensities of two multiplets

* Goudsmit, *PR*, 1930, **35** 1325.

† Laporte and Inglis, *PR*, 1930, **35** 1337. Pogány, *ZP*, 1933, **86** 729, confirms the predicted values in Kr I.

‡ Details on p. 106.

§ Langstroth, *PRS*, 1933, **142** 286.

arising in different configurations; but this may be determined experimentally provided some lines occur in each multiplet which are only slightly perturbed. A study of perturbed terms leads naturally to the identification of lines previously regarded as anomalous, for a weak multiplet may have one or more of its lines intensified by perturbation until they become strong enough to stand out in the spectrum. Several fragments of multiplets, whose other components are so weak that they escape observation, have in fact been found.

In barium no attempt was made to calculate the precise intensities of individual lines, for this is only possible when the parent configurations of all perturbing terms are known, and as yet the analysis of Ba II is insufficiently advanced. Indeed individual perturbed intensities seem to have been calculated and compared with experiment only once, by Kast* working on Sr I. He found satisfactory agreement, but the perturbation in the two terms considered was not as large as one could wish considering the magnitude of the experimental error.

BIBLIOGRAPHY

Pauling and Goudsmit, *The structure of line spectra*, 1930.

* Kast, *ZP*, 1932, **79** 731.

CHAPTER XIX

SERIES LIMIT

1. *J* values

Evidence has already been adduced to show that in the spectra of N, A II and Ne where the limit is a multiplet, some series converge to one component and some to another. To recall the point at issue consider the displaced terms of calcium; five terms of the

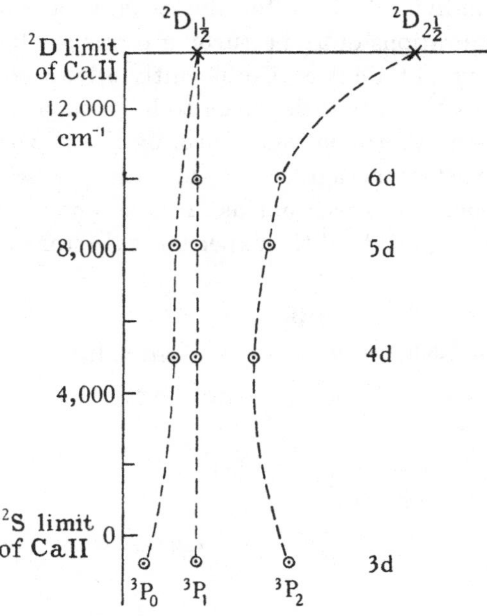

Fig. 19·1. Splitting of the 3d.*n*d ^{3}P terms of Ca I showing that all three do not tend to the same limit.

d.*n*p ^{3}P series are known and the separations are plotted in Fig. 19·1, showing quite clearly that $^3P_2$ converges to an upper limit, while $^3P_1$ and $^3P_0$ converge to a lower; these limits are naturally interpreted as the $^2D_{2\frac{1}{2}}$ and $^2D_{1\frac{1}{2}}$ terms of Ca II.

All these terms have been named because they approximate to the Russell-Saunders ideal; but there are other terms such as the

(²P) p and (²P) d terms of neon which cannot be named though their series have been worked out, and the limits they approach. Naturally then the question arises: cannot theory predict which terms will approach each limit? This was the problem which Hund set himself, and his solution will be considered in due course, but in that J is more easily determined than the orbital and spin vectors, it must take precedence.

Empirically the J values of unknown terms are always more easily determined than the multiplet structure. And theoretically the J values retain their significance even when the coupling is most irregular and L and S have no meaning. In particular, if both the series limit and the orbital vector of the series electron

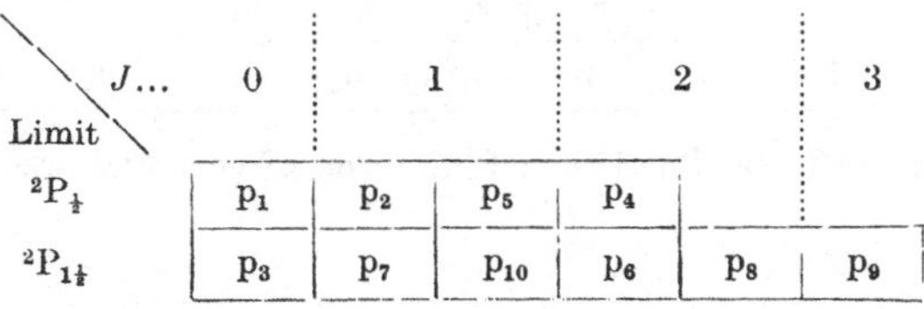

Fig. 19·2. Paschen's p series in neon and the limits to which they tend.

are known, then the J values of the terms approaching that limit are readily determined.

Ask, for example, which of the p terms of neon approach the lower $^2P_{1\frac{1}{2}}$ limit of Ne II. When the chief quantum number of the series electron is sufficiently large the coupling is roughly (jj) and the problem is simply to find what J values can result from a term having $j_1 = 1\frac{1}{2}$ and a p electron having $j_2 = \frac{1}{2}$ or $1\frac{1}{2}$. When $j_2 = \frac{1}{2}$ the permitted values of J are 1 and 2; while when $j_2 = 1\frac{1}{2}$, J will assume the values 0, 1, 2 or 3. Thus the J values of the terms approaching the lower limit should be 0, 1, 1, 2, 2, 3; while those approaching the upper $^2P_{\frac{1}{2}}$ limit will have J values of 0, 1, 1, 2; and this is in fact the division actually observed.

And these facts may be simply summarised in a cell diagram, constructed to show the limits from top to bottom and the J values from left to right.

This diagram shows that the structure of certain terms is determined uniquely, thus the empirical p_1 term is the only term arising by the addition of a p electron to a $^2P_{\frac{1}{2}}$ limit and having $J=0$; and it is thus the only term whose structure can be expressed symbolically as $(^2P_{\frac{1}{2}}.\mathrm{p})_0$. But unfortunately no method is known of distinguishing two terms which have the same limit and the same value of J, though theory shows clearly that the one arises when the p electron has $j_2=\frac{1}{2}$ and the other when $j_2=1\frac{1}{2}$. p_6 and p_8 must both be written $(^2P_{1\frac{1}{2}}.\mathrm{p})_2$, for they cannot be distinguished as $(^2P_{1\frac{1}{2}}.\mathrm{p}_{\frac{1}{2}})_2$ and $(^2P_{1\frac{1}{2}}.\mathrm{p}_{1\frac{1}{2}})_2$.

The same method when applied to the (^{2}P) d terms of neon or

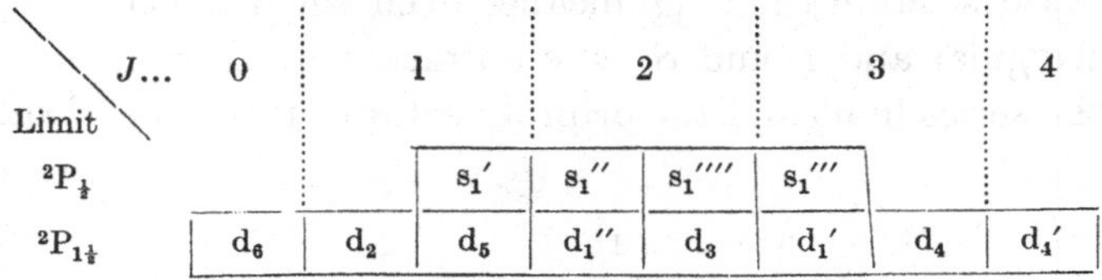

Fig. 19·3. Limits of the (^{2}P)nd terms of neon and argon.

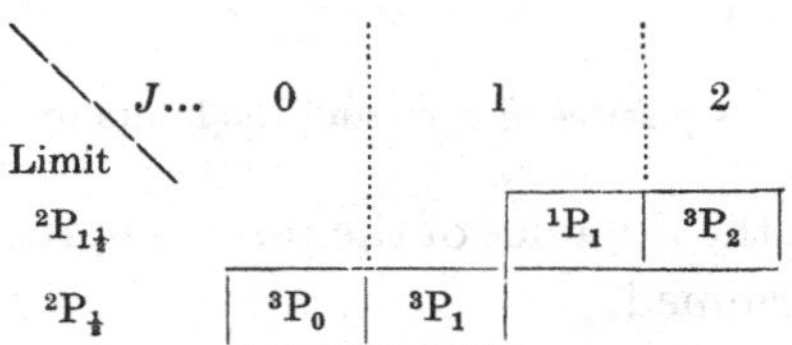

Fig. 19·4. Limits of the (^{2}P)ns terms of silicon.

argon gives Fig. 19·3, while applied to the 3p.ns configuration of silicon it gives Fig. 19·4.

The J values of all known series fit in with these requirements, and the agreement is equally satisfactory whether the low terms are roughly Russell-Saunders as they are in nitrogen and the s terms of neon, or whether a large gap separates the terms tending to the upper and lower limits as in the p and d terms of argon.

2. Hund's theories

The real problem, however, is not to account for the empirical terms by assuming a (jj) coupling, but to name them with the Russell-Saunders symbols.

Towards the solution of this problem Hund has made two

pronouncements. In the first* he fixed the limit towards which a term series must converge, and fixed it given only the ion, the series electron and the symbol of the term; but though this theory is successful enough when both terms and limit are erect it fails badly when the limit is inverted as it is in neon.† Thus the $(^2P)\,ns$ configuration yields two terms having $J=1$, $^3P_1$ and $^1P_1$; Hund's early theory states that of these $^3P_1$ will always approach the $^2P_{\frac{1}{2}}$ limit, and $^1P_1$ the $^2P_{2\frac{1}{2}}$ limit; but though this prediction is correct in silicon it is incorrect in neon.

Hund's revised prediction‡ is less precise and therefore less useful; it has usually been taken to mean that terms arising from

Term	$n=5$	$\Delta\nu$	$n=6$	$\Delta\nu$	Limit
$^3D_3$	19533·2		108313·0		195,195
		730·7		496·0	
$^3D_2$	18802·5		107817·0		194,824
		403·9		507·7	
$^3D_1$	18398·6		107309·3		194,268
		2276·1		−322·2	
$^1D_2$	16122·5		107631·5		195,978

Fig. 19·5. Values of the limits to which the four terms of the 4d.*ns* configuration of Zr III tend.

the same configuration and having the same J do not cross as they approach the limit; and in this form the theory is certainly in closer agreement with experiment. Thus in silicon the $^3P_1$ term lies below the $^1P_1$ term and accordingly $^3P_1$ will approach the lower limit, namely $^2P_{\frac{1}{2}}$ of Si II; in neon $^3P_1$, being s_4 in Paschen's notation, again lies below $^1P_1$ or s_2, so that again $^3P_1$ approaches the lower limit, but this time the limit is inverted and the lower component is $^2P_{1\frac{1}{2}}$.

There are many similar successes to the credit of the revised theory, but it is not without its own problems. Consider the 4d.*ns* configuration of Zr III for example;§ the levels of this configuration are shown in Fig. 19·5, measured upwards from the $^3F_2$ ground term. Each of these pairs of terms is treated as part of a Rydberg sequence, and the limits calculated.

These seem to show that $^1D_2$, though below $^3D_2$ in the low

* Hund, *Linienspektren*, 1927, 184 f.
† Shenstone, *N*, 1928, **121** 619; **122** 727.
‡ Hund, *ZP*, 1929, **52** 601.
§ Kiess, C. C. and Lang, *BSJ*, 1930, **5** 321.

terms, yet tends to a higher limit, for the separation of the limits of $^1D_2$ and $^3D_2$ works out as 1154 cm.$^{-1}$, which compares well with the interval of the ^2D term of Zr IV, which is 1250 cm.$^{-1}$ That the $^1D_2$ sequence should converge to the upper limit is in accord with Hund's early theory, but contrary to the usual interpretation of his later pronouncement.

A letter, however, written by Hund to Mack* shows that his statement has been interpreted too rigidly, for he specifically limited his prediction to terms which showed no symmetry property, and the letter states that any experimental evidence of crossing is to be taken as evidence of a symmetry property of the system. Accordingly the crossing of levels having the same J can never be in disagreement with Hund's conjecture, and the latter is in fact only a convention for naming levels where experiment does not distinguish between them. Mack pointed this out.

But even in this very limited form one may question whether the convention is useful, for if the Russell-Saunders notation is not to have any of its usual implications, it is surely better abandoned, and one which has no implications at all adopted in its place. And in fact Russell has advised that terms whose configuration and coupling are still undetermined shall be specified simply by numbers.

Present theory thus appears unable to make any general pronouncements, but it is not therefore valueless; experiment shows that in a large number of spectra, terms having the same J do not cross, and accordingly the consequences of this hypothesis are worthy of study, if only as an ideal with which the empirical may be compared. This ideal is the more useful in that if all terms obeyed Hund's energy rules only two types of convergence would be found, one when the limit is erect, the other when the limit is inverted. The erect type of convergence is identical with that dictated by Hund's first theory.

Take as an example the convergence of the terms of a (^{3}P) nd configuration when the limit is first erect and then inverted (Figs. 19·6, 19·7). When the limit is inverted all components of a multiplet tend to approach the same limit, but when the limit is

* Mack, *PR*, 1929, **34** 34.

erect they tend to approach different limits; and in this the $({}^3P)\,d$ terms are only one example of a general trend.

A word should be added on the filling up of these figures; the number of cells in each row and their J values have been con-

Limit \ J	$\frac{1}{2}$		$1\frac{1}{2}$		$2\frac{1}{2}$		$3\frac{1}{2}$		$4\frac{1}{2}$
${}^3P_2$	${}^2P_{\frac12}$	${}^4P_{\frac12}$	${}^2P_{1\frac12}$	${}^2D_{1\frac12}$	${}^2D_{2\frac12}$	${}^2F_{2\frac12}$	${}^2F_{3\frac12}$	${}^4D_{3\frac12}$	${}^4F_{4\frac12}$
${}^3P_1$		${}^4D_{\frac12}$	${}^4P_{1\frac12}$	${}^4D_{1\frac12}$	${}^4P_{2\frac12}$	${}^4D_{2\frac12}$	${}^4F_{3\frac12}$		
${}^3P_0$				${}^4F_{1\frac12}$	${}^4F_{2\frac12}$				

Fig. 19·6. Convergence of $({}^3P)nd$ terms when the limit is erect; the low terms are assumed to obey Hund's energy rules and terms with the same J do not cross as they tend to the limit.

Limit \ J	$\frac{1}{2}$		$1\frac{1}{2}$		$2\frac{1}{2}$		$3\frac{1}{2}$		$4\frac{1}{2}$
${}^3P_0$				${}^2D_{1\frac12}$	${}^2D_{2\frac12}$				
${}^3P_1$		${}^2P_{\frac12}$	${}^2P_{1\frac12}$	${}^4P_{1\frac12}$	${}^2F_{2\frac12}$	${}^4P_{2\frac12}$	${}^2F_{3\frac12}$		
${}^3P_2$	${}^4P_{\frac12}$	${}^4D_{\frac12}$	${}^4D_{1\frac12}$	${}^4F_{1\frac12}$	${}^4D_{2\frac12}$	${}^4F_{2\frac12}$	${}^4D_{3\frac12}$	${}^4F_{3\frac12}$	${}^4F_{4\frac12}$

Fig. 19·7. Convergence of $({}^3P)nd$ terms when the limit is inverted, the assumptions being the same as in the previous figure.

sidered earlier; the terms arising from $({}^3P)\,d$ are ${}^4F$, ${}^4D$, ${}^4P$, ${}^2F$, ${}^2D$, ${}^2P$ in that order from below upwards; accordingly, starting with ${}^4F$, insert each of the four components in the lowest cell which its J value permits; having finished ${}^4F$ continue to ${}^4D$ and so through ${}^4P$ and ${}^2(FDP)$ until all seventeen cells are full.

A number of these convergence types are given in Fig. 19·8.

BIBLIOGRAPHY

Chapter XIX of White's *Introduction to atomic spectra* may be usefully read, though its subject "Series Perturbations" is parallel to rather than a development of series limits.

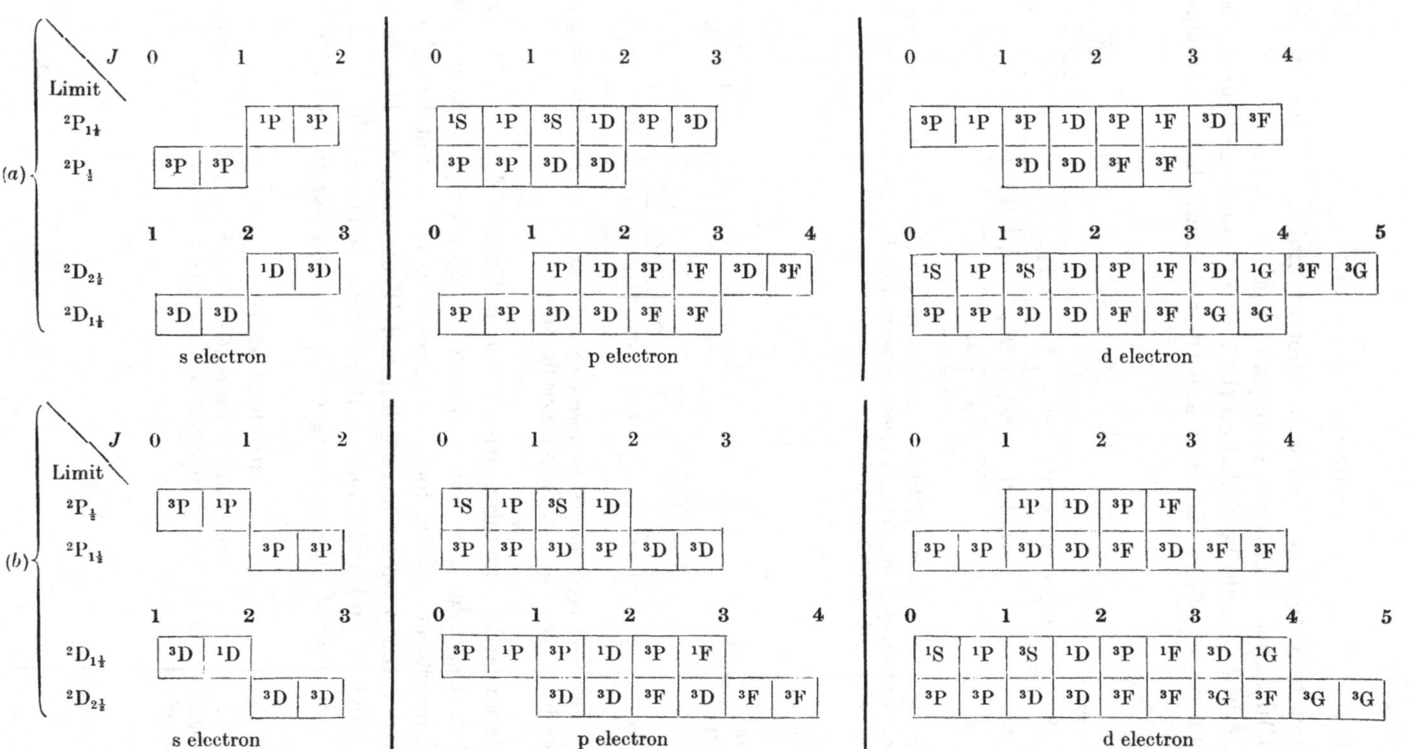

(a)
(b)
J
Limit
$^2P_{1\frac{1}{2}}$
$^2P_{\frac{1}{2}}$
$^2D_{2\frac{1}{2}}$
$^2D_{1\frac{1}{2}}$
s electron
p electron
d electron

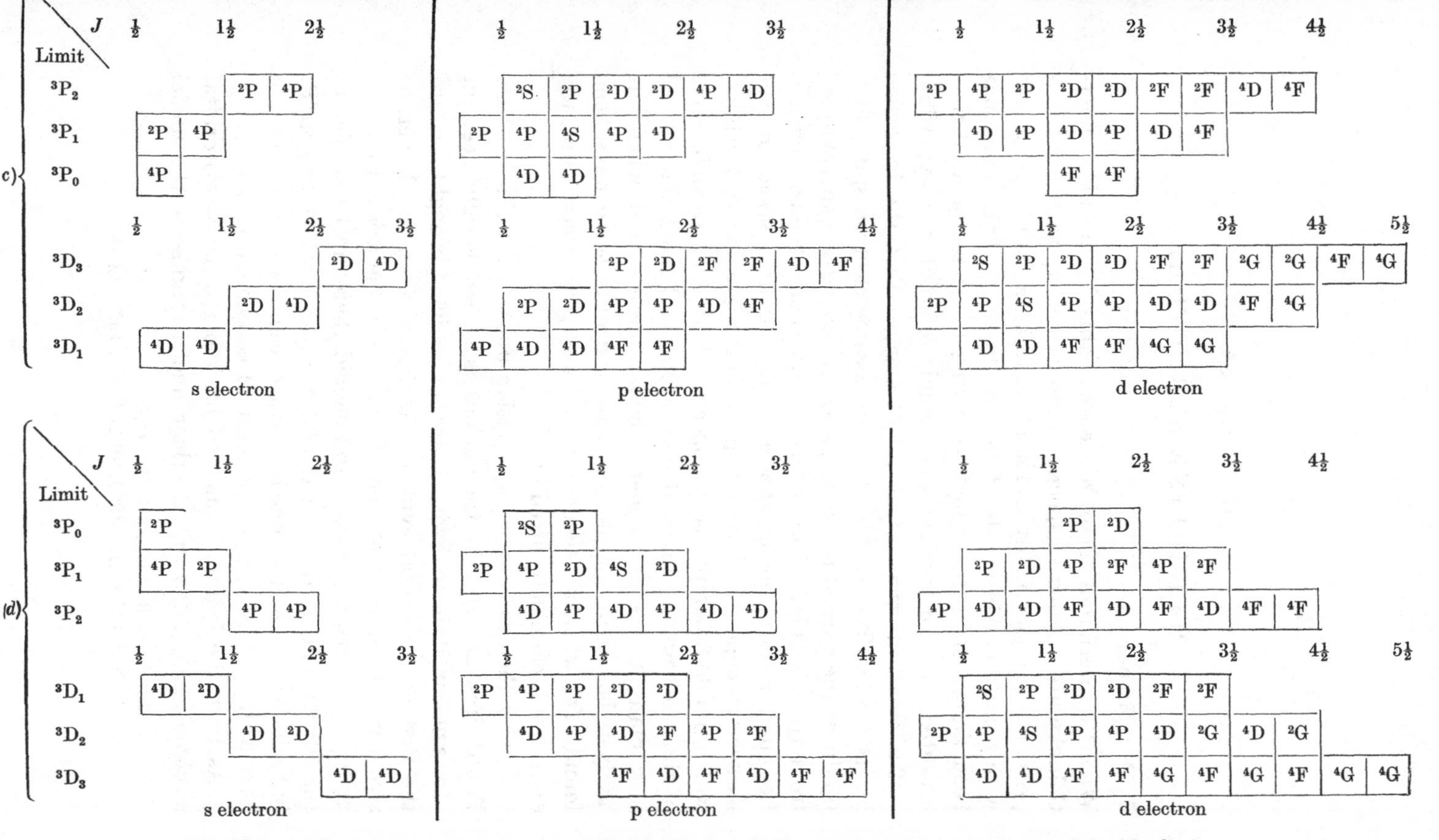

Fig. 19·8. (*a*) Convergence of terms to erect doublet limits. (*b*) Convergence of terms to inverted doublet limits. (*c*) Convergence of terms to erect triplet limits. (*d*) Convergence of terms to inverted triplet limits.

CHAPTER XX

HYPERFINE STRUCTURE

1. Empirical

When the interferometers of Michelson, Fabry-Perot and Lummer-Gehrcke are applied to spectral lines, which appear single in a prism spectroscope, many are found to exhibit a complex structure with component intervals of 0·1 to 1·0 cm.$^{-1}$ This much has been known since these high resolving instruments were first invented, the structure being commonly referred to as hyperfine.

The small intervals alone might suggest that the hyperfine levels can hardly be attributed to the same electron spin which produces the normal multiplet; and this suspicion gains support from the difficulty in fitting them into Hund's term scheme. Caesium, for example, must surely have a doublet structure, yet if a spectroscope of high resolving power is applied to certain lines, which should be simple, they appear as doublets with an interval of about 0·3 cm.$^{-1}$ Time and again Hund's scheme has predicted how many terms an element should possess, and when the analysis has been completed, these and only these have been found. Yet when an element exhibits hyperfine structure, the extra levels cannot be fitted in.

Clearly then some new explanation must be sought. In 1924 Pauli* drew attention to the nucleus as a possible influence; but it was not until the following year that Schüler† brought forward the first experimental evidence in his work on the hyperfine structure of the 5485 A. line of Li II. This line arises from the 2p $^3P_{0,1,2} \rightarrow$ 2s $^3S_1$ transition, and should therefore be similar to the 7065 A. line of He I, for the latter arises from 3s $^3S \rightarrow$ 2p ^3P; but whereas the helium line is a triplet, Schüler showed that the 5485 A. line of lithium has at least 14 components.

As both the helium atom and lithium ion have two orbital electrons, they differ only in their nuclei; and so to the nuclei

* Pauli, *Nw*, 1924, **12** 741.

† Schüler, *AP*, 1925, **76** 292; *ZP*, 1927, **42** 487.

must be attributed the hyperfine structure. Further, lithium has only two isotopes, Li^6 and Li^7, so that even if these did produce separate triplets they would account for only 6 out of the 14 components; the nuclear property responsible must therefore be something other than a simple mass effect.

2. Influence of nuclear mass

In both hydrogen and ionised helium, the series are governed by the relation

$$\nu = Z^2 R \left\{ \frac{1}{n_1^2} - \frac{1}{n_2^2} \right\},$$

but the Rydberg constant, R, has a slightly different value in the two spectra, 109,678 in hydrogen and 109,722 in helium. This small difference Bohr explained as due to the different mass of the two nuclei; for if the mass is infinite

$$R = \frac{2\pi^2 e^4 m_e}{ch^3}. \qquad \ldots\ldots(2{\cdot}9)$$

But when the mass of the nucleus is finite, the electron and nucleus both revolve about the common centre of gravity; to correct for this one substitutes for the mass of the electron, m_e, the quantity

$$\frac{M}{M + m_e} m_e.$$

This theory clearly shows that the spectral lines of the heavy isotope of hydrogen will be displaced, each line of the Balmer series having a weak component on its short wave-length side; moreover, as the mass of the isotope is 2, the intervals of the first four lines, H_α to H_δ, should be 4·16, 5·61, 6·29 and 6·65 $cm.^{-1}$* Photographs show that in fact a weak satellite does occur in this position, and that the satellite is stronger when the concentration of H^2 is increased.

When there is more than one electron, the theory is much more complicated, but Hughes and Eckart† have provided a solution for systems of two and three electrons. The separations of the

* Urey, Brickwedde and Murphy, *PR*, 1932, **40** 1.

† Hughes and Eckart, *PR*, 1930, **36** 694.

lithium isotopes, Li^6 and Li^7, is in reasonable agreement with theory.* The measured intervals are:

Li II	$2p\,^3P \rightarrow 2s\,^3S$	5485 A.	$1{\cdot}06$ cm.$^{-1}$
Li I	$2p\,^2P \rightarrow 2s\,^2S$	6708 A.	$0{\cdot}345$ cm.$^{-1}$
	$3p\,^2P \rightarrow 2s\,^2S$	3233 A.	$0{\cdot}56$ cm.$^{-1}$

For atoms with more than three electrons, no theory has been evolved, and in fact a displacement due to mass alone seems to have been demonstrated only in neon. Each arc line of this element is accompanied by a faint companion of shorter wavelength. If the single electron theory was not invalid, it would suggest that the intervals should be given by $\frac{\delta\nu}{\nu} = 247 . 10^{-8}$, while experiment shows that this ratio assumes the values $437 . 10^{-8}$ in the lines $2p_m \rightarrow 1s_2$, and $368 . 10^{-8}$ in the lines $2p_m \rightarrow 1s_{3,4,5}$. Thus the qualitative agreement is satisfactory.†

3. The extended vector model

The first theory put forward to explain hyperfine structure‡ assigned to the nucleus an angular momentum **I**, and made this combine with the electronic moment **J** to form a resultant atomic moment commonly written **F**. Then **F** must be quantised as well as **I**, and

$$\mathbf{J} + \mathbf{I} = \mathbf{F}.$$

Previous experience with the similar linking of **L** and **S** to form **J** suggests that possibly F will change only by ± 1 or 0, and that the transition from $F = 0$ to $F = 0$ will be forbidden; further, one may hope that the multiplet intervals will satisfy the Landé interval rule. This hypothesis was at first only a guess, but evidence drawn both from experiment and from the calculations of the wave mechanics shows that the guess is fortunate.

Evidence will be adduced first in support of the interval and selection rules; thus certain lines of the bismuth spectrum were

* Hughes, *PR*, 1931, **38** 857. Cf. Granath, *PR*, 1932, **42** 44.

† Hansen, *N*, 1927, **119** 237; Nagaoka and Mishima, *Imp. Acad. Tokyo, Proc.* 1929, **5** 200, 1930, **6** 143; Thomas and Evans, E. J., *PM*, 1930, **10** 128.

‡ Pauli, *Nw*, 1924, **12** 721.

PLATE VII. HYPERFINE STRUCTURE

1. *Potassium resonance lines*, 7699 and 7665 A., $^2P \to {}^2S$. Light from a potassium lamp was examined with a Fabry-Perot étalon, after passing through a beam of potassium travelling at right angles to the line of sight; the absorption pattern possesses a fine doublet structure, the Doppler width in absorption being much less than in emission.

2. *Rubidium resonance lines*, 7800 and 7948 A., $^2P \to {}^2S$. Photographed with a reflection échelon, each line consists of four components; the weak outer components are due to Rb_{87}, and the strong inner components to Rb_{85}. As the hfs is the same for both lines, it arises in the common level ^2S.

3. *Caesium line*, 4555 A., $^2P_{1\frac{1}{2}} \to {}^2S$. The hfs revealed by a Lummer-Gehrcke plate consists of doublets, arising in the ^2S level; the intensity ratio of 1·27:1 shows that $I = 3\frac{1}{2}$. The fringes of 4593 A., $^2P_{\frac{1}{2}} \to {}^2S$, appear very faint.

4. *Gallium resonance line*, 4033 A., $^2S \to {}^2P_{\frac{1}{2}}$. A reflecting échelon grating reveals three lines with intensity ratio of 4·9:6·1:5. The ^2S and ^2P terms have nearly the same interval so that two of the four components overlap; the theoretical patterns are the same for $I = \frac{1}{2}$ and $I = 1\frac{1}{2}$, but the intensity ratios are 1:2:1, and 5:6:5 respectively, thus $I = 1\frac{1}{2}$.

5. *Indium line*, 4101 A., $^2S \to {}^2P_{\frac{1}{2}}$. The photograph on the left is taken with a reflecting échelon grating, while on the right this has been crossed with a Fabry-Perot étalon. Since the four streaks A, B, C, D lie on a diagonal, the pattern consists of these and not of D', A, B, C as the left-hand figure might suggest. The intensity ratio of 2·72:1·82; 1·00:2·74 shows that $I = 4\frac{1}{2}$.

6. *Thallium line*, 5351 A., $^2S \to {}^2P_{1\frac{1}{2}}$. The left-hand photograph was taken with a reflecting échelon, in the right hand this has been crossed with a Fabry-Perot étalon. The line consists of two close doublets, for the components b' and B' can only be b and B appearing in the next order, since on the right the étalon fringes appear at the same height. The small interval is due to isotope displacement, a and b being due to Tl_{203} and A and B to Tl_{205}; the larger interval is due to the hfs of the ^2S term.

7. *Thallium line*, 3776 A., $^2S \to {}^2P_{\frac{1}{2}}$. The six components are arranged in three close doublets, but the pair B and b in the nth order overlap the pair C and c in the $(n+1)^{th}$ order, for when the échelon is crossed with a Fabry-Perot étalon, the fringes in these lines are double.

All photographs were lent by Dr D. A. Jackson.

Plate VII

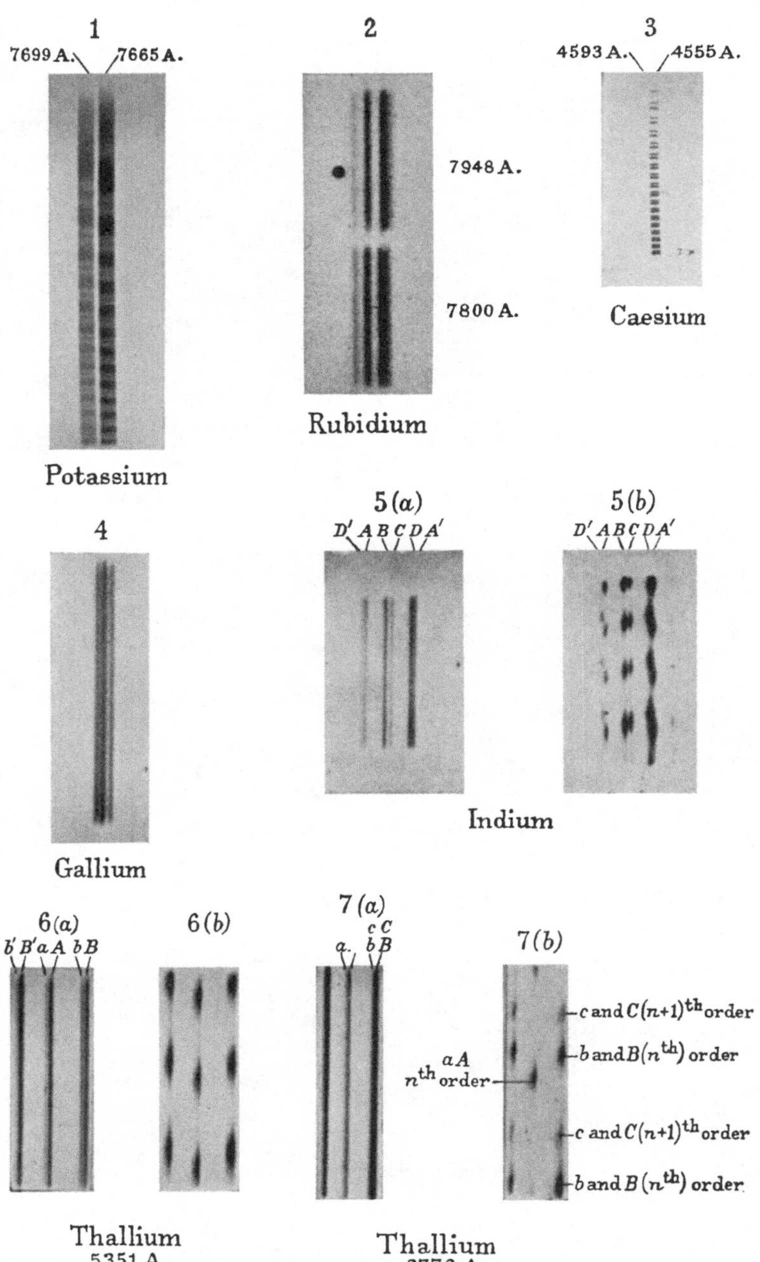

Hyperfine structure

early analysed* and the 4722 A. line, arising from the $p^2.s1_{\frac{1}{2}} \rightarrow p^3.{}^2D_{1\frac{1}{2}}$ transition may be quoted as an example. The observed pattern is shown in Fig. 20·1 and analysis of this shows that the lower term, ${}^2D_{1\frac{1}{2}}$, has intervals of 0·152, 0·198, and 0·255; these best satisfy the interval rule if F is assigned the values 3 to 6; and the vector model allows just these values if I is $4\frac{1}{2}$; moreover, the model predicts always $(2J+1)$ or $(2I+1)$ components, according as I or J is the larger, and in fact the $1_{\frac{1}{2}}$, $8_{1\frac{1}{2}}$ and ${}^2D_{2\frac{1}{2}}$ terms do

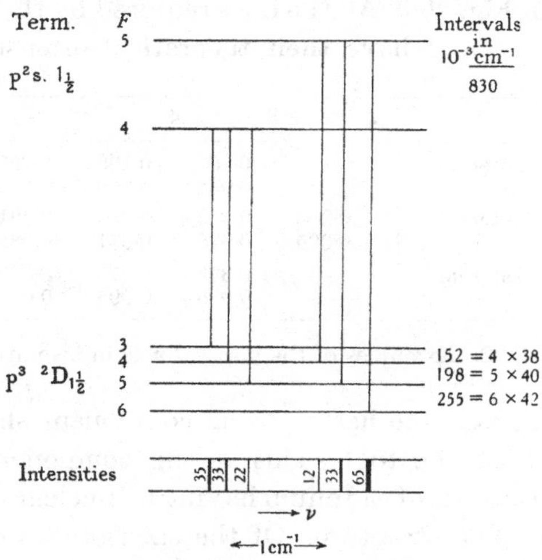

Fig. 20·1. Level diagram showing the structure of the 4722 A. line of Bi I; the intervals given are empirical, the intensities theortical. (After Goudsmit and Back, *ZP*, 1927, **43** 321.)

split into 2, 4 and 6 levels respectively. Of these terms the two last obey the interval rule even better than the ${}^2D_{1\frac{1}{2}}$ term cited above (Fig. 20·2).

The analysis of bismuth by the interval rule has been followed up by work on Mn, Pr, La and Cs. That bismuth was the first element successfully analysed was not due only to chance, but to the existence of only one isotope. For where several isotopes occur, each may have a different nuclear spin, and each spin then produces its own hyperfine pattern.

* Goudsmit and Back, *ZP*, 1927, **43** 321. Zeeman, Back and Goudsmit, *ZP*, 1930, **66** 1.

Thus cadmium possesses six isotopes and these manifest themselves in the structure of the triplet 4678, 4800, 5086 A., which arises from the transition $2\,^3S_1 \rightarrow 2\,^3P_{0,1,2}$.* The three hyperfine patterns have only one interval in common, namely 0·396 cm.$^{-1}$, and the $2\,^3S_1$ term must split therefore to two and only two components; but as the number of components is always $(2J+1)$ or $(2I+1)$, I must be $\frac{1}{2}$; and since this value must hold for all three P levels the term scheme and line pattern should be those shown in Fig. 20·3. All the lines required by this scheme are in fact observed and have their theoretical intensities, but in

Term	*F*	2	3	4	5	6	7
$6p^3.\,^2D_{1\frac{1}{2}}$	Interval		—	−0·152	−0·198	−0·255	—
	A		—	−0·038	−0·040	−0·042	—
$6p^3.\,^2D_{2\frac{1}{2}}$	Interval		0·256	0·312	0·385	0·491	0·563
	A		0·085	0·078	0·077	0·082	0·080
$6p^2.7s.\,8_{1\frac{1}{2}}$	Interval		—	0·379	0·473	0·563	—
	A		—	0·095	0·095	0·094	—

Fig. 20·2. Examples of the interval rule in bismuth.

addition each gross line has a strong component shown by the dotted line A at the foot. This strong component has been attributed to isotopes of cadmium having no nuclear moment and hence no hyperfine structure. Of the six isotopes of cadmium which Aston has identified therefore some have a nuclear moment of $\frac{1}{2}$ and some of zero; and the intensity of A relative to the other components makes it necessary to allocate the value $\frac{1}{2}$ to the odd isotopes 111 and 113, while all the even isotopes 110, 112, 114 and 116 have $I=0$.

Cadmium is peculiar in that the hyperfine levels are inverted, the levels with the largest values of F lying lowest; this arrangement is rather rare among hyperfine structures thus far analysed, though the $^2D_{1\frac{1}{2}}$ level of bismuth examined above happens also to be inverted.

The results thus far cited may be taken to prove that the hyperfine intervals satisfy the interval rule and that the selection rule is $\Delta F = \pm 1$ or 0; but no evidence has been adduced to show

* Schüler and Brück, *ZP*, 1929, **56** 291; **58** 735.

whether the transition from $F=0$ to $F=0$ is specifically forbidden. Consider therefore the 3776 A. line of Tl I, a line which arises from the jump $2\,^2S_{\frac{1}{2}} \rightarrow 2\,^2P_{\frac{1}{2}}$; examined for hyperfine structure this line

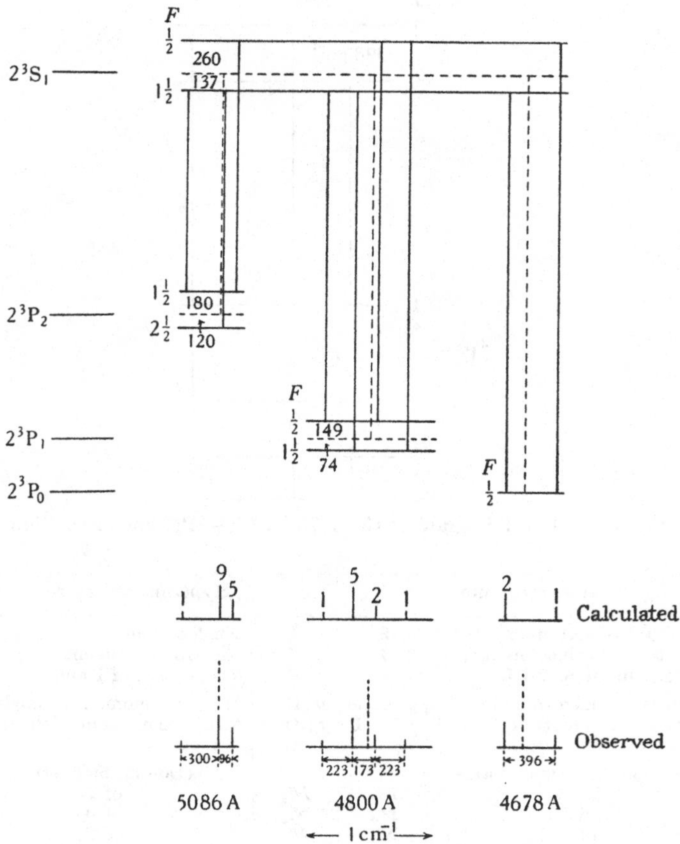

Fig. 20·3. Level diagram showing the structure of three cadmium lines.

reveals an unsymmetrical triplet (Fig. 20·4).* As the initial and final terms have the same value of J, this structure can be explained only if I is $\frac{1}{2}$, and if the jump from $F=0$ to $F=0$ is specifically forbidden; any other value of I would give rise to

* Schüler and Brück, *ZP*, 1929, **55** 575. Schüler and Keyston, *ZP*, 1931, **70** 1.

four components, and should two of these coincide then the triplet would be symmetrical.

That the same selection and interval rules are valid for both

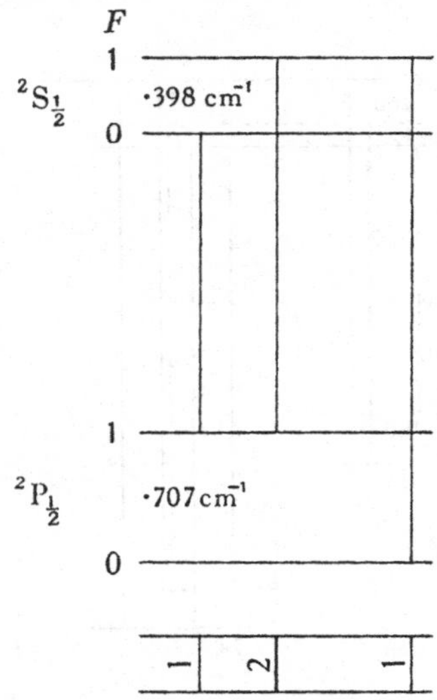

Fig. 20·4. Level diagram of the 3776 A., $^2S_{\frac{1}{2}} \rightarrow {}^2P_{\frac{1}{2}}$, line of thallium.

Gross structure		Hyperfine structure	
Resultant of spin moments	**S**	**I**	Nuclear spin
Resultant of orbital moments	**L**	**J**	Electronic moment
Vector sum of **S** and **L**	**J**	**F**	Vector sum of **I** and **J**
Magnetic moment of spin	$g(\mathbf{S}).\mathbf{S}$	$g(\mathbf{I}).\mathbf{I}$	Magnetic moment of nucleus
Magnetic moment of orbital vector	$g(\mathbf{L}).\mathbf{L}$	$g(\mathbf{J}).\mathbf{J}$	Magnetic moment of electronic vector
Projection on magnetic axis:			Projection on magnetic axis:
of **S**	$\mathbf{M_S}$	$\mathbf{M_I}$	of **I**
of **L**	$\mathbf{M_L}$	$\mathbf{M_J}$	of **J**
of **J**	$\mathbf{M_J}$	$\mathbf{M_F}$	of **F**

Fig. 20·5. Comparison of vectors used to explain gross and hyperfine structures.

gross and hyperfine structures suggests a far-reaching analogy, which is summarised and extended in Fig. 20·5, so as to suggest rules for the Zeeman and Paschen-Back splitting. The angular momenta are given in units $h/2\pi$ and the magnetic moments in Bohr magnetons

$$\frac{h}{2\pi} \cdot \frac{e}{2m_e c}.$$

That the interval rule is valid shows that if the interaction energy of **I** and **J** is E, then

$$\begin{aligned} E/ch &= AIJ\cos(\mathbf{IJ}) \\ &= \tfrac{1}{2}A\,.\{F(F+1)-J(J+1)-I(I+1)\}. \end{aligned} \qquad \text{......(20·1)}$$

The separation of the two levels F and $(F-1)$ is then

$$\tfrac{1}{2}A\,.\{F(F+1)-(F-1)F\}=AF;$$

so that successive intervals in a multiplet term are proportional to the greater of the adjacent quantum numbers.

Hyperfine multiplets usually obey the interval rule so well that exceptions attract an attention, which similarly deviating gross multiplets never obtain. These irregularities are to-day recognised as arising from at least two distinct causes, perturbation and absence of spherical symmetry in the electric field of the nucleus. As perturbation, however, produces also abnormal isotope displacements, and isotope displacements have not yet been considered, these irregularities are better postponed to a later section.

4. Zeeman and Paschen–Back effects

In work on the splitting of lines in the magnetic field, theory has long travelled ahead of experiment, so consider what splitting the vector model predicts.*

In a weak field **J** and **I** combine to form **F**, and the projection of **F** on the magnetic axis **H** is quantised. According to the theory of the Zeeman effect, the increment of energy will be ΔE, where

$$\Delta E/ch = M_F g(F)\,o_m. \qquad \text{......(20·2)}$$

o_m is here an abbreviation for the Lorentz unit

$$\frac{eH}{4\pi m_e c^2} = 4{\cdot}698\,.\,10^{-5}H\ \text{cm.}^{-1},$$

H being measured in gauss; while M_F is restricted by the condition

$$F \geqslant M_F \geqslant -F.$$

This energy equation gives the displacement of a level from the position it occupies when the magnetic field is zero; to obtain the

* Goudsmit and Bacher, *ZP*, 1930, **66** 13.

displacement from the centroid a second term must be added so that the energy becomes

$$E/ch = \tfrac{1}{2}A\,.\{F(F+1) - I(I+1) - J(J+1)\} + M_F g(F)\,o_m. \qquad \text{......(20·3)}$$

A is here the hyperfine interval quotient.

In an earlier chapter $g(J)$ was derived on the assumption that if two vectors **X** and **Y** combine to form a third vector **Z**, then

$$g(Z) = g(X)\frac{X}{Z}\cos(\mathbf{XZ}) + g(Y)\frac{Y}{Z}\cos(\mathbf{YZ}),$$

this expression being itself derived from the wave mechanics. Now $\mathbf{F} = \mathbf{J} + \mathbf{I}$, so that

$$g(F) = g(J)\frac{J}{F}\cos(\mathbf{JF}) + g(I)\frac{I}{F}\cos(\mathbf{IF}). \qquad \text{......(20·4)}$$

The magnitude of the hyperfine splitting itself and the theory of its cause both suggest that $g(I)$ will be very small, about 1/2000 say, so that in all fields which can be applied to the atom $g(I)$ will be negligible, and $g(F)$ may be written as

$$g(F) = g(J)\frac{F(F+1) + J(J+1) - I(I+1)}{2F(F+1)}. \qquad \text{......(20·5)}$$

This completes the description of the weak field. In a strong field **J** and **I** are no longer linked together, but precess independently round the magnetic axis, so that their projections on the magnetic axis have to be quantised separately. As in the theory of the Paschen-Back effect, so here

$$E/ch = AM_I M_J + M_J g(J)\,o_m, \qquad \text{......(20·6)}$$

where $g(I)$ has been treated as negligibly small.

This simple theory should give good agreement in fields so weak that $o_m \ll A$, and in fields so strong that $o_m \gg A$; but in intermediate fields resort must be had to the quantum mechanics. Moreover, in all fields the quantum mechanics should give more accurate numerical values.

A complex theory such as this is most satisfactorily tested, if applied, first to a transition producing a simple pattern, and afterwards to patterns of growing complexity. Accordingly,

Green and Wulff first examined the 3092 A. line of Tl II, which arises from 6s.7s $^1S_0 \rightarrow$ 6s.6p $^1P_1$,* and then turned to more complex lines in thallium and bismuth.† They worked with the first and second spark spectra, because hyperfine intervals are there larger than in the arc; and they examined each line in three

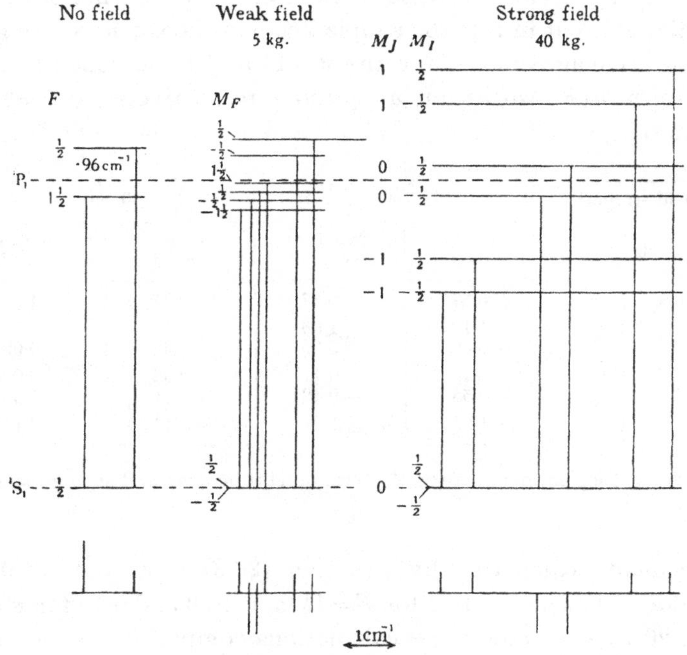

Fig. 20·6. Level diagram showing the splitting of the 3092 A., $^1S_0 \rightarrow {}^1P_1$, line of Tl II in weak and strong magnetic fields. The displacements shown are calculated from the vector model.

fields. In these heavy elements, the Doppler effect does not produce so wide a line as when the atom is light; moreover, thallium and bismuth provide a convenient contrast, for their nuclear moments are $\frac{1}{2}$ and $4\frac{1}{2}$ respectively, so that in a strong field the thallium lines split to two components while the bismuth split to ten.

Consider then the 3092 A. line of Tl II; of the two terms, $^1S_0$ and

* Green and Wulff, *ZP*, 1931, **71** 593.

† Green and Wulff, *PR*, 1931, **38** 2176, 2186.

$^1P_1$ (Fig. 20·6), from which it arises, the first does not split but the second has a hyperfine interval of $-0{\cdot}96$ cm.$^{-1}$, so that the interval quotient A is $-0{\cdot}64$ cm.$^{-1}$; while in the weakest field used, namely 14,700 gauss, the Lorentz unit o_m is 0·69 cm.$^{-1}$ In this field then o_m is not 'much smaller' than $\Delta\nu$ and the splitting is likely to be characteristic of an intermediate rather than of a weak field; but in fact the simple Zeeman theory does give good qualitative agreement. For the $F = 1\frac{1}{2}$ and $\frac{1}{2}$ components of the $^1P_1$ term $g(F)$ works out at $\frac{2}{3}$ and $\frac{4}{3}$ respectively, so that the patterns predicted are $\pm$ (1) 1 3/3 and $\pm$ (2) 2/3, while the dis-

Field strength	14,700 g		43,350 g	
M_F	Vector model	Quantum mechanics	Vector model	Quantum mechanics
$1\frac{1}{2}$	0·37	0·37	1·51	1·51
$\frac{1}{2}$	1·10 −0·09	1·06 −0·05	2·15 0·32	2·17 −0·025
$-\frac{1}{2}$	0·18 −0·55	0·11 −0·48	−0·32 −1·51	0·005 −1·50
$-1\frac{1}{2}$	−1·01	−1·01	−2·15	−2·15

Fig. 20·7. Displacements of the $^1P_1$ term of Tl II in weak and strong magnetic fields.

placements calculated by equation (20·3) work out as 0·37, $-0{\cdot}09$, $-0{\cdot}55$ and $-1{\cdot}01$ for $F = 1\frac{1}{2}$, and 1·10 and 0·18 for $F = \frac{1}{2}$. Fig. 20·7 shows how these calculations compare with the more accurate predictions of the quantum mechanics.

In contrast a field of 43,350 gauss produces a Lorentz unit of 2·02 cm.$^{-1}$, so that splitting characteristic of a strong field is to be expected. In fact theory shows that the 3092 A. line splits to six components, whose displacements are $\pm$ (0·32), 0·32, 1·51, 2·15. The pattern thus consists of three pairs of lines, one at the centroid of the hyperfine doublet and the other two arranged symmetrically on either side; and in general terms this is what experiment reveals, though the numerical agreement is not close.

Better numerical agreement is, however, obtained if the more precise theory of the wave mechanics is substituted for the vector model (Fig. 20·8, 9); the energies of the six $^1P_1$ terms are then

$$\left.\begin{array}{ll} M_F = 1\tfrac{1}{2} & E_\nu = \tfrac{1}{2}A + go_m, \\ M_F = \tfrac{1}{2} & E_\nu^2 + E_\nu(\tfrac{1}{2}A - go_m) - \tfrac{1}{2}A^2 = 0, \\ M_F = -\tfrac{1}{2} & E_\nu^2 + E_\nu(\tfrac{1}{2}A + go_m) - \tfrac{1}{2}A^2 = 0, \\ M_F = -1\tfrac{1}{2} & E_\nu = \tfrac{1}{2}A - go_m. \end{array}\right\} \quad \ldots\ldots(20{\cdot}7)$$

E_ν is here to be measured in wave-numbers, and g is an abbreviation for $g(J)$ not $g(F)$; moreover, g must be assigned the value

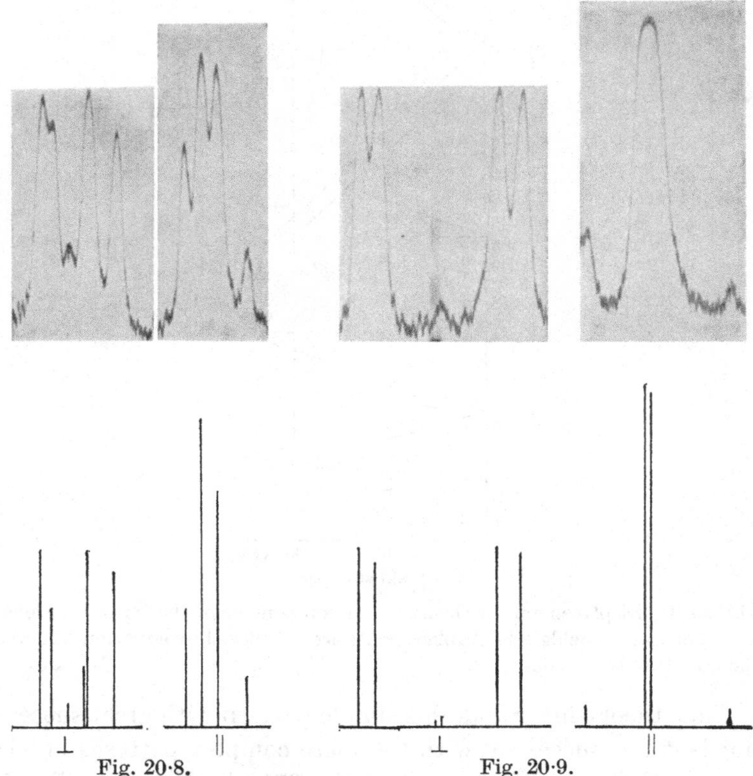

Fig. 20·8. Fig. 20·9.

Fig. 20·8. Splitting of the 3092 A., $^1P_1 \rightarrow {}^1S_0$, line of Tl II in a field of 14,700 gauss. The upper curve is the microphotometer trace taken from a photograph; below is the theoretical pattern. (After Green and Wulff, *ZP*, 1931, **71** 597.)

Fig. 20·9. Splitting of the 3092 A. line of Tl II in a field of 43,350 gauss. (After Green and Wulff, *ZP*, 1931, **71** 599.)

1·025 instead of 1 since the coupling is not pure (**LS**), but shows some signs of the (**jj**) type so that the g factor of the $^1P_1$ term is influenced by other terms which lie near; the numerical value of the correction is obtained by the method of Houston.*

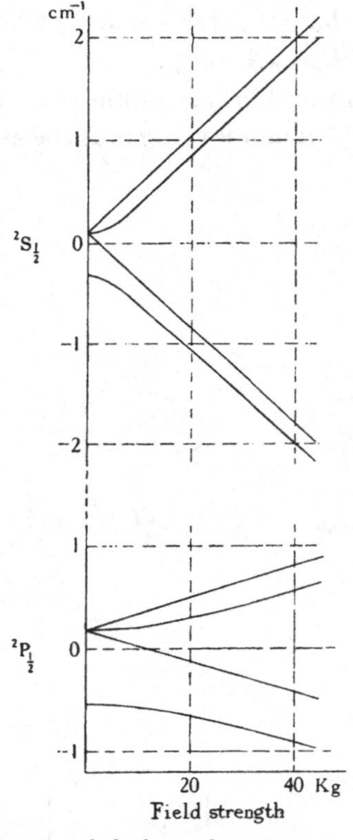

Fig. 20·10. Displacements of the hyperfine components of the $^2S_{\frac{1}{2}}$ and $^2P_{\frac{1}{2}}$ levels of Tl I in various fields. The displacements are calculated by equations (20·8) of the quantum mechanics.

Thus theory interprets this simple pattern with great success; nor is it less successful with the more complex patterns of Tl I. Back and Wulff† photographed the 3776 A., $^2S_{\frac{1}{2}} \rightarrow {}^2P_{\frac{1}{2}}$, line in fields of 17,050, 29,700 and 43,350 gauss. In these fields the Lorentz unit assumes the values 0·80, 1·40 and 2·08 cm.$^{-1}$, while

* Houston, *PR*, 1929, **33** 297.

† Back and Wulff, *ZP*, 1930, **66** 31.

the hyperfine intervals of the $^2S_{\frac{1}{2}}$ and $^2P_{\frac{1}{2}}$ terms are 0·40 and 0·71 respectively, so that all three fields are technically 'strong'. The magnetic field and nuclear moment would thus separately split each term in two; the Zeeman components being determined by $M_J = \pm \frac{1}{2}$, and the hyperfine components by $M_I = \pm \frac{1}{2}$. According to the wave mechanics the displacements of these four levels from the centroid are given by the equations:

$$\begin{array}{llll} M_J = \frac{1}{2} & M_I = \frac{1}{2} & M_F = 1 & E/ch = A/4 + \frac{1}{2} g o_m \\ \frac{1}{2} & -\frac{1}{2} & 0 & = -A/4 + \frac{1}{2}\sqrt{A^2 + g^2 o_m{}^2} \\ -\frac{1}{2} & \frac{1}{2} & 0 & = -A/4 - \frac{1}{2}\sqrt{A^2 + g^2 o_m{}^2} \\ -\frac{1}{2} & -\frac{1}{2} & -1 & = A/4 - \frac{1}{2} g o_m \quad \ldots (20 \cdot 8) \end{array}$$

and shown graphically in Fig. 20·10. As previously g is here an abbreviation for $g(J)$.

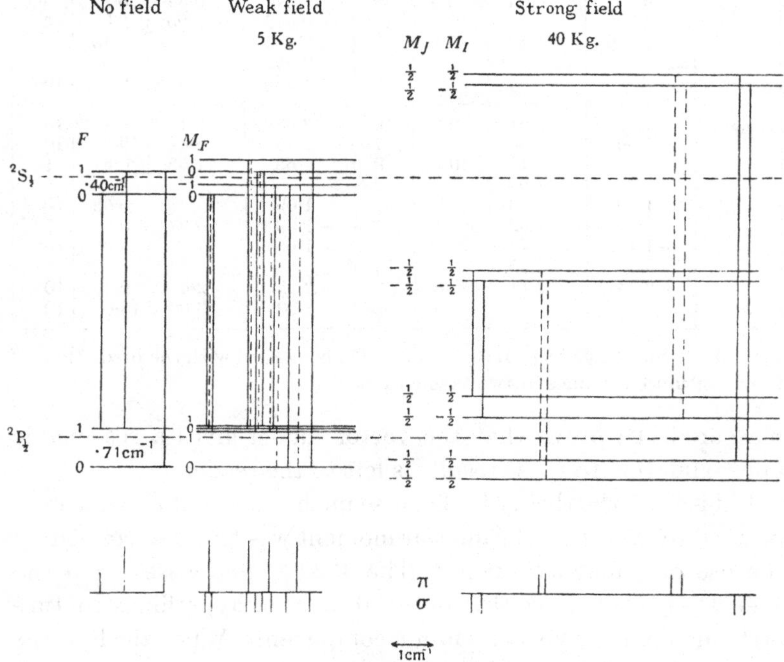

Fig. 20·11. Level diagram to show the structure of the 3776 A., $^2S_{\frac{1}{2}} \rightarrow {}^2P_{\frac{1}{2}}$, line of Tl I in weak and strong magnetic fields. The diagram assumes that the laws of the vector model are valid.

When the four levels of each gross term combine, they would give rise to six π and eight σ components in a weak field, but in a strong field two π and four σ components fade, for in a strong field transitions in which M_I changes its value are no longer allowed (Fig. 20·11). The empirical displacements are compared with this theory in Fig. 20·12, the centroid of the empirical pattern being adjusted so that the line of lowest frequency has the value dictated by theory. The empirical intensities are estimated, not

Field in gauss	Parallel components				Perpendicular components			
	Position (cm.$^{-1}$)		Intensity		Position (cm.$^{-1}$)		Intensity	
	Calc.	Obs.	Calc.	Obs.	Calc.	Obs.	Calc.	Obs.
17,050	1·344	1·39	3·1	3	1·521	1·62	8·0	9
	0·456 }	0·50	6·9 }	10	0·810	0·93	9·85	9
	0·455 }		10·0 }		0·633	0·76	2·0	1
	−0·302	−0·24	6·9	8	0·278	0·34	0·15	0·5
	−0·609	−0·57	10	10	−0·075	−0·05	2·0	3
	−1·190	−1·19	3·1	3	−0·863	—	0·15	—
					−0·963	−0·94	8·0	6
					−1·368	−1·368	9·85	10
43,350	2·89	2·91	0·8	6	3·08	3·11	9·4	10
	1·36	1·50	9·2	7	2·45	2·46	10·0	10
	1·28	1·17	10	9	1·55	1·59	0·6	1
		−1·06		2				
	−1·20	(1·19)	9·2		1·09	—	0·02	—
		−1·32		7	−1·00	−0·97	0·6	1
	−1·44	(1·43)	10		−1·65	—	0·02	—
		−1·55		10				
	−2·74	−2·74	0·8	6	−2·52	−2·52	9·4	10
					−3·00	−3·00	10·0	10

Fig. 20·12. Splitting of the 3776 A. line of Tl I compared with the predictions of the quantum mechanics in two magnetic fields.

measured. To prove that the vector model gives a reasonable approximation to these results is left to the reader.

In bismuth the theory is of course unchanged, but one line may be cited to show how the nuclear moment was first determined by the use of a magnetic field.* The 4722 A. line arises from the $1_{\frac{1}{2}} \rightarrow {}^2D_{1\frac{1}{2}}$ transition, so that in the absence of hyperfine structure it should split to two π and four σ components. When the line was examined in a field of 43,340 gauss, these components were found,

* Back and Goudsmit, *ZP*, 1928, **47** 174. Zeeman, Back and Goudsmit, *ZP*, 1930, **66** 1.

but all were exceptionally wide, and four were clearly resolved into ten components (Fig. 20·13). Now the energy of a magnetic level is

$$E/ch = AM_I M_J + M_J g(J) o_m,$$

where the second term determines the gross Zeeman level and the first term its hyperfine structure; as M_I can assume $(2I+1)$

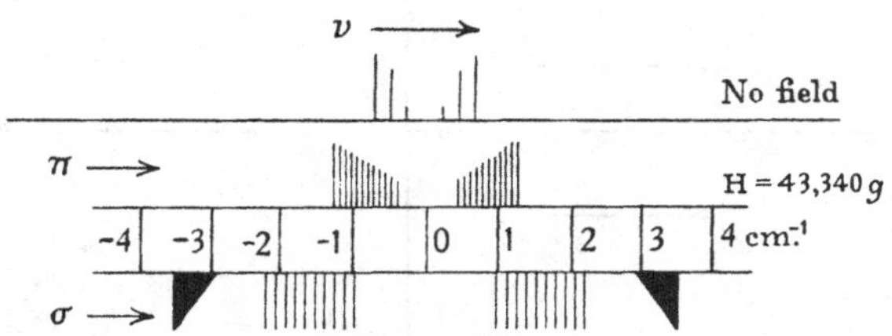

Fig. 20·13. Empirical structure of the 4722 A., $1_{\frac{1}{2}} \rightarrow {}^2D^\circ_{1\frac{1}{2}}$, line of Bi I in a strong magnetic field. (After Back and Goudsmit, *ZP*, 1928, **47** 179.)

values, each gross term will have $(2I+1)$ components (Fig. 20·14). Moreover, as M_I is not allowed to change in a transition the gross Zeeman lines will also split to $(2I+1)$ components, the interval being $A'M_J' - A''M_J''$ (Fig. 20·15). Thus, if in fact the lines split to ten components the nuclear moment must be $4\frac{1}{2}$. And it is of some interest to note that the value of the interval was also confirmed; the hyperfine interval quotients of the ${}^2D_{1\frac{1}{2}}$ and $1_{\frac{1}{2}}$ terms are known to be $-0{\cdot}0403$ and $0{\cdot}166$ cm.$^{-1}$ respectively, so that the interval of the π components should be

$$\begin{aligned} A'M_J' - A''M_J'' &= -0{\cdot}0403\,(-\tfrac{1}{2}) - 0{\cdot}166\,(-\tfrac{1}{2}) \\ &= 0{\cdot}1031 \text{ cm.}^{-1}, \end{aligned}$$

and of the outer σ components

$$\begin{aligned} A'M_J' - A''M_J'' &= -0{\cdot}0403 \times \tfrac{1}{2} - 0{\cdot}166\,(-\tfrac{1}{2}) \\ &= 0{\cdot}0628 \text{ cm.}^{-1} \end{aligned}$$

Now the width of the π components was measured as $0{\cdot}880$ cm.$^{-1}$, so that the interval between successive components must be $0{\cdot}880/9$ or $0{\cdot}0977$ cm.$^{-1}$ The width of the outer σ components is $0{\cdot}520$ cm.$^{-1}$, and the interval accordingly $0{\cdot}0577$ cm.$^{-1}$ Both are in satisfactory agreement with theory.

In Figs. 20·16–17 evidence is presented for two still more complex lines, 2298 A., $2\,{}^3S_1 \rightarrow 2\,{}^3P_2$ of Tl II and 5719 A., ${}^3P_0 \rightarrow {}^3P_1^\circ$ of

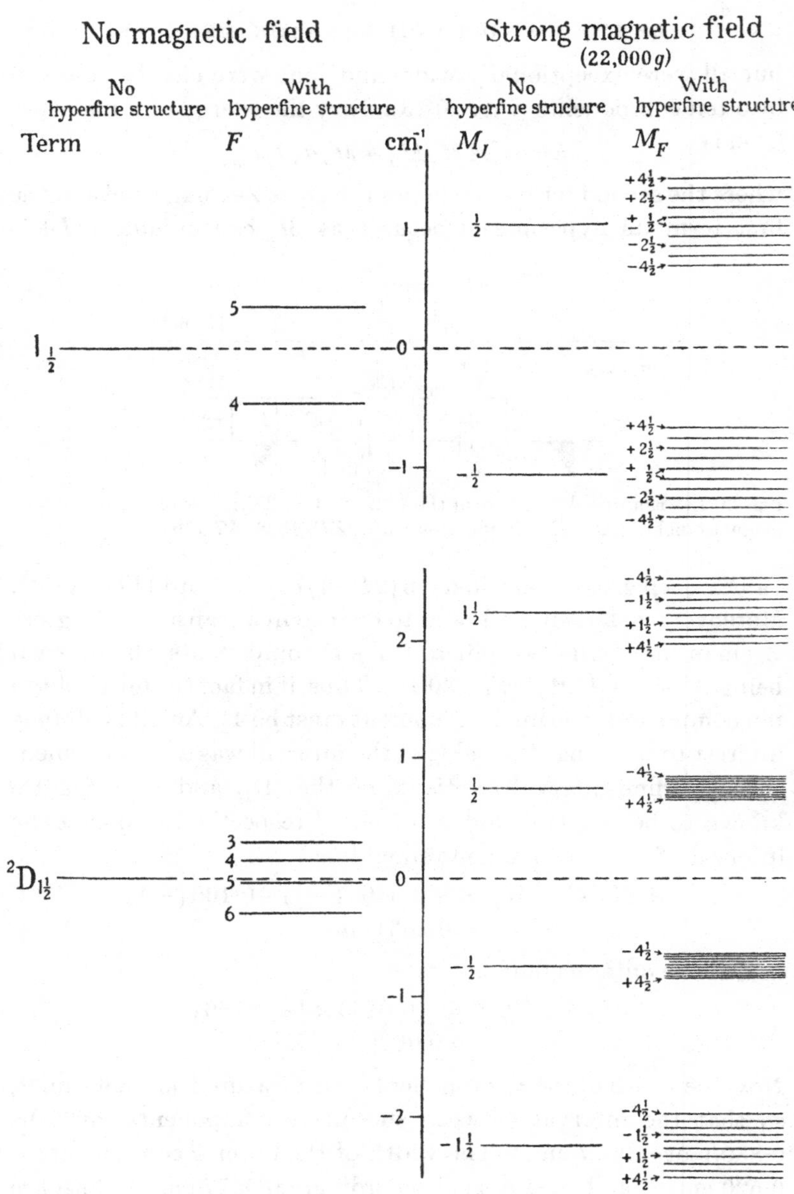

Fig. 20·14. Structure of the $1_{\frac{1}{2}}$ and $^2D^\circ_{1\frac{1}{2}}$ levels, which produce the 4722 A. line of Bi I. The laws of the vector model are assumed, and the field strength has been reduced to half that actually used to keep the figure within reasonable limits; the g factors are abnormal being 1·225 and 2·088 in the two terms.

Bi II;* in each figure the photometer curve appears above and the theoretical pattern below, and in both the agreement is all that can be desired.

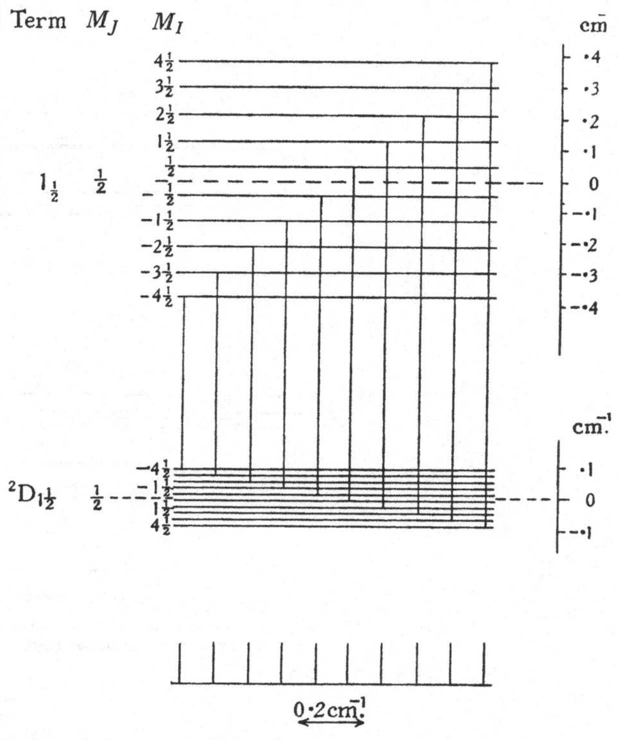

Fig. 20·15. Level diagram to show the hyperfine structure of one gross σ component of the 4722 A. line of Bi I.

5. Intensities

No absolute measurements of intensity seem to have been made, but the photometric curves obtained by Zeeman, Back and Goudsmit in 1930 showed that the intensity formulae deduced from the wave mechanics are qualitatively correct; while more recently these rules have been so widely applied to the analysis of the complex structures described in the next section and the results have been so satisfactory, that few serious deviations can exist.

* Green and Wulff, *PR*, 1931, **38** 2182, 2186.

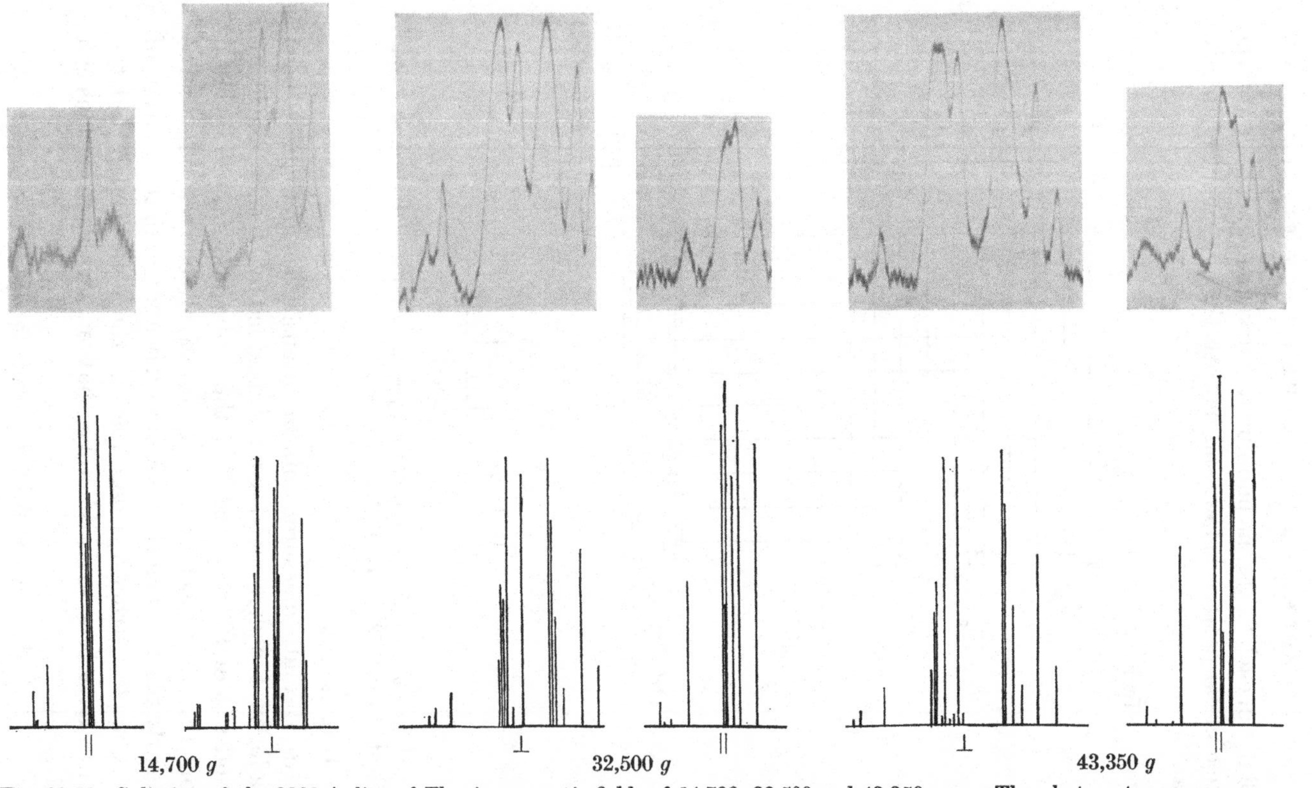

Fig. 20·16. Splitting of the 2298 A. line of Tl II in magnetic fields of 14,700, 32,500 and 43,350 gauss. The photometer curve appears above and the theoretical pattern below. The line arises as 6s.7s $^3S_1 \rightarrow$ 6s 6p $^3P_2$. (After Green and Wulff, *PR*, 1931, **38** 2182.)

In order that the hyperfine components of a line may be observed with their theoretical intensities, however, the exciting source must fulfil certain conditions. Thus the temperature of the source must be as low as possible so as to reduce the Doppler broadening of the lines; the source is therefore worked in liquid air or liquid hydrogen. And secondly, the vapour pressure of the

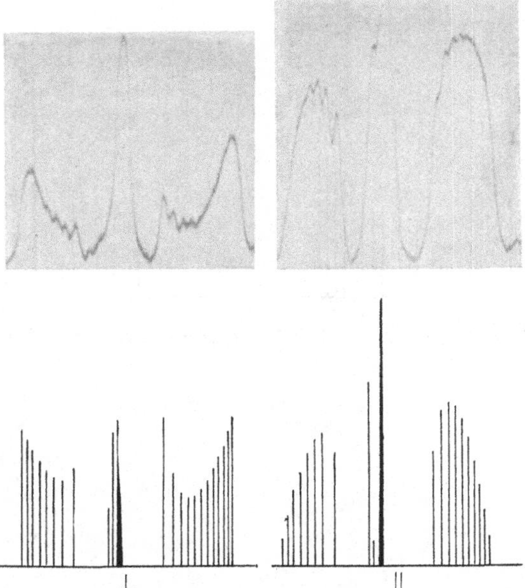

Fig. 20·17*a*. Splitting of the 5719 A. line of Bi II in a magnetic field of 14,700 gauss; again the photometer curve is above and theory below. The line arises as $6p.7p\ ^3P_0 \rightarrow 6p.7s\ ^3P_1$. (After Green and Wulff, *PR*, 1931, **38** 2186.)

element in the source must be kept low, for high pressure leads to self-absorption; besides hyperfine intensities seem in some way disturbed by inter-atomic fields, though why they should be is not yet clear.*

The intensity formulae, deduced from the wave mechanics by Hill,† correspond to the multiplet intensity formulae with the transformation (Fig. 20·5) already used in the interval rule.

* Schüler and Keyston, *ZP*, 1931, **71** 413.

† Hill, *Nat. Acad. Sci., Proc.* 1929, **15** 779.

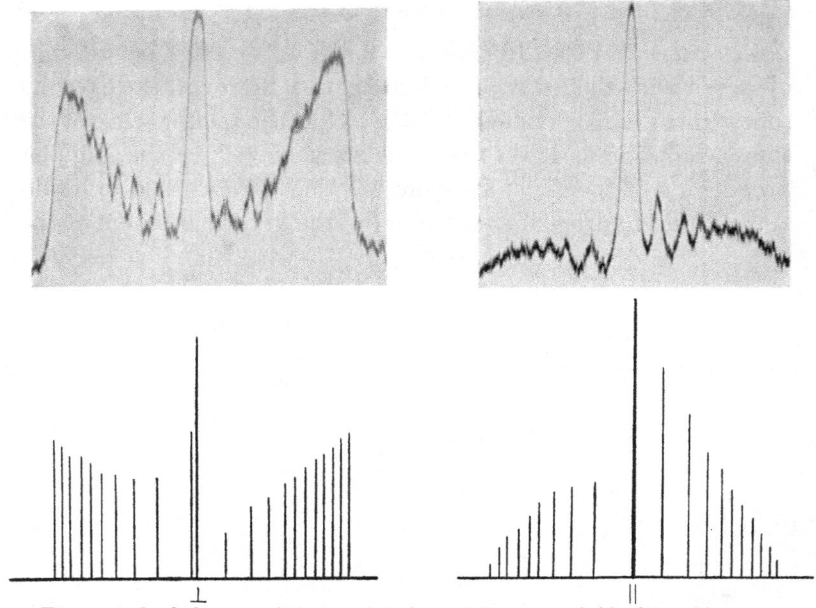

Fig. 20·17*b*. Splitting of the 5719 A. line of Bi II in a field of 32,500 gauss.

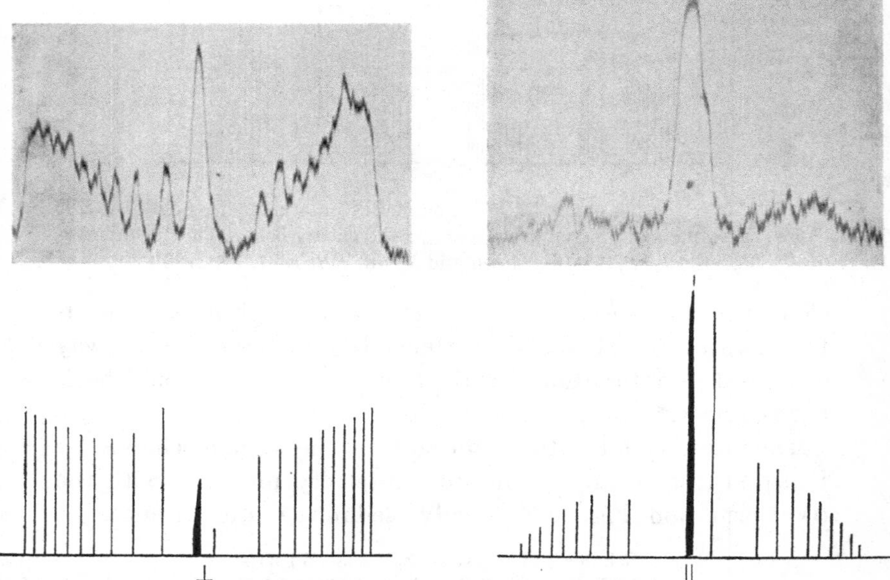

Fig. 20·17*c*. Splitting of the 5719 A. line of Bi II in a field of 43,350 gauss.

In the jump $J \to (J-1)$ and

$$F \to (F-1) \qquad I_- = \frac{1}{F} P(F) . P(F-1),$$

$$F \to F \qquad I_0 = \frac{(2F+1)}{F(F+1)} . P(F) . Q(F),$$

$$(F-1) \to F \qquad I_+ = \frac{1}{F} . Q(F) . Q(F-1),$$

while in the jump $J \to J$, and

$$F \to F \qquad I_0 = \frac{(2F+1)}{F(F+1)} . R^2(F),$$

$$\begin{matrix} F \to (F-1) \\ \text{or } (F-1) \to F \end{matrix} \qquad I_\pm = \frac{1}{F} P(F) . Q(F-1),$$

where

$$P(F) = (F+J)(F+J+1) - I(I+1),$$
$$Q(F) = I(I+1) - (F-J)(F-J+1),$$
$$R(F) = F(F+1) + J(J+1) - I(I+1).$$

In all these formulae the transitions are so chosen that J and F are the larger of the two quantum numbers involved. Moreover, a term containing only J and I has been dropped since in fact the formulae are applied to the components of a single gross line.

The Zeeman intensity formulae may be transformed in the same way and compared with the visual estimates of Back and Wulff.*

6. Isotope displacement

The vector model, based on the hypothesis of nuclear spin, solves many of the problems of hyperfine structure, but there is clear evidence that when several isotopes exist, the vector model does not suffice.

Consider, for example, the Tl I line, 5351 A.,† which arises as $7s\,^2S_{\frac{1}{2}} \to 6p\,^2P_{1\frac{1}{2}}$; work on the 3776 A. line of Tl has already been adduced to show that the nuclear moment of Tl is $\frac{1}{2}$, so that the 5351 A. line might be expected to split to three components with an intensity ratio of 5 : 2 : 1 as shown in Fig. 20·18; but in fact the

* Back and Wulff, *ZP*, 1930, **66** 31.

† Schüler and Keyston, *ZP*, 1931, **70** 1.

line splits to four components, and these have intensities roughly in the ratio of 6·9 : 3 : 2·3 : 1, that is two pairs with an intensity ratio of 3 : 1; now the three lines of theory can be reduced to two with an intensity ratio of 3 : 1 if the two components of the $^2P_{1\frac{1}{2}}$ term are not resolved. Accordingly, the empirical facts can be explained if there are two isotopes of Tl with an abundance ratio of 2·3 : 1, and if the term schemes of these two isotopes are displaced slightly relative to one another; in short that is if the term

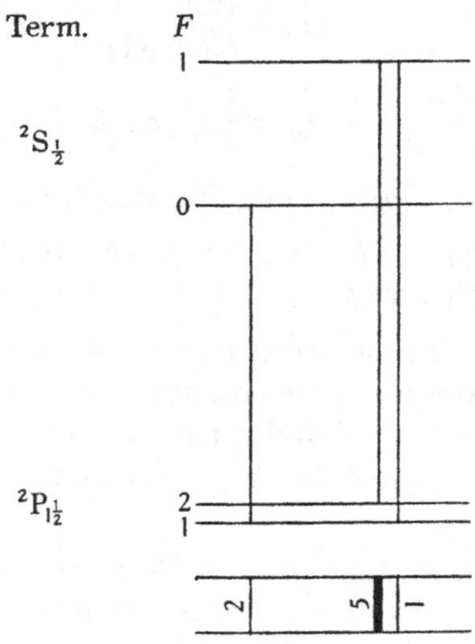

Fig. 20·18. Structure of a normal $^2S_{\frac{1}{2}} \rightarrow {}^2P_{1\frac{1}{2}}$ line.

scheme of Fig. 20·19 is postulated. In this diagram the energy of the $^2S_{\frac{1}{2}}$ term is supposed the same in both isotopes, though in fact only the displacements of the one term relative to the other can be measured; experiment then shows that the $^2P_{1\frac{1}{2}}$ term is displaced through 55 X. in one of the isotopes 203 and 205. The existence of the isotopes 203 and 205 in the abundance ratio of 1 : 2·3 has been confirmed by the analysis of many Tl II lines, while more recently Aston* has found a ratio of 1 : 2·40.

Isotopes have been observed displaced in the lines of many elements, but of all mercury is the most complex, for it consists of

* Aston, *PRS*, 1932, **134** 571.

six isotopes;* of these the four of even atomic weight have zero moment, but the two odd isotopes have different nuclear moments, namely $\frac{1}{2}$ and $1\frac{1}{2}$, and to make confusion worse confounded the terms of one are inverted.

To illustrate the methods by which such a complicated line

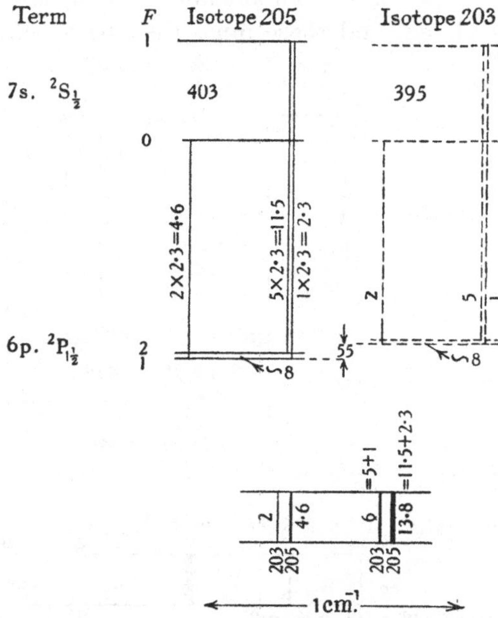

Fig. 20·19. Structure of the 5351 A., $^2S_{\frac{1}{2}} \rightarrow {}^2P_{1\frac{1}{2}}$, line of Tl I. (After Schüler and Keyston, *ZP*, 1931, **70** 3.)

structure can be analysed, consider two mercury arc lines, 2536 and 4078 A.,† which arise as $2\,^3P_1 \rightarrow 1\,^1S_0$ and $2\,^1S_0 \rightarrow 2\,^3P_1$; these lines are in fact particularly simple, for in one term J is zero.

In the analyses described heretofore, three criteria have been used: first, the picking of a common hyperfine interval from lines with a common gross structure level; second, the interval rule; and third the intensity rules. Of these the first two suffice for the

* Recent work has revealed more than six isotopes, but in work on hyperfine structure only those present in a proportion of more than 1 per cent. need usually be considered.

† Schüler and Keyston, *ZP*, 1931, **72** 423.

analysis, when the structure is simple, but the last is the most valuable when the structure is complex. That the method has been so successful is perhaps the best evidence of the wide validity of the intensity formulae.

To obtain the theoretical intensities of the components of a line, the intensity ratios of the components due to a single isotope must first be calculated, and these must then be weighted with the

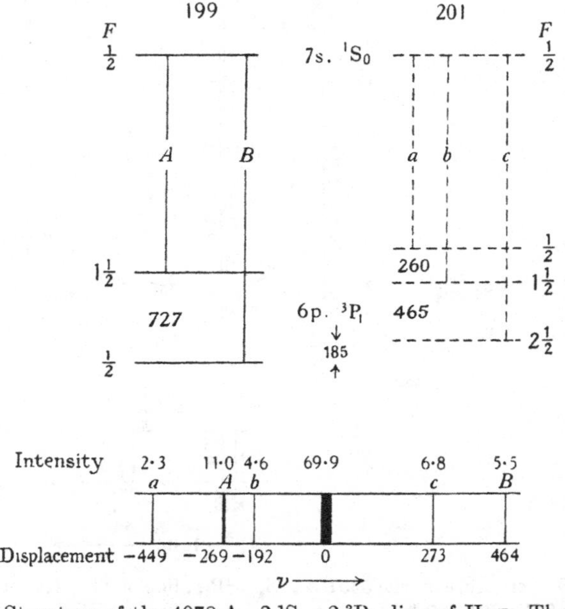

Fig. 20·20. Structure of the 4078 A., 2 $^1S_0 \rightarrow 2\ ^3P_1$, line of Hg I. The displacements are given in 10^{-3} cm.$^{-1}$ (After Schüler and Keyston, *ZP*, 1931, **72** 428.)

isotope abundance. Thus the even isotopes produce only one component each, but the isotopes 199 and 201, having nuclear moments of $\frac{1}{2}$ and $1\frac{1}{2}$, produce two and three components respectively. These are shown as A, B and a, b, c in Figs. 20·20–21, and the theoretical intensity ratios are calculated as 2 : 1 and 1 : 2 : 3. But the mass spectrograph has shown that mercury contains 16·44 per cent. of isotope 199, so that the absolute intensity of the A component should be $\frac{2}{3}$ of 16·44 or 10·96; and similarly since isotope 201 occurs to the extent of 13·68 per cent., the line b

should be of absolute intensity $\frac{1}{3}$ of 13·68 or 4·56. The intensities calculated in this way are summarised in Fig. 20·22.

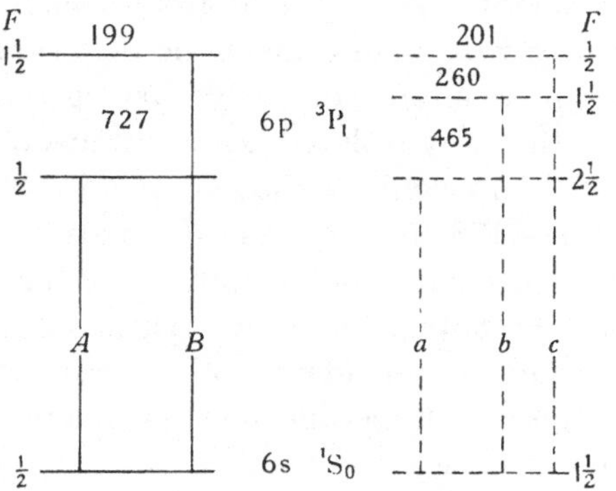

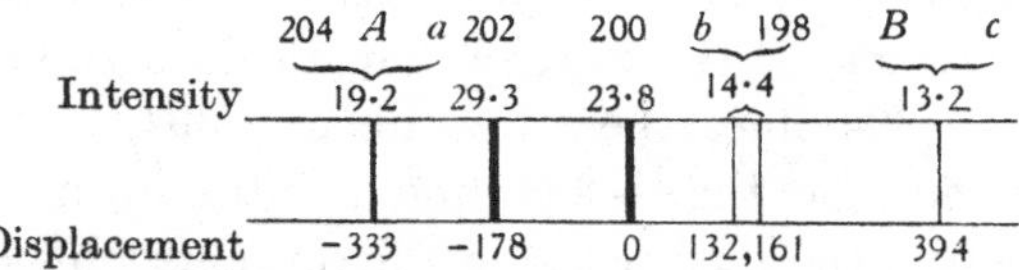

Fig. 20·21. Structure of the 2536 A., 2 $^3P_1 \to 1\ ^1S_0$, line of Hg I. The displacements are again in 10^{-3} cm.$^{-1}$ (After Schüler and Keyston, *ZP*, 1931, **72** 434.)

Isotope	Abundance ratio	Component	Relative intensity	Absolute intensity
199	16·44	*A*	2	10·96
		B	1	5·48
201	13·68	*a*	1	2·28
		b	2	4·56
		c	3	6·84
198	9·89	No hyperfine structure		9·89
200	23·77			23·77
202	29·27			29·27
204	6·85			6·85
	99·90			99·90

Fig. 20·22. Intensities in a $^3P_1 \to {}^1S_0$ or $^1S_0 \to {}^3P_1$ line of Hg I.

With these predictions in mind, the analysis of the structures shown in Figs. 20·20 and 21 presents no great difficulty, provided one understands that components separated by less than 0·030 cm.$^{-1}$ are not resolved.

This somewhat complicated analysis of the 2536 A. line has

been beautifully confirmed by Mrozowski,* who has shown that when separately excited the various isotopes behave like a mixture of independent gases. If light from a mercury arc is sent through mercury vapour in a magnetic field, various hyperfine components can be filtered out. By varying the field strength Mrozowski in this way isolated three groups of lines, (*a*) the 394.10^{-3} cm.$^{-1}$ component, (*b*) the $+161$ and -333 components, (*c*) the 0·0 and -178 components. No matter which of these groups is used to excite cold mercury vapour, the resonance radiation contains only that group; when only the $+161$ and -333 components are incident, only these two components are emitted. Thus each hyperfine line is itself a resonance line and there is no fluorescence.

Fluorescence appears, however, when a little helium is added to the tube,† for the increased pressure means inelastic collisions for the mercury atoms. If the cold vapour mixed with helium is excited by the $+394$ line, the three lines $+394$, $+161$ and -333 are emitted, for the $+394$ line excites both the 199 and 201 isotopes and these after losing a little energy in colliding with a helium atom can return to the ground state while emitting the $+161$ or -333 line. The details are easily followed in Fig. 20·21. On the other hand, irradiate the vapour with the 0·0 and -178 lines, and only these lines are emitted, for these lines excite only the 200 and 202 isotopes, and as these have only one upper level they must radiate what they absorbed.

The analysis of a single line (Fig. 20·23) does not determine the absolute isotope displacements of either term; it determines only the displacements of the isotopes in one term relative to their displacements in the other; and what is true of the analysis of a single line is true of the analysis of any number of lines, so that apparently the absolute displacements can never be determined. Many lines, however, which arise between high levels of the mercury atom show no hyperfine structure at all, so that all the terms concerned must have identical isotope displacements, and

* Mrozowski, *Sci. Bull. Acad. Pol.* 1930, 464; 1931, 489. The lines with displacements of 132 and 161.10^{-3} cm.$^{-1}$ in Fig. 2021 were not resolved in these experiments; they are referred to on this page as the 161 line.

† Mrozowski, *ZP*, 1932, **78** 826.

this is extremely improbable unless the displacements of all are zero. In support of this thesis, theory and experiment show that isotope displacements must and do decrease, as the chief quantum number increases (Figs. 20·24–26).

The hyperfine levels thus obtained may be simplified by substituting for the odd components a single level at the centroid of the isotope, and when this is done an interesting regularity

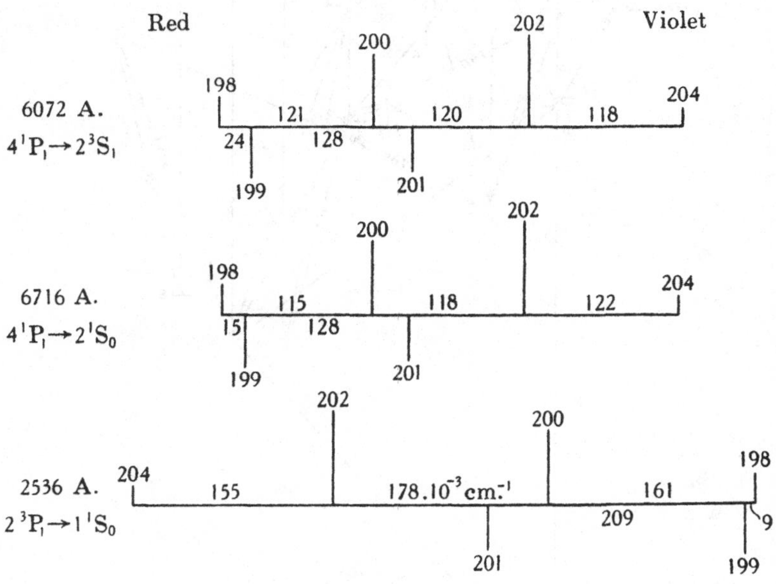

Fig. 20·23. Centroids of the isotopes in three lines of Hg I. (After Schüler and Keyston, *ZP*, 1931, **72** 423.)

emerges; the interval separating each even isotope from its neighbour is roughly constant and equal to the interval separating the odd isotopes; but the odd isotopes are not spaced mid-way between the even isotopes; this is clearly shown in Fig. 20·23 and is true not only in Hg I, but also in Pb I and Pb II. As the isotopes are not displaced in the upper levels of many lines, this property may well be demonstrated in a single line.

The relative energies of the isotopes also need to be related to their masses, for there has never been any doubt that the sequence of the isotopes is the sequence of their masses, nor that in the arc and spark spectra of mercury the isotope of lowest atomic

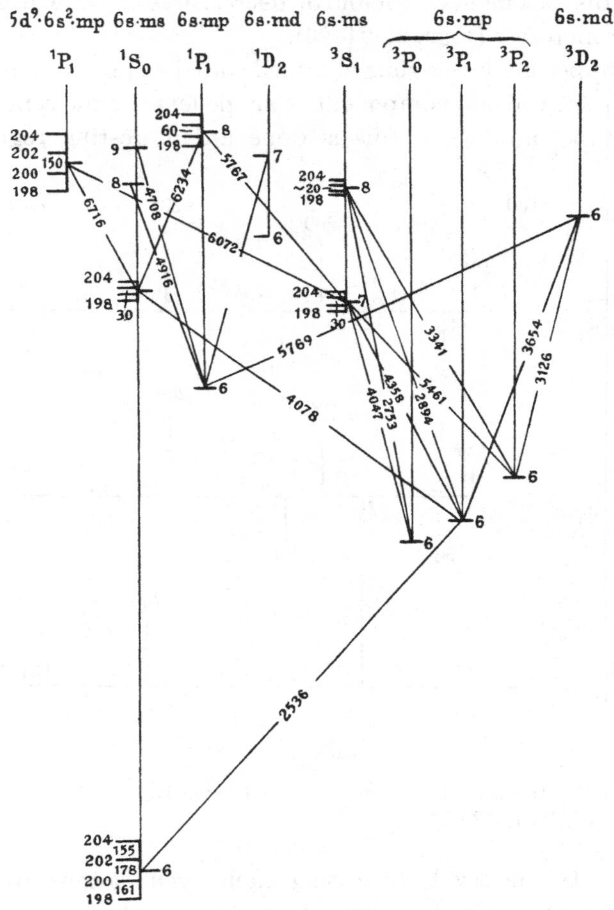

Fig. 20·24. Isotope displacements in Hg I. The wide splitting of the 8p.$^1P_1$ term is anomalous. (After Kallmann and Schüler, *EEN*, 1931, **11** 165.)

weight lies lowest. In the light of recent theory this order seems to be general, but several spectra have been a little difficult to bring into line; thus the displacements in thallium seemed to be in different directions in the arc and spark spectra, but Breit* has

* Breit, *PR*, 1932, **42** 348.

since shown that this was due to a misinterpretation of certain lines of Tl II and that in fact the isotopes of thallium are arranged like the isotopes of mercury, the smallest at the bottom. The

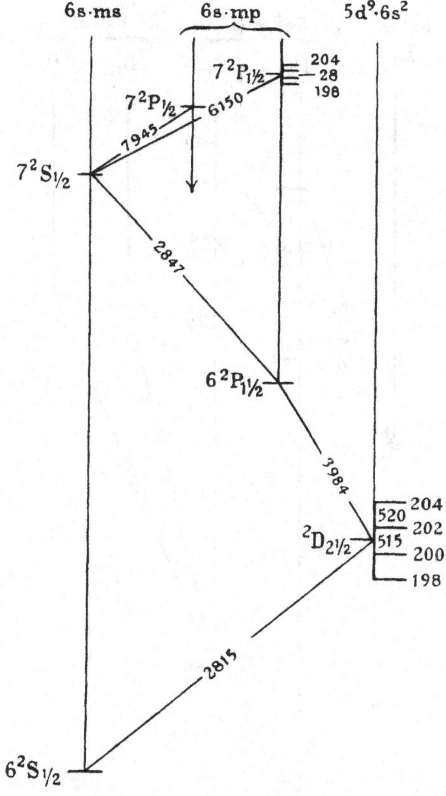

Fig. 20·25. Isotope displacements in Hg II. (After Kallmann and Schüler, *EEN*, 1932, **11** 167.)

theoretical explanation is interesting; isotope displacements are due to the increase in the volume of the nucleus resulting on the addition of a neutron;* this causes the same charge to occupy a larger volume and so reduces the electrostatic energy of the system; as the electrostatic energy of a condenser is negative,

* Schüler and Schmidt, *ZP*, 1935, **94** 463.

this means that the terms of the larger isotope lie higher than the smaller.

Turning to the quantitative side, the displacement of one isotope relative to another seems to be the same for all terms of a

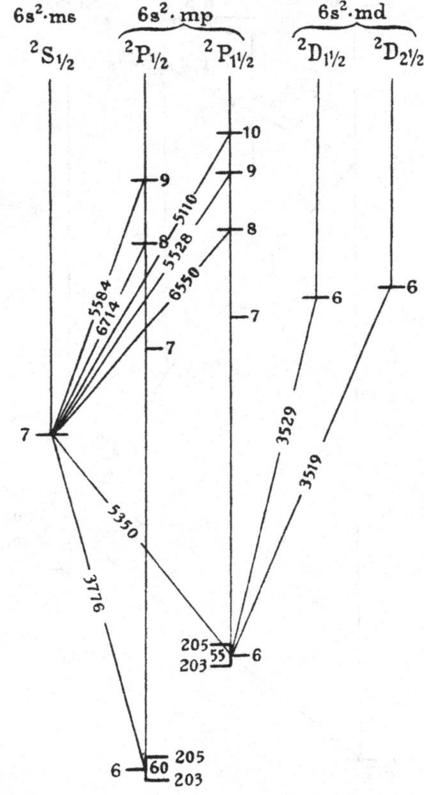

Fig. 20·26. Isotope displacements in Tl I. (After Kallmann and Schüler, *EEN*, 1932, **11** 167.)

configuration; Fig. 20·27 shows this regularity in the $6s^2 . 6p^2$ configuration of Pb I. Attempts actually to calculate the displacement have not yet met with complete success, because large corrections have to be made for the screening of one configuration by another, and these cannot be accurately estimated. Of partial solutions the most important is that produced by Breit,* who has

* Breit, *PR*, 1932, **42** 348.

related the displacement to the nuclear radius, the latter being assumed proportional to the cube root of the mass, but there is little doubt that other influences contribute.*

	$^1S_0$	$^1D_2$	$^3P_2$	$^3P_1$	$^3P_0$	
204	—	—	—	—	—	204
	65	84	?	?	?	
206	—	—	—	—	—	206
	72	88	90	85	83	
208	—	—	—	—	—	208

Fig. 20·27. Isotope displacements in the $6s^2.6p^2$ configuration of Pb I.

7. Deviations from the interval rules

In general hyperfine multiplets obey the interval rule better than gross multiplets, but this very regularity makes exceptions the more striking. When two gross levels lie near one another and satisfy certain conditions they perturb one another; one may expect the same phenomenon among hyperfine levels, and in fact Schüler and Jones† have shown that when in Hg I the 6d $^1D_2$ and 6d $^3D_1$ levels lie at a distance apart comparable with their hyperfine intervals, the interval rule is no longer valid and the isotope displacements are anomalous. In the Hg^{201} isotope the hyperfine intervals of the 6d $^1D_2$ term are 181, 301 and 313.10^{-3} cm.$^{-1}$, numbers which are in the ratio of 3·0 : 5·0 : 5·2 instead of in the normal ratio of 3 : 5 : 7. Moreover, in these terms the even isotopes are all coincident, showing that there is no isotope displacement, but the centroids of the odd isotopes instead of coinciding with the even isotopes lie farther apart. This suggests at once that the reason why both intervals and isotope displacements are abnormal, is that the two terms perturb one another, for the characteristic sign of perturbation is repulsion. Among gross multiplets two components can perturb one another only if they have the same J, and so among hyperfine multiplets we may expect components to perturb one another only if they have the same F; if this is true the even isotopes will not be displaced, for in them I is zero, and the two levels will have different values

* Breit, *PR*, 1933, **44** 418*a*; Bartlett and Gibbons, *PR*, 1933, **44** 538; Grace and More, *PR*, 1934, **45** 169.

† Schüler and Jones, *ZP*, 1932, **77** 801.

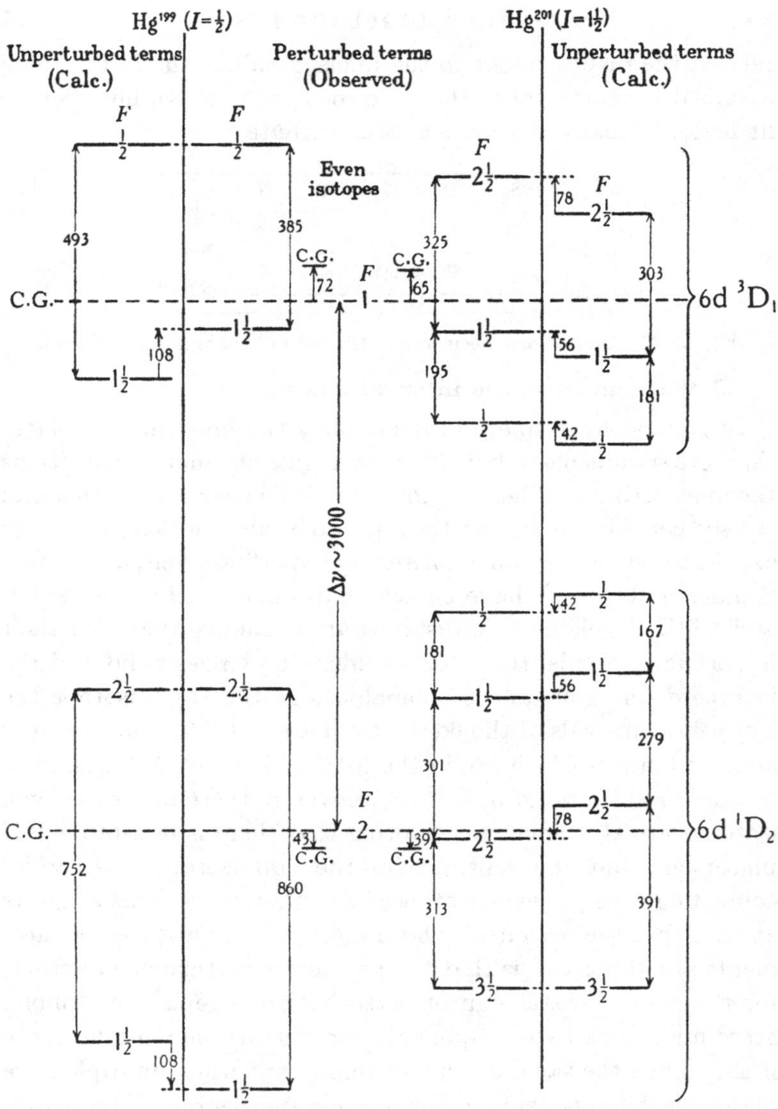

Fig. 20·28. Two perturbed terms of mercury. In the middle between the two uprights is the observed term system, while to left and right the terms are shown in the positions they would occupy if they were not perturbed. The argument by which the various displacements are obtained is fully given by Schüler and Jones, *ZP*, 1932, **77** 806.

of F, namely 1 and 2. This makes the investigation much simpler than the analogous investigation of two gross multiplets, for the even isotopes provide a fixed scale of reference.

A bird's eye view of the changes produced by perturbation is provided by Fig. 20·28; in the 199 isotope system, the levels in which F is $1\frac{1}{2}$ perturb one another, while the levels in which F is $\frac{1}{2}$ above and $2\frac{1}{2}$ below remain undisturbed. As the two levels in which F is $1\frac{1}{2}$ repel one another, the $^1D_2$ term has an enlarged and the $^3D_1$ a reduced interval. In the isotope 201 on the other hand three components wander, for there are levels with F values of $\frac{1}{2}$, $1\frac{1}{2}$ and $2\frac{1}{2}$ in both gross terms; only the $F = 3\frac{1}{2}$ level of the $^1D_2$ term remains undisturbed. The repulsion of the levels with $F = \frac{1}{2}$, $1\frac{1}{2}$ and $2\frac{1}{2}$ leads to a great reduction of the $2\frac{1}{2}$–$3\frac{1}{2}$ interval, and a consequent departure from Landé's interval rule.

When the interval rule is not obeyed the interaction between the nucleus and the electron shells is no longer proportional to the scalar product of the two vectors, and the vector model fails;* the quantum mechanics shows, however, that two states lying close to one another perturb one another and gives a satisfactory account of the displacements.†

As perturbation is a phenomenon well known when two gross multiplets lie close to one another, physicists have naturally tried to use it to explain all deviations from the interval rule. But whereas two gross levels may perturb one another when 2000 cm.$^{-1}$ apart, two hyperfine levels must lie within a few units of one another, and so it is much less probable that an unknown level lies close enough to produce the observed distortion. And in fact Schüler and Schmidt have shown that some multiplets in Eu^{151}, Eu^{153},‡ Lu^{175}§ and Hg^{201}‖ do not satisfy the interval rule, and that perturbation is an inadequate explanation. Instead of the displacements of the levels from the centroid being proportional simply to $\cos(\mathbf{IJ})$, a second term proportional to $\cos^2(\mathbf{IJ})$

* Casimir, *ZP*, 1932, **77** 811.

† Goudsmit and Bacher, *ZP*, 1933, **43** 894.

‡ Schüler and Schmidt, *ZP*, 1935, **94** 457.

§ Schüler and Schmidt, *ZP*, 1935, **95** 265.

‖ Schüler and Schmidt, *ZP*, 1935, **98** 239.

appears; this second term has a coefficient small compared with that of the normal cos (**IJ**) term, but though the deviations are small, they are greater than the experimental error. Whether this second term affects the extreme interval is doubtful, its obvious result is to make the levels crowd towards the levels of large or small F, the one when the term is positive the other when it is negative.

The appearance of a $\cos^2$ term means that the interaction of the nucleus and electrons cannot be wholly explained as due to the action of a magnetic nuclear dipole on the electrons; the form the deviation takes suggests that the new force is electrostatic in origin, and is due to a lack of spherical symmetry in the electric field of the nucleus.*

8. The spin of the nucleus

The spin of an atomic nucleus may be obtained from the hyperfine structure of a line, from the deflection of an atomic ray in a magnetic field and from the alternating intensities of band spectra. All three methods have been applied to sodium, and all give the same nuclear spin; and more generally where two methods have been used the results are consistent. Besides these three methods there are others which are permissible in theory, but which do not seem to have been used as yet; these are the specific heat at low temperatures† and scattering.‡

An atomic ray can be used only if the resolution is increased by including a magnetic velocity selector;§ this consists of an inhomogeneous magnetic field, which spreads the beam out into a velocity spectrum. A movable slit selects a part of the beam homogeneous in velocity to about 10 per cent., and this then passes through a weak inhomogeneous field of the Stern-Gerlach type; thereafter a third magnetic field is used to focus the beam on the detector wire. This technique has only recently been perfected, but already in Rabi's hands it has disclosed the spin of

* The equations for the interaction of such a core with the nucleus have been given by Casimir, *Physica*, 1935, **2** 719.

† Dennison, *PRS*, 1927, **115** 483.

‡ Sexl, *Nw*, 1934, **22** 205.

§ Rabi and Cohen, *PR*, 1933, **43** 582*a*.

potassium and in Stern's the magnetic moments of the proton and deuton,* important quantities which could not have been easily measured by other methods. When applied to sodium the

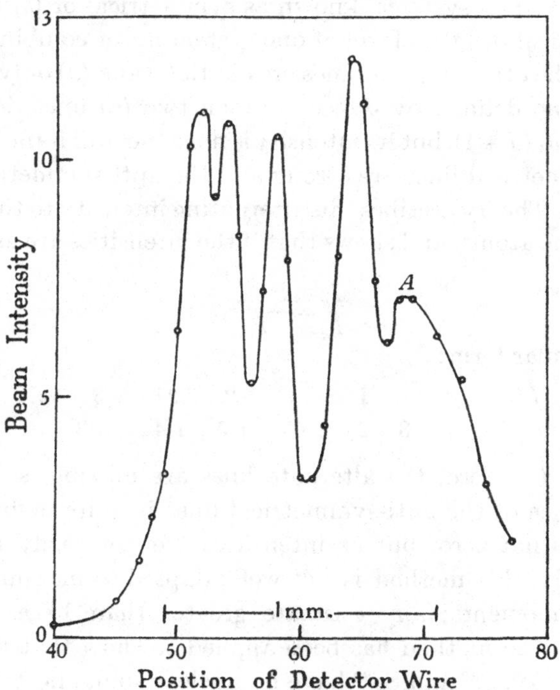

Fig. 20·29. An atomic ray of sodium split into four components in passing through a magnetic field. The high resolution is obtained by using an inhomogeneous magnetic field and a slit to select a beam homogeneous in velocity to 10 per cent. The figure shows the points obtained in one particular run. The peak marked *A* is due to atoms other than those selected. (After Rabi and Cohen, *PR*, 1933, **43** 582.)

beam splits into four components as shown in Fig. 20·29; the spin of sodium is therefore $1\frac{1}{2}$.†

In the application of band spectra‡ to this purpose, the relative intensities of the alternating bands of symmetrical molecules

* References to particular elements appear in the bibliography of h.f.s.

† Breit and Rabi, *PR*, 1931, **38** 2082*a*; Rabi and Cohen, *PR*, 1933, **43** 582*a*; Rabi, Kellogg and Zacharias, *PR*, 1934, **46** 157; Dickinson, *PR*, 1934, **46** 598.

‡ Jevons, *Band spectra of diatomic molecules*, 1932, 140.

are measured; $Cl^{35}Cl^{35}$ is a symmetrical molecule, $Cl^{35}Cl^{37}$ is not; there are no alternating intensities in the bands of unsymmetrical molecules. The rotational levels of a symmetrical molecule divide into two systems, known as symmetrical or (s) and anti-symmetrical or (a); a level of one system never combines with a level of the other; (s) combines only with (s) and (a) only with (a). An (s) line defined by J lies between two (a) lines defined by $(J-1)$ and $(J+1)$, but its intensity is not a mean of its neighbours; the symmetrical lines are strong and the anti-symmetrical lines are weak. Theory ascribes this alternating intensity to the nuclear spin of the atoms, and shows that if the intensities are as I_s to I_a, then

$$\frac{I_s}{I_a}=\frac{I+1}{I},$$

or in tabular form:

$I=$	0	$\frac{1}{2}$	1	$1\frac{1}{2}$	2	$2\frac{1}{2}$	3	...
$I_s/I_a=$	∞	3	2	1·67	1·5	1·4	1·33	...

When I is zero, the alternate lines are missing, so that the appearance of the anti-symmetrical lines is quite definite proof that I is not zero, but as intensities are not easily measured accurately, this method is not well adapted to measuring large nuclear moments; any estimate greater than $1\frac{1}{2}$ is certainly dubious. The method has been applied to show that the nuclei He^4, C^{12}, O^{16}, S^{32} and Se^{80} have no nuclear spin, and to estimate the nuclear spin of ten other elements; in the table of values these are indicated by a B.

An examination of the spins obtained by hyperfine structure and band spectra (Figs. 20·30–32) shows that they divide nuclei into two groups, in one of which the mass is odd and in the other even. Other evidence suggests that nuclei should be further divided according as the charge is odd or even. When the mass and charge are both even, the spin is always zero; this statement is supported by measurements on more than 30 nuclei. When mass is even and the charge odd, the spin is certainly 1 in H^2 and N^{14}; of this the alternating intensities of band spectra leave no doubt; in Li^6 the absence of hyperfine structure has been taken to indicate zero spin, but a narrow hyperfine structure can easily be missed. Nuclei of this type are so unstable that only four examples are known. When the mass is odd, the spin is always half an odd integer; from time to time, zero moment has been ascribed to a nucleus of this type, but in view of the 42 nuclei which conform and the fact that hyperfine structure can always be missed if the lines are too blurred or the resolving power too low, they need not be further considered; no exception has been established.

How these regularities are related to be the structure of the nucleus can be most profitably considered after the magnetic moments have been discussed.

9. The magnetic moment of the nucleus

The simplest theory of hyperfine structure ascribes a magnetic moment to the nucleus, and argues that when this acts upon an electron in its orbit, the electronic term splits. If the coupling of the nucleus and electron obeys a cosine law, this hypothesis can be shown to predict the doublet formulae, the interval rule and the usual magnetic splitting.*

In particular the displacement of a particular level from the centroid will be given by the formula

$$\begin{aligned}\nu - \nu_G &= AIJ\cos(IJ)\\ &= \tfrac{1}{2}A\{F(F+1) - J(J+1) - I(I+1)\},\end{aligned}$$

where A is the hyperfine interval quotient.* When only one electron is active A is identical with a, the electronic interval quotient, which is determined by the formula

$$a = \frac{R\alpha^2 Z^3}{n^3(l+\frac{1}{2})j(j+1)}\frac{g(I)}{1838}\,\text{cm.}^{-1}$$

for a hydrogen-like orbit, or by

$$a = \frac{R\alpha^2 Z_i^2 Z_a^2}{n^{\star 3}(l+\frac{1}{2})j(j+1)}\frac{g(I)}{1838}\,\text{cm.}^{-1}$$

for a penetrating orbit. The latter is considered as composed of two parts, an inner in which the electron moves as if under the influence of a charge Z_i, and an outer where the effective charge is Z_a. $n^\star$ is the effective quantum number. In theory these

* Goudsmit, *PR*, 1931, **37** 663; Fermi and Segrè, *ZP*, 1933, **82** 729.

Element	Charge	Mass	Spin	Magnetic moments
H	1	1	$\frac{1}{2}$ *B*	
Li	3	7	$1\frac{1}{2}$ *H*, *B*, *R*	3·29 S, G; 3·20 F; 3·20
B	5	11		
N	7	15		
F	9	19	$\frac{1}{2}$ *H*, *B*	3
Na	11	23	$1\frac{1}{2}$ *H*, *B*, *R*	2·14 S; 2·09 F; 2·08
Al	13	27	$\frac{1}{2}$ *H*	1·93 S; 2·1 G; 2·4
P	15	31	$\frac{1}{2}$ *B*	
Cl	17	35	$2\frac{1}{2}$ *B*	
		37		
K	19	39	$1\frac{1}{2}$ *R*	0·38, 0·43, ·397
		41	$>\frac{1}{2}$ and $\leqslant 2\frac{1}{2}$ *R*	
Sc	21	45	$3\frac{1}{2}$ *H*	3·6
V	23	51	$3\frac{1}{2}$ *H*	
Mn	25	55	$2\frac{1}{2}$ *H*	
Co	27	59	$3\frac{1}{2}$ *H*	2·5?
Cu	29	63	$1\frac{1}{2}$ *H*	2·74 S
		65	$1\frac{1}{2}$ *H*	2·74 S
Ga	31	69	$1\frac{1}{2}$ *H*	2·14 S, F; 2·01 G
		71	$1\frac{1}{2}$ *H*	2·74 S; 1·49 F; 2·55 G
As	33	75	$1\frac{1}{2}$ *H*	0·90 S, G
Br	35	79	$1\frac{1}{2}$ *H*, *B*?	
		81	$1\frac{1}{2}$ *H*, *B*?	
Rb	37	85	$2\frac{1}{2}$ *H*	1·49 S; 1·36 F; 1·3 G
		87	$1\frac{1}{2}$ *H*	3·06 S; 2·79 F; 2·7 G
Y	39	89		
Cb	41	93	$4\frac{1}{2}$ *H*	3·7
Ma	43	—		
Rh	45	—		
Ag	47	107		} <0·24 S
		109		
In	49	115	$4\frac{1}{2}$ *H*	5·25 S; 5·3 F; 5·4 G
Sb	51	121	$2\frac{1}{2}$ *H*	2·70 S; 2·7 G; 2·75
		123	$2\frac{1}{2}$ *H*	2·10 S; 2·1 G; 2·0
I	53	127	$4\frac{1}{2}$ *H*, *B*?	
Cs	55	133	$3\frac{1}{2}$ *H*	2·91 S; 2·63 F; 2·50
La	57	139	$3\frac{1}{2}$ *H*	2·5
Pr	59	141	$2\frac{1}{2}$ *H*	
Il	61			
Eu	63			
Tb	65			
Ho	67	165		
Tu	69	169		
Lu	71	175	$3\frac{1}{2}$ *H*	2·5
Ta	73	181	$3\frac{1}{2}$ *H*	
Re	75	185	$2\frac{1}{2}$ *H*	
		187	$2\frac{1}{2}$ *H*	
Ir	77	191	$\frac{1}{2}$? *H*	
		193	$\frac{1}{2}$? *H*	
Au	79	197	$1\frac{1}{2}$ *H*	0·23 S; 1·67 F
Tl	81	203	$\frac{1}{2}$ *H*	1·47 S; 1·42 F; 1·8 G
		205	$\frac{1}{2}$ *H*	1·47 S; 1·42 F; 1·8 G
Bi	83	209	$4\frac{1}{2}$ *H*	3·60 S; 3·54 F; 4·0 G

Fig. 20·30. Protonic nuclei with charge and mass both odd. In the spin column, the letters *H*, *B*, and *R* show the method used, hyperfine structure, band spectra or atomic rays. In the column of magnetic moments the letters denote the authority, S for Schüler, *ZP*, 1933, **88** 324, F for Fermi and Segrè, *ZP*, 1933, **82** 748, and G for Goudsmit, *PR*, 1933, **43** 638. Figures given without a letter are from recent papers which can be traced in the bibliography.

Element	Charge	Mass	Spin	Magnetic moments
H	1	2	1 *B*	0·7–0·8 *R*
Li	3	6	0 *H*, $\geqslant$ 1 *R*	0·7–0·8 *R*
B	5	10		
N	7	14	1 *B*	$\leqslant$ 0·2 *H*

Fig. 20·31. Deutonic nuclei with charge odd and mass even. For key to letters see Fig. 20·30.

Element	Charge	Mass	Spin	Magnetic moments
Be	4	9	½? *H*	
C	6	13		
O	8	17		
Ne	10	21		
Mg	12	25		
Si	14	29		
S	16	33		
Cr	24	53		
Zn	30	67	1½ *H*	
Ge	32	73		
Se	34	77		
Kr	36	83	$\geqslant$ 3½ *H*	
Sr	38	87	$\geqslant$ 1½ *H*	−0·86 S
Mo	42	95		
		97		
Ru	44	99		
		101		
Pd	46	?		
Cd	48	111	½ *H*	−0·63 S; −0·53 F; −0·67 G
		113	½ *H*	−0·63 S; −0·53 F; −0·67 G
Sn	50	115		
		117	½ *H*	−0·95 S; −0·89
		119	½ *H*	−0·95 S; −0·89
		121		
Te	52	123		
		125		
Xe	54	129	½ *H*	−ve } $\mu_{129}/\mu_{131} = -1{\cdot}1$
		131	1½ *H*	+ve }
Ba	56	135	2½ *H*	
		137	2½ *H*	0·94 S; 1·05 F
Sm	62	?		
Gd	64	?		
Dy	66	161		
		163		
Er	68	167		
Yb	70	171		
		173		
Hf	72	?		
W	74	183		
Os	76	187		
		189		
Pt	78	195	½ *H*	
Hg	80	199	½ *H*	0·55 S; 0·46 F; 0·55 G
		201	1½ *H*	−0·62 S; −0·51 F; −0·62 G
Pb	82	207	½ *H*	0·60 S; 0·67 F; 0·60 G
		209		

Fig. 20·32. Neutronic nuclei with charge even and mass odd. For key to letters see Fig. 20·30.

equations offer a simple method of determining the magnetic splitting factor $g(I)$, and hence the magnetic moment of the nucleus; but in practice a large relativity correction has to be made before these formulae can be applied.* A more general formula is thus of greater value.

When several electrons are active the interval quotient can be shown to have the value

$$A=\frac{\mu H}{IJ},$$

where μ is the magnetic moment of the nucleus, and H is the magnetic field produced by the electrons at the nucleus. When the eigenfunctions of the electrons are known, the field H can be calculated, and then the experimental values of the interval quotient determine the magnetic moment of the nucleus.

Whether this theory can be considered satisfactory depends on how well nuclear moments agree when calculated from different terms of the same spectrum, and from different spectra with the same nucleus. That 20 terms distributed among the first three spectra of thallium do all give the same magnetic moment is therefore encouraging.

The magnetic moments thus obtained are summarised in Figs. 20·30–32. In these figures the nuclear moment is given in nuclear magnetons, μ_N, where

$$\mu_N=\frac{eh}{4\pi Mc}=\frac{\mu_B}{1838},$$

M is the mass of the nucleus and μ_B is the Bohr magneton.

Of the isotopes of even mass only two magnetic moments are known, H^2 and N^{14}; in all the other isotopes the mechanical moment is zero, so that there is no hyperfine structure and therefore no evidence. An inspection of the table of isotopes of odd mass, however, shows a clear division between those of even and odd charge; among the former positive and negative moments are equally frequent, the latter being shown empirically by inverted hyperfine terms; moreover, the moment is always less

* Goudsmit, *PR*, 1933, **43** 636. The precise form the formulae should take seems also in doubt. Pauling and Goudsmit, *Structure of line spectra*, 1930, 60 and 209, note 1.

than 1; on the other hand, when the charge is odd, the moment is always positive, and with two or three exceptions it is greater than 1.*

Though this last statement certainly indicates a tendency, it must not be taken quite at its face value, for there are many spectra in which spectroscopists have commonly failed to find any structure at all; among these are N I, P II, Cl II, K I, K II, Ag I and Cd II. Band spectra dispose of the hypothesis that the mechanical moments of these atoms are all zero; nitrogen, phosphorus and potassium all spin. A second alternative is that the magnetic field produced by the electrons at the nucleus is very weak, but as some of the atoms possess a penetrating s electron such as usually produces wide hyperfine intervals, this explanation is improbable. Better is the suggestion that the magnetic moments of these nuclei are small; in two elements certainly recent work supports this; thus Jones† has shown that in Cd II, though 12 out of the 13 lines which he examined have no structure, the thirteenth reveals the ground, ^{2}S, term as double and indicates a nuclear magnetic moment of $0{\cdot}625\mu_N$. In K I too Jackson and Kuhn‡ have recently shown that if sufficient care is taken, the resonance lines may be resolved to a narrow doublet; the lines were obtained from a discharge tube containing neon at a pressure of a few mm. and potassium at a pressure of less than one two-thousandth of a mm.; before entering the spectrograph, the light passed through a ray of potassium atoms travelling at right angles across its path. This atomic ray passed through a cool tube whose length was twenty times its diameter, so that the Doppler width of absorption lines was only 1/20 of that obtained with a random distribution of velocities. Resolved with a Fabry-Perot étalon, the lines appeared as a doublet with an interval of 0·015 cm.$^{-1}$

* The magnetic moments have been tabulated by: Fermi and Segrè, *ZP* 1933, **82** 748; Goudsmit, *PR*, 1933, **43** 638; Schüler, *ZP*, 1933, **88** 324. Schüler gives a critical survey of the data and recalculates the values.

† Jones, *Phys. Soc. Proc.* 1933, **45** 625.

‡ Jackson and Kuhn, *PRS*, 1934, **148** 335.

10. The structure of the nucleus

At the present day the evidence of hyperfine structure suffices to divide the nuclei into four types, though history relates that Aston* recognised these types long before hyperfine structure had advanced far enough to be of much service. The nuclear type is most simply determined by the mass and charge, the crucial question being whether each of these is odd or even (Fig. 20·33).

Nuclear mass	Even	Even	Odd	Odd
Nuclear charge	Even	Odd	Even	Odd
No. of isotopes known	114	4	44	45
No. of nuclear moments known	>30	3	12	30
Value of I	0	1 or ? 0	Half integral	Half integral
No. of magnetic moments known	0	2	10	20
Values and sign of μ	No evidence, probably 0	—	As often −ve as +ve; $\mu<1$	Always +ve; in all but 4, $\mu>1$
Structure of nucleus suggested	Core with $I=\mu=0$	Core + 1 deuton	Core + 1 neutron	Core + 1 proton

Fig. 20·33. Frequency of occurrence of the four nuclear types and their properties.

When both mass and charge are even, the nuclear moment is always zero; nothing is then known of the magnetic moment, though it is commonly assumed zero. This structure is particularly stable, Aston's figures showing that more than half the known isotopes are built to this plan.

When the mass is even and the charge is odd, the nucleus is usually unstable, though four light isotopes do occur; the heaviest of these is N^{14}. Of the known mechanical moments, two are of magnitude 1 and a third is possibly zero.

When the mass is odd and the charge even, the mechanical moment is half an odd integer; the magnetic moment is as often negative as positive, and is always less than one. In terms of the number of isotopes known, this nuclear type is less than half as stable as the even-even type.

* Aston, *Mass-spectra and isotopes*, 1933, 175.

Finally, when the mass and charge are both odd, the mechanical moment is again half an odd integer, but the magnetic moment is always positive and usually greater than 1. This type of nucleus is about as stable as the preceding type. Further, of ten of these elements, which have two isotopes, nine pairs have the same spin; the one exception, rubidium, is exceptional also in being radioactive. When two isotopes are of even charge, they do not obey this law.*

These facts are independent of any theory of the structure of the nucleus; but to proceed further some simplifying assumptions are necessary. A very few years ago, the nucleus was supposed constructed of protons and electrons,† and there seemed no alternative to allotting the proton a spin of $\frac{1}{2}h/2\pi$ and depriving the electron of the spin it has outside the nucleus. But to-day this awkward trick of giving the electron different properties in different places can be avoided by introducing the neutron, and supposing the nucleus built of three bricks only, the proton, the neutron and the α-particle. The odd-even rule of mechanical moments is then easily explained if both the proton and neutron have a spin of $\frac{1}{2}h/2\pi$, while the α-particle has no spin at all. The band spectra of H^1 and He^4 confirm this assignment for they show that these nuclei have respectively moments of $\frac{1}{2}$ and 0; while the moment assigned to the neutron is that suggested by the quantum mechanics.‡ Becoming still more definite, if a nucleus of mass M and charge Z is composed of N neutrons, P protons and A α-particles, then obviously

$$M = 4A + N + P,$$
$$Z = 2A + P.$$

As these equations are indeterminate, there are current at the present time two simplifying assumptions; the first due to Heisenberg§ eliminates the α-particles altogether and leaves the

* Schüler and Schmidt, *ZP*, 1935, **94** 460.

† Bryden, *PR*, 1931, **38** 1989; White, *PR*, 1931, **38** 2073*a*.

‡ Temple, *PRS*, 1934, **145** 344.

§ Heisenberg, *ZP*, 1932, **78** 156. This theory has been developed by Eastman, *PR*, 1934, **46** 1.

nucleus composed of neutrons and protons only; the second proposed by Iwanenko and used by Landé* does not allow the presence of more than one proton, this one being present whenever the charge is odd.

For the present purpose, however, the respective merits of these two hypotheses are of little account; for both are consistent with the assumption that the even-even nucleus is the core from which the other three types are formed by the addition of a proton, a neutron or a deuton. For convenience we may then describe a nucleus with odd mass and odd charge as 'protonic', one with odd mass and even charge as 'neutronic', and one with even mass and odd charge as 'deutonic'. This hypothesis in its simplest form has not thus far been stretched to cover all the facts of hyperfine structure, but it has succeeded so well that recent papers have agreed that some variation is indicated.

Consider first the protonic nuclei.† If the mechanical and magnetic moments are to be ascribed to a single proton moving outside a core in which both moments are zero, the proton is best assumed to move in an orbit and to have on this account an angular momentum **l**, which is always integral; the inherent spin of the proton is then written **s** with value $\frac{1}{2}$, and this is made to set parallel or anti-parallel to the orbital vector, so that the resultant moment **I** is half-integral. That the magnetic moments are often large suggests that the magnetic moment of the proton is itself large; and in fact Stern‡ has arrived at the same conclusion from an entirely independent experiment. Working on the deflection of an atomic ray in a magnetic field, he concluded that the magnetic moment of the proton is $2\frac{1}{2}$ nuclear magnetons, the experimental error being given as 10 per cent.; Landé, however, in applying this observation to hyperfine structure has found that a g factor of 4 fits the facts better than one of 5.

If the proton revolves about a core, the magnetic moment may be calculated by the methods developed by Goudsmit for an

* Iwanenko, *Comptes Rendus*, 1932, **195** 439; Landé, *PR*, 1933, **43** 620; Inglis and Landé, 1934, **45** 842*a*.

† Landé, *PR*, 1933, **44** 1028*a*; Kallmann and Schüler, *ZP*, 1934, **88** 210.

‡ Stern, *Helv. Phys. Acta*, 1933, **6** 426.

electron; the moment will be gI, where the splitting factor g is given by the equation

$$g=\frac{I(I+1)+l(l+1)-s(s+1)}{2I(I+1)}g(l)+\frac{I(I+1)+s(s+1)-l(l+1)}{2I(I+1)}g(s).$$

As $g(l)$ is the splitting factor of a revolving particle, its value must be 1; $g(s)$ is assigned the value 4 for reasons already given. The nuclear moments resulting from this system are shown in Fig. 20·34.

The experimental moments do not agree very well with this theory, but as no observation can be trusted to less than 10 per cent. and the calculations of different authors from the same data

l \ I	$\frac{1}{2}$	$1\frac{1}{2}$	$2\frac{1}{2}$	$3\frac{1}{2}$	$4\frac{1}{2}$
0	2				
1	0	3			
2		$\frac{3}{5}$	4		
3			$\frac{10}{7}$	5	
4				$\frac{21}{9}$	6
5					$\frac{36}{11}$

Fig. 20·34. Protonic nuclei. Magnetic moments dictated by theory.

show even wider discrepancies, comparison is not easy (Fig. 20·35). The general tendency of the magnetic moments to increase with the spin is well covered by the theory; while if we concentrate attention on the nuclei with a spin of $1\frac{1}{2}$, the division into seven elements with magnetic moments greater than 2 and three elements with moments less than 1 is striking and in accord with theory; on the other hand the prediction that for any value of I there can be only two values of μ seems definitely at variance with experiment; the range from sodium with an estimated moment of 2·14 to lithium with a moment of 3·29 seems to exceed the possible error; moreover, there is no doubt that the two gallium isotopes have different moments. Ways of circumventing this difficulty will, however, be better taken up when the neutronic nuclei have been considered.

Landé* has pictured the neutronic nuclei built to a similar plan; the mechanical and magnetic moments of the nucleus are due wholly to a single neutron, moving with angular moment **l** and inherent spin **s** round an inert core; the inherent spin is again

Spin	Element	Mass	Magnetic moment	
			Obs.	Theory
½	F	19	3·?	2 or 0
	Al	27	1·93	
	Tl	203	1·47	
		205	1·47	
1½	Li	7	3·29	3 or 0·60
	Na	23	2·14	
	K	39	0·38	
	Cu	63	2·74	
		65	2·74	
	Ga	69	2·14	
		71	2·74	
	As	75	0·90	
	Rb	87	3·06	
	Au	197	0·23	
2½	Rb	85	1·49	4 or 1·43
	Sb	121	2·70	
3½	Sb	123	2·10	5 or 2·33
	Cs	133	2·91	
4½	In	115	5·25	6 or 3·27
	Bi	209	3·60	

Fig. 20·35. Protonic nuclei. Observed magnetic moments compared with theory.

$\frac{1}{2}h/2\pi$, but the magnetic splitting factors are different; an uncharged particle will create no magnetic field in its motion so that $g(l)$ will be zero, and the nuclear splitting factor g will be simply

$$g=\frac{I(I+1)+s(s+1)-l(l+1)}{2I(I+1)}g(s).$$

That the magnetic moment of a neutronic nucleus is as often negative as positive has been taken to mean that the inherent magnetic moment of a neutron is negative, but it would appear that equal numbers of positive and negative moments appear provided that $g(l)$ is equated to zero; the sign of $g(s)$ is irrelevant.

* Inglis and Landé, *PR*, 1934, **45** 842*a*.

Thus if $g(s)$ is assigned the value 1, so that the magnetic moment of the neutron is $\frac{1}{2}$ a nuclear magneton, the permitted nuclear moments are those shown in Fig. 20·36, while if $g(s)$ is given the value -1, the numerical values will be unaltered, but the + and − signs must be interchanged.

l \ I	$\frac{1}{2}$	$1\frac{1}{2}$	$2\frac{1}{2}$
0	$+\frac{1}{2}$		
1	$-\frac{1}{6}$	$+\frac{1}{2}$	
2		$-\frac{3}{10}$	$+\frac{1}{2}$
3			$-\frac{5}{14}$

Fig. 20·36. Neutronic nuclei. Magnetic moments dictated by theory.

That the nuclear moments thus calculated are rather smaller than those observed is of no account, for the neutron may easily be assigned a larger magnetic moment; but the inequality of the positive and negative values is serious, for experiment suggests that positive and negative values occur over the same numerical range; indeed, all observed moments lie between $+\frac{1}{2}$ and $+1$ or between $-\frac{1}{2}$ and -1. Clearly these figures are more in conformity with the suggestion that the magnetic moment of the neutron is $-0{\cdot}6\,\mu$ and can set itself parallel or anti-parallel to the field, there being no orbital motion. Whether the observed moments cover too wide a range actually to invalidate this suggestion is probably still a matter of opinion.

In the hope of obtaining better agreement in both protonic and neutronic nuclei, Schüler* has proposed the introduction of another vector **r**, which shall combine with the resultant of **l** and **s** to produce **I**; while Tamm and Altschuler† have proposed that in some nuclei two neutrons uncouple from the core, and combine freely with the external proton or neutron. Both authors can claim that the magnetic moments thus calculated nowhere clash with experiment, but both theories give a choice of values, which must be considered rather wide when compared with the paucity

* Schüler, *ZP*, 1934, **88** 323.

† Tamm and Altschuler, *Comptes Rendus de l'Acad. de Sci. de l'U.S.S.R.*, 1934, **1** 458.

and inaccuracy of the experimental figures. Inglis and Landé* have expressed their preference for the solution of Tamm and Altschuler, but where rapid advance is likely in the next few years, a detailed analysis of what is tentative seems hardly justified.

Thus the deutonic nuclei alone remain to be considered. In these the mass is even and the charge odd, so that both a neutron and a proton must be added to the core of even mass and even charge. If these always combine in the same way to form the deuton nucleus of heavy hydrogen, nuclei of this type should always have the unit spin of the deuton itself; N^{14} actually has this spin, but Li^6 has always been assumed to have zero spin. Whereas the spins of H^2 and N^{14} were obtained from band spectra, however, the spin of Li^6 has been obtained only from hyperfine structure and signifies therefore only that if the nucleus has a spin moment its magnetic moment is very small. Indeed the lines of N^{14} are also narrow, so that all we know of its magnetic moment is that it is certainly less than 0·2 nuclear magneton. Clearly band spectra observations which will determine the nuclear moments of the two deutonic nuclei Li^6 and B^{10} are much to be desired.

BIBLIOGRAPHY

Much the most thorough article is one by Kallmann and Schüler in *Erg. d. Exact. Naturwiss.* **1932**, **11** 134. The treatment is exhaustive, but much has been learnt since 1932.

* Inglis and Landé, *PR*, 1934, **46** 76*a*.

CHAPTER XXI

QUADRIPOLE RADIATION

1. Forbidden lines

The transition producing a normal line must satisfy two selection rules; Laporte's rule forbids a jump from even term to even term and from odd to odd, and a second limits the changes in the angular momentum of the atom, J. On occasion, however, lines appear which violate these rules; in an electric field, for example, both rules are violated, the phenomena being referred to as the 'completion of the multiplet'. As this has been known for twenty years the more recently discovered forbidden lines have been frequently ascribed to an unsuspected electric field; the precise conditions under which forbidden lines appear are therefore important.

For this purpose the forbidden lines may be classified in three divisions: (1) the alkali D→S doublets; (2) the green aurora line 5577 A. and some nebular lines; (3) the mercury line $^3P_2 \rightarrow {}^1S_0$, 2270 A. Of these the last may be dismissed as probably due to the interaction between the electrons and the nucleus, the J selection rule applying rigidly only to the total angular momentum of the atom F.* The two other divisions merit fuller discussion.

The D→S lines of sodium and potassium seem to have been studied for the first time in 1922, when Datta† showed that they are absorbed by a tube of potassium vapour; the lines appeared at all pressures used from 2·5 up to 46 mm., while the potassium bands, which are known to be sensitive to electrostatic fields, did not blur until the pressure rose to 30 mm. At about the same time Foote, Mohler and Meggers‡ showed that the D→S lines are emitted in a space shielded from the applied electrostatic field, even when the field itself is very weak; though the presence of ions in the vapour may be an essential condition of the experi-

* Rayleigh, *PRS*, 1927, **117** 294; Huff and Houston, *PR*, 1930, **36** 842.

† Datta, *PRS*, 1922, **101** 539.

‡ Foote, Mohler and Meggers, *PM*, 1922, **43** 659.

ment, it seems more probable that the lines are both radiated and absorbed in a field-free space. Later the lines were shown to occur also in a newly struck arc;* the arc was struck and run until the tips of the electrodes were red hot, when caesium carbonate was fused on to both electrodes; thereafter whenever the arc was struck, the D → S lines appeared strong for a few seconds and as they faded, the ordinary caesium lines shone forth.

The next forbidden line to attract attention was the auroral line 5577 A., which McLennan† examined and traced to a forbidden transition in the O I spectrum. As the spectrum of the night sky does not contain any nitrogen bands, the potential necessary to produce the green auroral line is presumably less than the 11·5 volts required to produce the most easily excited of these. Experiment in the laboratory confirms this hypothesis; in a discharge tube containing pure oxygen the 5577 A. line is swamped by the band spectrum, but if some neon or argon is introduced so that when the total pressure is 3 cm. of mercury the partial pressure of the oxygen is only 3 mm., the green line comes out strongly. Now the inert gases are known to have a very small potential drop throughout the discharge, so that the oxygen lines produced when the inert gas is in large excess will be limited to those of low excitation energy.

But if the green line is produced by a potential of less than 11·5 electron volts, the energy which has to be given to the oxygen atom must be less than 4·4 volts, for oxygen can be dissociated into neutral atoms by light of wave-length 1750 A. corresponding to a dissociation potential of 7·1 volts, and other evidence serves to confirm this value. The resonance potentials of atomic oxygen are, however, 9·11 volts for $^5S \leftarrow {}^3P$ and 9·48 for $^3S \leftarrow {}^3P$, so that an excitation energy of 4·4 volts can hardly cause the atom to emit any lines of the triplet or quintet systems. In fact with only this potential available, the terms which can be concerned in the production of the 5577 A. line seem to be limited to the five members of the p^4 configuration, $^3P_{2,1,0}$, $^1D_2$ and $^1S_0$.

* Shrum, Carter and Fowler, H. W., *PM*, 1927, **3** 27.

† McLennan and Shrum, *PRS*, 1925, **108** 501; McLennan, *PRS*, 1928, **120** 327.

Of these low levels $^1D_2$ and $^1S_0$ are metastable, and if one of them is the initial state, the final state must be $^1S_0$ or one of the ^3P components, but even this leaves seven possible transitions. To distinguish between them McLennan and his co-workers examined the longitudinal Zeeman effect, and found the two circularly polarised lines characteristic of a singlet line, a result which seems to reduce the seven alternatives to three, namely $^1S_0 \rightarrow {}^1D_2$, $^1S_0 \rightarrow {}^3P_0$, $^1D_2 \rightarrow {}^3P_0$. Of these the last may be rejected because in nebular spectra which show the transitions $^1D_2 \rightarrow {}^3P_{2,1}$ occurring strongly, the $^1D_2 \rightarrow {}^3P_0$ line is always so weak that it has so far escaped detection; accordingly if the line 5577 A. line is $^1D_2 \rightarrow {}^3P_0$, two other lines $^1D_2 \rightarrow {}^3P_{2,1}$ should occur even more brightly in positions which can be calculated; but in fact these lines donot appear either in the aurora or in oxygen gas excited to produce the 5577 A. line. A similar argument applied to $^1S \rightarrow {}^3P_0$ transition compels us to reject it too; so that the green auroral line may be taken as $^1D_2 \rightarrow {}^1S_0$ without any reference to the theory of quadripole radiation; history shows indeed that this evidence was accepted in 1928, though only later was the corner stone added in an observation of the transverse Zeeman effect, work which will be discussed in a later section.

While McLennan was developing this work on the auroral line, Bowen* showed that eight strong nebular lines could be ascribed to transitions between various metastable states of oxygen and nitrogen (Fig. 21·1). Whereas stellar spectra contain lines characteristic of almost all elements, nebulae emit only a few lines and of these all that have been assigned arise from the six elements H, He, C, O, N and A; thus of a list of 79 lines between 3300 and 7300 A., 57 have been assigned to these elements.† The lines due to hydrogen and helium are of no particular interest, for they are simply the lines normally produced in the laboratory, but of the lines due to the other elements many are normally forbidden.

That lines forbidden in the laboratory should appear in nebulae

* Bowen, *N*, 1927, **120** 473.

† Becker and Grotrian, *EEN*, 1928, **7** 56. A IV lines from Boyce, *PR*, 1935, **48** 401.

is not difficult to explain. The intensity of a line depends both on the transition probability and the concentration of atoms in the excited state; in nebulae a low probability may be balanced by a high concentration, but this is not usually possible in the labora-

Spectrum	Wave-length	Transition	Allowed or forbidden
O I	6302	$2p^4\,{}^1D_2 \rightarrow 2p^4\,{}^3P_2$	f.
	6364	$2p^4\,{}^1D_2 \rightarrow 2p^4\,{}^3P_1$	f.
O II	3726	$2p^3\,{}^2D_{1\frac{1}{2}} \rightarrow 2p^3\,{}^4S_{1\frac{1}{2}}$	f.
	3729	$2p^3\,{}^2D_{2\frac{1}{2}} \rightarrow 2p^3\,{}^4S_{1\frac{1}{2}}$	f.
	4076	$2p^2.3d\,{}^4F_{4\frac{1}{2}} \rightarrow 2p^2.3p\,{}^4D_{3\frac{1}{2}}$?	a.
	4416 {	$2p^2.3p\,{}^2D_{1\frac{1}{2}} \rightarrow 2p^2.3s\,{}^2P_{\frac{1}{2}}$	a.
		$2p^2.3p\,{}^2D_{2\frac{1}{2}} \rightarrow 2p^2.3s\,{}^2P_{1\frac{1}{2}}$	a.
	4649	$2p^2.3p\,{}^4D_{3\frac{1}{2}} \rightarrow 2p^2.3s\,{}^4P_{2\frac{1}{2}}$	a.
	7325	$2p^3\,{}^2P \rightarrow 2p^3\,{}^2D$	f.
O III	3313	$2p.3p\,{}^3S_1 \rightarrow 2p.3s\,{}^3P_1$	a.
	3342	$2p.3p\,{}^3S_1 \rightarrow 2p.3s\,{}^3P_2$	a.
	3445	$2p.3p\,{}^3P_2 \rightarrow 2p.3p\,{}^3P_2$	a.
	3759	$2p.3p\,{}^3D_3 \rightarrow 2p.3s\,{}^3P_2$	a.
	4363	$2p^2\,{}^1S_0 \rightarrow 2p^2\,{}^1D_2$	f.
	4959	$2p^2\,{}^1D_2 \rightarrow 2p^2\,{}^3P_1$	f.
	5007	$2p^2\,{}^1D_2 \rightarrow 2p^2\,{}^3P_2$	f.
N II	5755	$2p^2\,{}^1S_0 \rightarrow 2p^2\,{}^1D_2$?	f.
	6548	$2p^2\,{}^1D_2 \rightarrow 2p^2\,{}^3P_1$	f.
	6584	$2p^2\,{}^1D_2 \rightarrow 2p^2\,{}^3P_2$	f.
N III	4097	$3p\,{}^2P_{1\frac{1}{2}} \rightarrow 3s\,{}^2S_{\frac{1}{2}}$	a.
	4102	$3p\,{}^2P_{\frac{1}{2}} \rightarrow 3s\,{}^2S_{\frac{1}{2}}$	a.
	4634	$3d\,{}^2D_{1\frac{1}{2}} \rightarrow 3p\,{}^2P_{\frac{1}{2}}$	a.
	4641	$3d\,{}^2D_{2\frac{1}{2}} \rightarrow 3p\,{}^2P_{1\frac{1}{2}}$	a.
A IV	4711	$3p^3\,{}^2D_{2\frac{1}{2}} \rightarrow 3p^3\,{}^4S_{1\frac{1}{2}}$	f.
	4740	$3p^3\,{}^2D_{1\frac{1}{2}} \rightarrow 3p^3\,{}^4S_{1\frac{1}{2}}$	f.

Fig. 21·1. Nebular lines. This list is abstracted from one given by Becker and Grotrian, *EEN*, 1928, **7** 56; the original gives lines of hydrogen and helium as well.

tory, because atoms leave the excited state by collision. This general explanation has received interesting confirmation in a comparison of the relative intensities of the magnesium lines 2852 and 4571 A.* The first is the line $2\,{}^1P \rightarrow 1\,{}^1S$, which is very intense in the arc, the second $2\,{}^3P_1 \rightarrow 1\,{}^1S$, which is very weak. The natural lives of the atoms in these initial states, calculated by the

* Frayne, *PR*, 1929, **34** 590.

methods of the quantum mechanics,* are $3 . 10^{-9}$ and $4 . 10^{-3}$ sec. respectively, so that if the concentration of the atoms in the two states are at all comparable, the $2\,^1P$ line will appear much the more intense. In the arc the number of atoms in the $2\,^3P$ state is kept down only by frequent collisions, but if the pressure is sufficiently reduced and an inert gas is introduced to prevent the magnesium atoms diffusing to the walls, the 4571 A. line should increase in intensity. Kinetic theory shows that if the time between successive collisions is to be reduced to 10^{-3} sec., the vapour pressure must be reduced to 10^{-4} mm.; and in fact the vapour pressure of magnesium is of this magnitude at 500° C. Working with an electrodeless discharge at this temperature Frayne found that the 4571 A. line was fairly prominent and that the introduction of 10 mm. of various inert gases increased the intensity 50–100 times. In agreement with this work the auroral line 5577 A. appears only in the presence of an inert gas, and the intensity increases with increase in the diameter of the tube.

As a general explanation this hypothesis of Bowen's appears satisfactory enough, but recent theory shows that it may be made much more precise.

2. Quantum mechanics

The quantum mechanics shows that the ordinary spectral lines arise from a dipole oscillation and that, usually, this alone is important; but mathematically it is only the first term in the series which arises when the vector potential is developed in powers of the atomic radius divided by the wave-length of the emitted light; the second and third terms indicate quadripole and octapole radiation of much lower intensity. Of these the quadripole oscillation will be shown responsible for the forbidden lines described in the last section; no line has yet been attributed to an octapole oscillation.

In agreement with this theory is the low intensity of the quadripole lines. Thus experiments in absorption suggest that 12,000 times as many atoms absorb the $3\,^2P \leftarrow 1\,^2S$ lines of K and Rb as absorb the $3\,^2D \leftarrow 1\,^2S$ lines, the ratio being roughly the

* Houston, *PR*, 1929, **33** 297.

same in the two metals;* while Rasetti,† using a more sensitive method depending on anomalous dispersion, found the intensity ratio of the emitted lines of K to be $1{\cdot}1 . 10^{-6}$; and as Rasetti estimates his accuracy as only 50 per cent. this may be considered in satisfactory agreement with the theoretical value of $1{\cdot}5 . 10^{-6}$.‡ That the absorption measurements show rather a large error is of little account, for the authors themselves insist that their estimate was intended to be little more than qualitative.

The selection rules observed in quadripole lines can be predicted from the simple hypothesis that a quadripole line results from two simultaneous dipole transitions, though this rough analogy is of course a poor substitute for the rigid argument of the quantum mechanics. Thus in a dipole jump the sum of the orbital vectors Σl_1 changes always from odd to even or even to odd, and accordingly in a quadripole jump the same quantity changes from even to even or odd to odd; in particular a quadripole line may be emitted when the jump is between two terms of the same configuration. Again in a dipole line

$$\Delta J = 0 \quad \text{or} \ \pm 1,$$

so that in a quadripole line we expect

$$\Delta J = 0 \quad \text{or} \ \pm 1 \quad \text{or} \ \pm 2.$$

And in fact all known lines do satisfy this condition.

In practice the quadripole selection rules are not used like the dipole rules to predict which lines will appear, since so few of the lines allowed have yet been produced; rather do they serve to determine whether an observed line arises from a dipole or a quadripole transition, and for this in a field-free space the Laporte rule is crucial; if a transition is known to start from one term and end in another, then it can be written down at once as a dipole or a quadripole. But so often one cannot be certain whether there is a stray electric field or not, and then the Laporte rule cannot be trusted; instead one has to examine the Zeeman effect and let it decide.

* Sowerby and Barratt, *PRS*, 1926, **110** 190.

† Rasetti, *Accad. Lincei, Atti*, 1927, **6** 54.

‡ Stevenson, *PRS*, 1930, **128** 591.

3. Zeeman effect

In the Zeeman splitting of quadripole, as of dipole, lines, the term displacements are simple fractions of the Lorentz unit, $o_m = \frac{eH}{4\pi m_e c}$; so that the change in energy ΔE of any state may be written

$$\Delta E/hc = Mgo_m,$$

where g has the normal Landé values given by

$$g = 1 + \frac{J(J+1) + S(S+1) - L(L+1)}{2J(J+1)}.$$

But whereas in a dipole line the transitions are governed by the selection rule

$$\Delta M \leqslant 1,$$

in quadripole lines $\Delta M \leqslant 2.$

And the polarisation rules too are different, as the summary given in Fig. 21·2 shows; in particular, when the pattern is viewed at an angle of 45° to the field, components appear which are invisible when the line of sight is parallel or perpendicular to the field.

Transition	Polarisation when viewed		
	Across field	Diagonal at 45°	Parallel to field
$\Delta M = -2$	σ	l. ell.	0
$\Delta M = +2$	σ	r. ell.	0
$\Delta M = -1$	π	σ	l. circ.
$\Delta M = +1$	π	σ	r. circ.
$\Delta M = 0$	0	π	0

Fig. 21·2. Quadripole radiation. Transitions permitted in a magnetic field and the polarisation of the line produced.

Applied to a singlet line, such as the green auroral line, this theory shows that the Zeeman pattern should be that shown in Fig. 21·3; thus when viewed parallel to the field the Zeeman pattern of a dipole and a quadripole line are identical, a fact which explains why McLennan* was able to confirm the identification of the 5577 A. line as due to $^1D_2 \rightarrow {}^1S_0$ before the quadripole theory was developed, though later Frerichs and Campbell† showed that

* McLennan, McLeod and Ruedy, *PM*, 1928, **6** 558; Sommer, *ZP*, 1928, **51** 451.

† Frerichs and Campbell, *PR*, 1930, **36** 1460.

it arises in a quadripole transition; for McLennan viewed the pattern along the field axis, while Frerichs and Campbell viewed it transversely. The latter obtained a beautiful confirmation of theory, finding two π components at a distance of one, and two σ components at a distance of two, Lorentz units, all four lines being of equal intensity (Fig. 21·3).

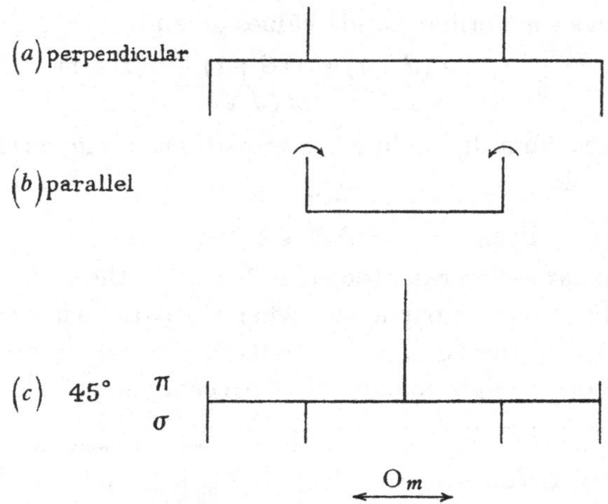

Fig. 21·3. Zeeman pattern of a quadripole singlet line, when viewed (*a*) perpendicular, (*b*) parallel, (*c*) at 45°, to the field. The intensities are proportional to the lengths of the verticals.

A more rigorous because more complex confirmation of theory has been obtained by Segrè and Bakker, who measured both the Zeeman and Paschen-Back splitting of SD alkali doublets,* and contrasted these with the splitting of the mercury line 3680 A.; this line arises as $7p\,^3P_2 \rightarrow 6p\,^3P_2$ and is normally forbidden, but it appears in an electric field.†

The Zeeman effect of the SD doublet was demonstrated with a field of 7500 gauss applied to the potassium lines

$$\left.\begin{array}{l} 4d\,^2D_{2\frac{1}{2}} \rightarrow 4s\,^2S_{\frac{1}{2}}\ 4642{\cdot}27\text{ A.} \\ 4d\,^2D_{1\frac{1}{2}} \rightarrow 4s\,^2S_{\frac{1}{2}}\ 4641{\cdot}77\text{ A.} \end{array}\right\}$$

with an interval of 2·325 cm.$^{-1}$

* Segrè and Bakker, *ZP*, 1931, **72** 724.
† Bakker and Segrè, *ZP*, 1932, **79** 655.

For this doublet the theoretical patterns are:

(1) When observed parallel to the field,

$^2D_{2\frac{1}{2}} \rightarrow {}^2S_{\frac{1}{2}}$ 4.8/5

$^2D_{1\frac{1}{2}} \rightarrow {}^2S_{\frac{1}{2}}$ 1.7/5

all components being circularly polarised.

(2) When observed transverse to the field,

$^2D_{2\frac{1}{2}} \rightarrow {}^2S_{\frac{1}{2}}$ (4).(8).10.14/5

$^2D_{1\frac{1}{2}} \rightarrow {}^2S_{\frac{1}{2}}$ (1).(7).11/5.

The figures in brackets measure π components, the other figures σ components.

(3) When observed at an angle of $\pi/4$ to the field,

$^2D_{2\frac{1}{2}} \rightarrow {}^2S_{\frac{1}{2}}$ (2).4.8.*10*.*14*/5

$^2D_{1\frac{1}{2}} \rightarrow {}^2S_{\frac{1}{2}}$ 1.(3).7.*11*/5.

The figures in italics are elliptically polarised with an axis ratio of $\sqrt{2}:1$ between the σ and π axes. The intensities in these simple transitions may be obtained from the polarisation rule and the sum rule of Burger and Dorgelo (Fig. 21·4). Fig. 21·5

		$^2D_{1\frac{1}{2}} \rightarrow {}^2S$		$^2D_{2\frac{1}{2}} \rightarrow {}^2S$		
	M	$\frac{1}{2}$	$1\frac{1}{2}$	$\frac{1}{2}$	$1\frac{1}{2}$	$2\frac{1}{2}$
M	gM	$\frac{2}{5}$	$\frac{6}{5}$	$\frac{3}{5}$	$\frac{9}{5}$	3
$\frac{1}{2}$	1	$-\frac{3}{5}$ (2)	$\frac{1}{5}$ (1)	$-\frac{2}{5}$ (3)	$\frac{4}{5}$ (4)	2 (5)
$-\frac{1}{2}$	-1	$\frac{7}{5}$ (3)	$\frac{11}{5}$ (4)	$\frac{8}{5}$ (2)	$\frac{14}{5}$ (1)	

Fig. 21·4. The displacements and in brackets the intensities of the Zeeman components of a D→S doublet.

shows that the experimental results are in complete agreement with theory.

The Zeeman effect thus satisfactorily confirmed, Segrè and Bakker changed from potassium to sodium, in which the two lines lie so close together that they were not resolved in the instrument used. The pattern should then be (1).2/1, when observed transverse to the field. Experiment confirmed this Paschen-Back effect (Fig. 21·6) though the photographs are somewhat

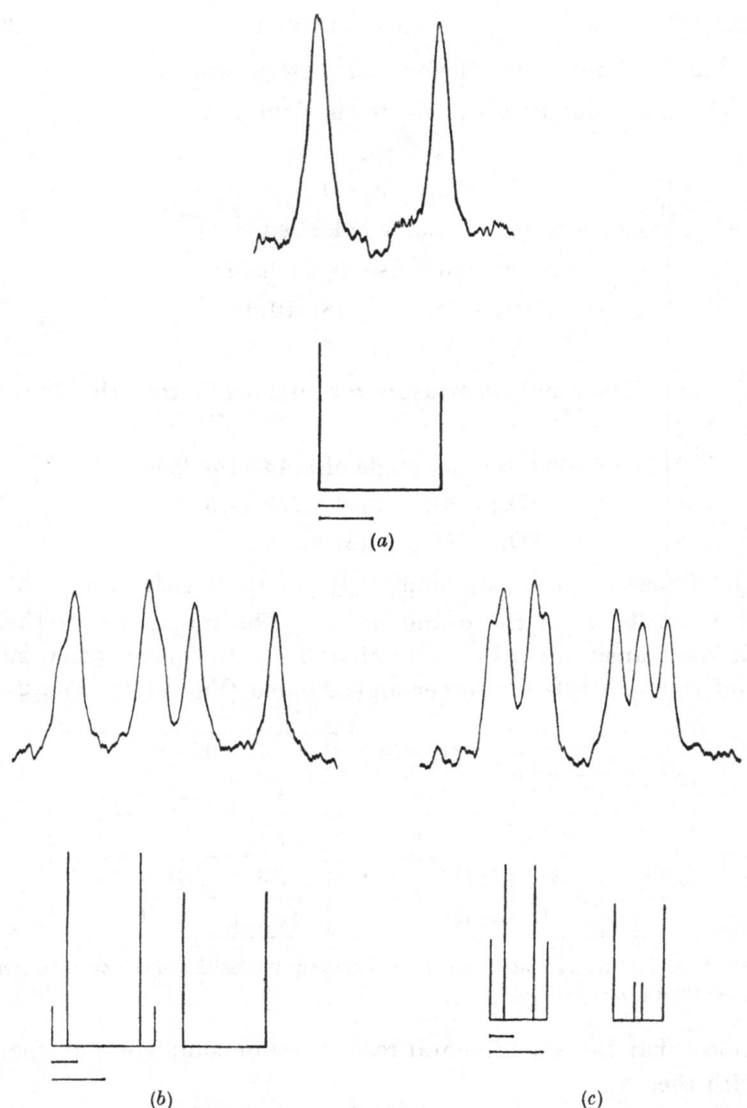

Fig. 21·5. The 4642–41 A., D→S, doublet of potassium. In each figure the photometer curve appears above and the theoretical splitting below. The intensity scale is arbitrary, but the same in all figures. The $^2D_{2\frac{1}{2}} \to {}^2S_{\frac{1}{2}}$ line is on the left, and the $^2D_{1\frac{1}{2}} \to {}^2S_{\frac{1}{2}}$ line on the right. At the foot the scale is shown by two lines of length 0·1 A. and 1 cm.$^{-1}$ The three figures show respectively (*a*) the doublet in the absence of a magnetic field, (*b*) the σ components viewed perpendicular to a field of 7500 gauss, and (*c*) the π components viewed from the same direction. (After Segrè and Bakker, *ZP*, 1931, **72** 728.)

distorted by the appearance of bands due to the sodium molecule; but if the intensities observed in the absence of a field are compared with those obtained with a field, the lines can be distinguished. Not content with these successes the authors measured

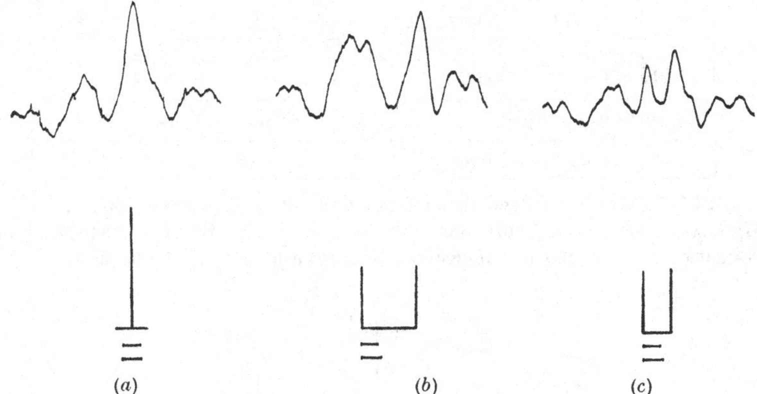

Fig. 21·6. Paschen-Back effect in the 3427 A. $3\,^2D \rightarrow 1\,^2S$, line of sodium. Figure *a* is taken in the absence of a magnetic field, *b* and *c* in a field of 16,100 gauss; *b* shows the σ components viewed perpendicular to the field, *c* the π components viewed in the same direction. The appearance of the Na_2 bands confuses the pictures a little, but it is the changes which occur between *a* on the one hand and *b* and *c* on the other which are of importance. (After Segrè and Bakker, *ZP*, 1931, **72** 731.)

the potassium doublet in an 'intermediate field', the splitting and intensity having been calculated by Millianzuk* with the help of Darwin's work on the gross Paschen-Back effect. A field of 17,800 gauss was used giving the ratio $o_m/\Delta\nu$ a value of 0·35, and again satisfactory agreement was obtained.

Finally Segrè and Bakker† examined the $7p\,^3P_2{}^\circ \rightarrow 6p\,^3P_2{}^\circ$ line of mercury in order to show that it arises, not as a quadripole transition, but as a dipole conditioned by the electric field. The quantum mechanics makes this probable, for a rough calculation shows that the quadripole term in the radiation of a forbidden line is usually larger than the dipole produced by an electric field, but this is not true when a state with which both levels may com-

* Milianczuk, *ZP*, 1932, **74** 825.

† Segrè and Bakker, *ZP*, 1932, **79** 655.

bine lies close to one of them;* and in mercury the 7p $^3P_2^\circ$ term lies at a distance of only 188 cm.$^{-1}$ from the 6d $^3D_2$ term.

An electric field modifies the ordinary selection rule; when the pattern is viewed transverse to the magnetic field, π components

M	-2	-1	0	1	2
Mg of 7p $^3P_2$	-3	$-1\frac{1}{2}$	0	$1\frac{1}{2}$	3
Mg of 6p $^3P_2$	-3	$-1\frac{1}{2}$	0	$1\frac{1}{2}$	3
Pattern if a dipole	(0) (3) 0 3 6/2				
Pattern if a quadripole	(3) 6/2				

Fig. 21·7. Theoretical Zeeman pattern of the 3680 A., 7p $^3P_2 \rightarrow$ 6p $^3P_2$, line of Hg I, when viewed perpendicular to the magnetic field, if it arises as (*a*) a dipole transition conditioned by an electric field; (*b*) a quadripole transition.

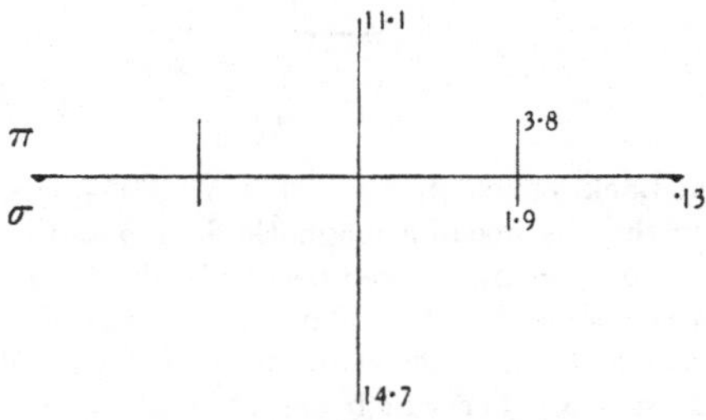

Fig. 21·8. Theoretical intensities of the Zeeman components of the 3680 A. line of Hg I, if it arises as a dipole in an electric field. The line is viewed transverse to the field. (After Segrè and Bakker, *ZP*, 1932, **79** 655.)

arise from transitions in which ΔM is 0 or ± 1 and σ components when ΔM is 0, ± 1 or ± 2; so that the pattern should be (0) (3) 0 3 6/2 (Fig. 21·7) and intensity calculations show that of these components the undisplaced line is far the strongest (Fig. 21·8). In contrast the selection rules of a quadripole line allow π components to arise only when ΔM is ± 1 and σ components only when ΔM is ± 2, so that the pattern predicted is (3) 6/2. The photometer curve reveals simply a strong central line, but as there seems no doubt that the resolving power was sufficient to have separated the side components of a quadripole line, this may be taken as proof that the transition is a dipole.

* Huff and Houston, *PR*, 1930, **36** 842.

4. Intensities

Theory has been able to predict the intensities of normal quadripole multiplets,* but very little has been confirmed by experiment. In part this is due to the difficulty of producing quadripole lines in the laboratory, and in part to the restriction of theory to multiplets arising from Russell-Saunders coupling, a restriction which implies the exclusion of all inter-system lines; no general theory of the latter has yet been developed, so that the intensity of each separate line has to be calculated.

BIBLIOGRAPHY

There are two important articles, the first by Becker and Grotrian "Über die galaktischen Nebel und den Ursprung der Nebellinien" in *Erg. d. Exact. Naturwiss.* 1928, **7**`8, and the second by Rubinowicz and Blaton on "Die Quadrupolstrahlung" in *Erg. d. Exact. Naturwiss.* 1932, **11** 170. The second of these deals with the theory more thoroughly than here.

* Rubinowicz and Blaton, *EEN*, 1932, **11** 176.

CHAPTER XXII

FLUORESCENT CRYSTALS

1. The energy levels of crystals

Though the energy levels of many gases and vapours are now well known, the energy levels of solids are the hills of an unexplored land; for only two routes have yet led to quantitative results, and both of these are of limited value. X-ray levels do change from one chemical compound to another,* but as 15,000 calories correspond to a shift of between 0·1 and 0·01 X. unit or 10^{-4} of the energy involved, the measurements have to be very accurate. The light scattered in the Raman effect,† on the other hand, is accurate enough, but it reveals only certain energy states out of the many which exist.

A third route runs through the absorption and fluorescent spectra of crystals, though as yet this has given hardly any quantitative results. At room temperature the lines are blurred by the agitation of the molecules, for the levels of one ion are split by the electric field of the ions which surround it, but the lines sharpen as the temperature is reduced. Much valuable work has been done in liquid air, but Spedding and his co-workers in California are rapidly making this of little more than historic interest by their work in liquid hydrogen.

Many absorption lines are of course due to molecular vibrations, but in the last five years evidence has accumulated to show that in the rare earths and the phosphors of chromium at least, many lines arise as electronic transitions within the atom. This evidence is both simpler and more conclusive in chromium.

2. Fluorescence and phosphorescence

When a solution of chlorophyll is illuminated with a beam of violet or ultra-violet light, a green band can be seen from all

* Siegbahn, *Spektroskopie der Röntgenstrahlen*, 1931, 278.

† Kohlrausch, *Der Smekal-Raman-Effekt*, 1931. Placzek, *Rayleigh Streuung und Raman-Effekt*, 1934. Symposium at Faraday Congress by Raman, Wood, Cabannes and others, *Trans. Far. Soc.* 1929, **25** 781. Dadieu and Kohlrausch, "Raman effect in organic chemistry", *JOSA*, 1931, **21** 286.

sides; this is an example of fluorescence. Again, in time past luminous paints consisted of impure barium sulphide; if first exposed to sunlight, this remains visible in the dark for some hours; it is said to phosphoresce. Fluorescence is distinguished from phosphorescence by the persistence of the latter after the exciting rays have been extinguished. To-day, however, this distinction does not seem of great importance; it is probably related to the fact that only solids phosphoresce, though solids, liquids and gases may all fluoresce.

In general fluorescence and phosphorescence may both be excited by cathode rays and X-rays as well as by light. When light is used, however, the wave-length of the exciting beam must be shorter than the wave-length of the light emitted; this law, discovered in 1853 by Sir George Stokes,* is simply explained by the quantum theory; energy may be lost between absorption and emission, it cannot be gained. This must not be taken to mean, however, that all wave-lengths shorter than those emitted are equally effective; on the contrary, a fluorescent solution or crystal absorbs only certain bands, and these are characteristic of the substance. The emission spectrum too is typically independent of the exciting source.

Many organic substances, such as quinine sulphate and eosin, fluoresce; but in these the cause is undoubtedly molecular. Of more interest here are the solid phosphors, a group of substances which are conveniently divided under five heads: the phosphors of chromium, the phosphors of the rare earths, the uranyl salts, the Lenard phosphors and the platino-cyanides. The phosphorescence of the first two has recently been shown to originate within the atom, and the evidence deserves detailed consideration. The phosphorescence of the other three may be atomic in origin, but no evidence is yet forthcoming; accordingly they are treated much more briefly.

3. Chromium phosphors

Many naturally occurring stones, such as the ruby, sapphire, red spinell, alexandrite and topaz, emit a red fluorescence. This

* Stokes, *Roy. Soc., Phil. Trans.*, 1853, **142** 463; 1853, **143** 385.

spectrum was thoroughly studied at a time when a ruby was supposed to consist of almost pure aluminium oxide, the odd 1 per cent. of chromium oxide being regarded as an unimportant impurity. Of recent years however the ruby has been imitated in the laboratory, and the chromium oxide, though present in such a small proportion, has been proved an essential constituent; for without it the aluminium oxide is not fluorescent, while with it the artificial ruby exhibits a spectrum identical with that of the natural stone.

In preparing one of the chromium phosphors in the laboratory a few drops of a chromium salt solution are added to the salt of some other metal, such as calcium nitrate, and heated twice to a temperature of about 1400° C. The spectrum of this phosphor does not depend on the amount of chromium, though it is brightest with a few parts per thousand present. But the oxide must be an intrusion in the regular lattice of the bedding; mixed crystals of $KAl(SO_4)_2.12H_2O$ and $KCr(SO_4)_2.12H_2O$ show no phosphorescence, whatever the proportions of chromium and aluminium;* moreover, the Cr_2O_3, or some molecule which contains it such as $MgCr_2O_4$, must be isomorphous with the material in which it is bedded, for Deutschbein has shown that chromium fluoresces in:

αAl_2O_3	which is isomorphous with	Cr_2O_3;	both	trigonal.
$MgAl_2O_4$	,, ,,	$MgCr_2O_4$;	,,	regular.
$ZnGa_2O_4$	,, ,,	$ZnCr_2O_4$;	,,	regular.
$BeAl_2O_4$	,, ,,	$BeCr_2O_4$;	,,	rhombic.
MgO	because it forms mixed crystals with	$MgCr_2O_4$;	,,	regular.

On the other hand, chromium does not fluoresce in:

γAl_2O_3 because it is not isomorphous with Cr_2O_3, the one being regular and the other trigonal;

BeO because it forms no mixed crystals with $BeCr_2O_4$, the former being hexagonal and the latter rhombic.†

That Cr_2O_3 must be isomorphous with the bedding means presumably that the outer electrons must not be much distorted;

* Deutschbein, *AP*, 1932, **14** 713.

† Deutschbein, *PZ*, 1932, **33** 874.

and in fact the spectrum of the ruby is very sharp and bright, for the Cr_2O_3 molecule fits easily into the αAl_2O_3 lattice. On the other hand, in Mg_2TiO_4 when the distortion is greater, the spectrum is more diffuse; moreover, the unaided eye notices that the phosphorescent light is much weaker.

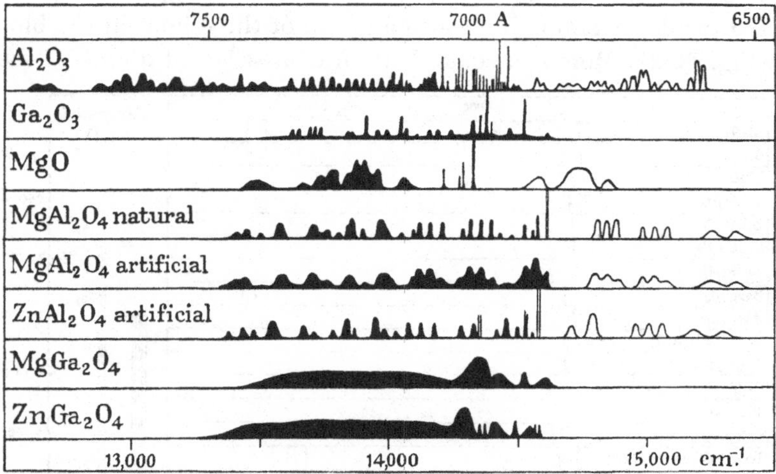

Fig. 22·1. Spectra of some chromium phosphors. The lines shown black are those emitted at $-186°$ C.; those shown in outline appear in absorption at low temperatures, but in emission only at high temperatures. Comparison of the spectra in Al_2O_3 and Ga_2O_3, crystals belonging to the same system, shows that the principal doublet appears in both, but with fivefold greater interval in Ga_2O_3; the two strongest subsidiary lines are also more widely separated in Ga_2O_3. The spectra in the regular crystal system—MgO, $MgAl_2O_4$, $ZnAl_2O_4$—show greater symmetry and a much smaller doublet interval, the doublet of MgO being still unresolved. (After Deutschbein, *PZ*, 1932, **33** 876.)

Whether the chromium phosphors are excited by cathode rays, X-rays or ultra-violet light, the spectrum produced is always the same; moreover, the emission spectrum is very like that produced in absorption. In both the lines are easily divided into three groups; first, principal lines, which are intense, sharp and only some half-dozen in number; second, subsidiary lines, which are weaker and more numerous; and third, bands, which are weak and often lie near the lattice bands of the pure bedding. Plate V shows these three in the emission spectrum of chromium bedded in aluminium oxide; at the short wave-length end of the spectrum lies the principal doublet, which is here some thousand times over-exposed. Then follow on the long wave-length side a few weaker but very sharp subsidiary lines, while beyond again lie some diffuse bands. That this type is general Fig. 22·1 shows.

All the chrome phosphors have one or two principal lines in the red, while several exhibit also a group of three lines in the blue (Fig. 22·2). Moreover, examination of a series of alums of the

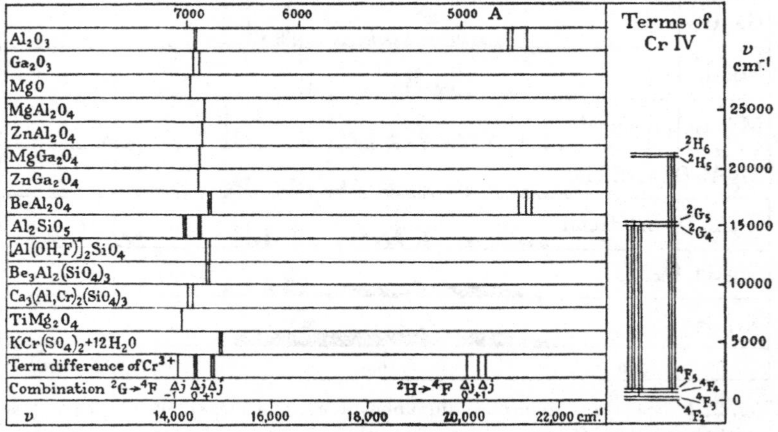

Fig. 22·2. Principal emission and absorption lines of some chrome phosphors compared with the term differences of Cr IV. (After Deutschbein, *PZ*, 1932, **33** 877.)

types $RCr(SO_4)_2.12H_2O$ and $RCr(SeO_4)_2.12H_2O$, where R is variously K, Rb, Tl and (NH_4), shows that in liquid air all absorb a strong doublet near 6700 A. (Fig. 22·3).* Other chromium compounds are not of course so regular as these, but most show line absorption.† The frequencies of these lines, and especially of the phosphorescent lines, are comparable with term differences found in an analysis of the Cr IV spectrum; in this the $3d^3$ configuration produces a ^4F ground term and above it metastable ^2G and ^2H

* Sauer, *AP*, 1928, **87** 197.

† Snow and Rawlins, *Camb. Phil. Soc. Proc.* 1932, **28** 522; Joos and Schnetzler, *Z. Phys. Chem.* 1933, B, **20** 1. A paper on $KCr(SO_4)_2.12H_2O$ by Spedding and Nutting, *J. Chem. Phys.* 1934, **2** 421, appeared after this section was written.

terms; in the vapour these terms do not combine, but in the phosphors the mechanism is probably not a pure dipole, so that

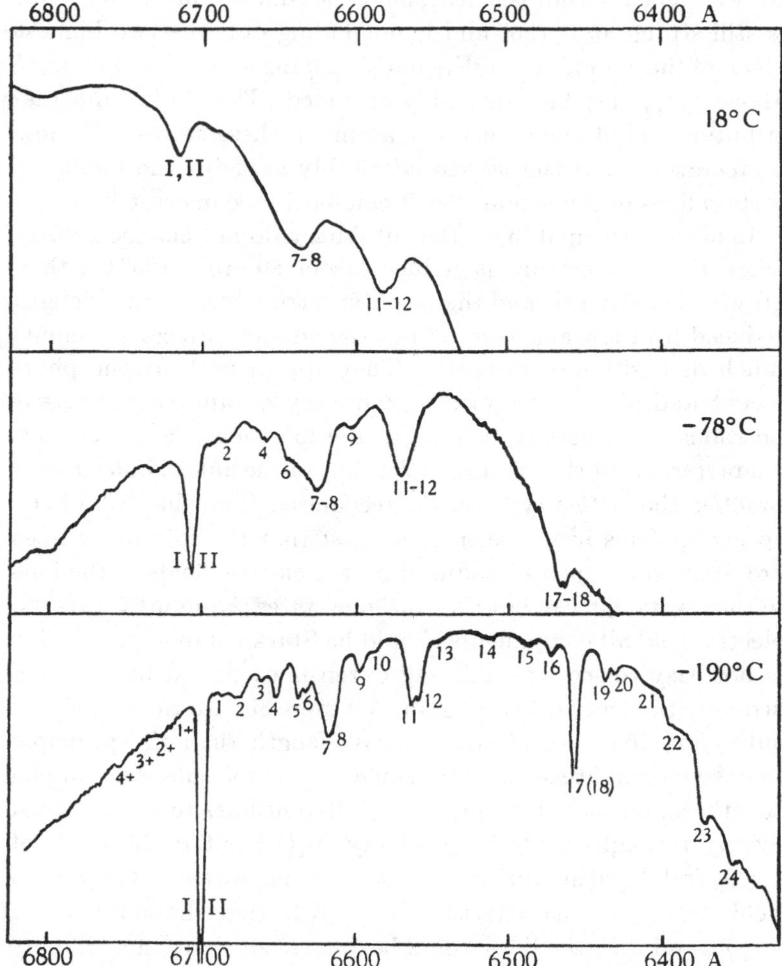

Fig. 22·3. Absorption curves of $RbCr(SeO_4)_2 . 12H_2O$ at 18, − 78 and −190° C. Crystal thickness about 4 mm. (After Sauer, *AP*, 1928, **87** 219.)

the transitions may reasonably be allowed. At any rate, Fig. 22·2 shows that the frequencies agree so remarkably that the identification can hardly be doubted; the red principal lines are there shown to arise in the $^2G \rightarrow {}^4F$ transition and the blue lines in $^2H \rightarrow {}^4F$;* the precise transitions to which the individual lines are to be assigned are not yet clear, but as the doublet of chrome alum is still strong at $-195°$ and is not fading out, the two highest levels of the ground term, $^4F_{4\frac{1}{2}}$ and $^4F_{3\frac{1}{2}}$, lying at 950 and 550 cm.$^{-1}$ above $^4F_{1\frac{1}{2}}$, may be reasonably excluded. The Boltzmann distribution would allow very few atoms in these states at so low temperature, and the law so admirably explains the fading of certain lines in samarium that it can hardly be in error.†

Like the principal lines, the subsidiary do not change greatly when the temperature is reduced from 20° to $-195°$ C.; they grow rather sharper, and the whole spectrum has its wave-length reduced by a few angstroms,‡ but they do not otherwise change much in position or intensity. They appear both in phosphorescent and absorption spectra, but they occupy very different positions in different beddings; the positions, however, are characteristic of the crystal lattice, not of the anion; indeed with practice the lattice type can be recognised from the look of the spectrum. This would seem to suggest that the subsidiary lines are Stark components produced by the electric fields of the ions in the crystal lattice; but if the atomic states are split by a strong electric field all components should be Stark components, and it is not easy to see why the half-dozen lines should be so much stronger than any others. Again, is it mere chance that nearly all subsidiary lines are of shorter wave-length than the principal lines? One might assume that some Cr_2O_3 molecules are lumped together, and that these produced different lines to those spread evenly through the bedding of say Al_2O_3; but if the fields of Cr_2O_3 and Al_2O_3 are different, as is probable, we should expect all fields from that characteristic of Cr_2O_3 to that characteristic of Al_2O_3, and then the lines would be diffuse not sharp; accordingly one must suppose the Cr_2O_3 molecules spread evenly through the Al_2O_3 lattice like currants in a cake. The electric field is then effectively that of the Al_2O_3 lattice.

* Deutschbein, *ZP*, 1932, **33** 877.

† Spedding, *PR*, 1933, **43** 143*a*.

‡ Deutschbein, *AP*, 1932, **14** 720.

The bands are considerably more complicated than either the principal or the subsidiary lines, for they are not the same in emission and absorption, and grow more complex if the temperature is raised. At low temperatures, however, the emission spectrum consists only of a long wave-length band lying in the red and infra-red; while the absorption spectrum consists of a short wave-length band lying in the blue. If the principal lines, which lie between these two bands, are unaltered electron transitions, and the bands conform to Stokes's law, these facts are easily explained; the absorption bands must lie on the high-frequency side of the principal lines, and the emission bands on the low-frequency side.* At room temperature emission still occurs chiefly in the red and infra-red, but some diffuse bands appear on the violet side of the principal lines; these short-wave or 'anti-Stokes' bands must be emitted by molecules passing from an excited state to one of low energy, so that their intensity should be proportional to the number of molecules in the excited state; this number is determined by the Boltzmann law and decreases rapidly with decrease of temperature. Again, in the absorption spectrum there appear at room temperature bands of wave-length longer than the principal lines; these anti-Stokes absorption bands can be explained in the same way as the anti-Stokes emission bands, they are due to absorption by molecules which are already in an excited state.†

The frequencies of some of the stronger bands are also of interest. The Raman spectrum of pure αAl_2O_3 reveals a lattice frequency of 417 $cm.^{-1}$ Now in the emission spectrum of a ruby, the principal doublet and a strong band lie at 14,416 and 14,006 $cm.^{-1}$ respectively; thus in the phosphor a band occurs with a frequency difference slightly less than that of pure Al_2O_3, 410 instead of 417 $cm.^{-1}$ This relationship is not uncommon in the chrome phosphors, and is accounted due to a decrease in the lattice frequency caused by the introduction of the chromium oxide molecules.*

* Deutschbein, *ZP*, 1932, **77** 490.

† For the effect of a magnetic field on the lines of a ruby: Du Bois and Elias, *AP*, 1908, **27** 233, 1911, **35** 617; Du Bois, *PZ*, 1912, **13** 128.

4. Rare earth phosphors

As long ago as the eighties of last century Crookes* showed, during the course of his pioneer work on high vacua, that when rare earth minerals are irradiated with cathode rays, they emit a strong phosphorescent spectrum; but his work is only of historic interest, because very few of his rare earth preparations were pure. More recent work shows that salts, which are colourless like those of lanthanum, gadolinium and ytterbium, exhibit no after-glow; but the smallest trace of active impurity makes the substance phosphorescent; 4.10^{-6} gm. of samarium in 1 gm. of calcium oxide is sufficient to produce a red after-glow, while the spectrum is brightest if the active earth is present in a proportion of only 1 per cent.; thus the smallest trace of dysprosium in yttrium, or of terbium in gadolinium, can easily be detected.†

Naturally occurring fluorites too have long been known to exhibit blue, yellow and green fluorescence, when excited with radium; but little progress was made until synthetic calcium fluoride was used and 1/10 per cent. of a rare earth added. Then it was shown that the blue band appears only when europium is present, and the yellow-green band only when ytterbium. The active agent of the red band has not yet been traced; all other rare earths are inactive. As europium and ytterbium are the two rare earths which most readily become divalent,‡ and the fluorescence may be excited by heating the activated fluorite in a reducing flame, the transition from tri- to divalent form is clearly linked with the fluorescence.§

Of all the rare earth phosphors, samarium has been most thoroughly studied (Fig. 22·4); as elsewhere the spectrum is much the same if excited by ultra-violet light instead of by cathode rays. The lines are somewhat sharper than those of the chromium phosphors, especially at room temperature; while at low temperatures even naturally occurring fluorspar emits lines which are as sharp as the D lines of a flame poor in sodium.‖ Again rare

* Crookes, *Chem. Soc. J.* 1889, **55** 255.

† Urbain, *Chem. Rev.* 1924, **1** 167.

‡ Jantsch and Klemm, *Z. f. anorg. und allgem. Chemie*, 1933, **216** 80.

§ Haberlandt, Karlik and Przibram, *Akad. Wiss. Wien, Ber.* 1934, **143** 151.

‖ Tomaschek, *AP,* 1927, **84** 329, 1047.

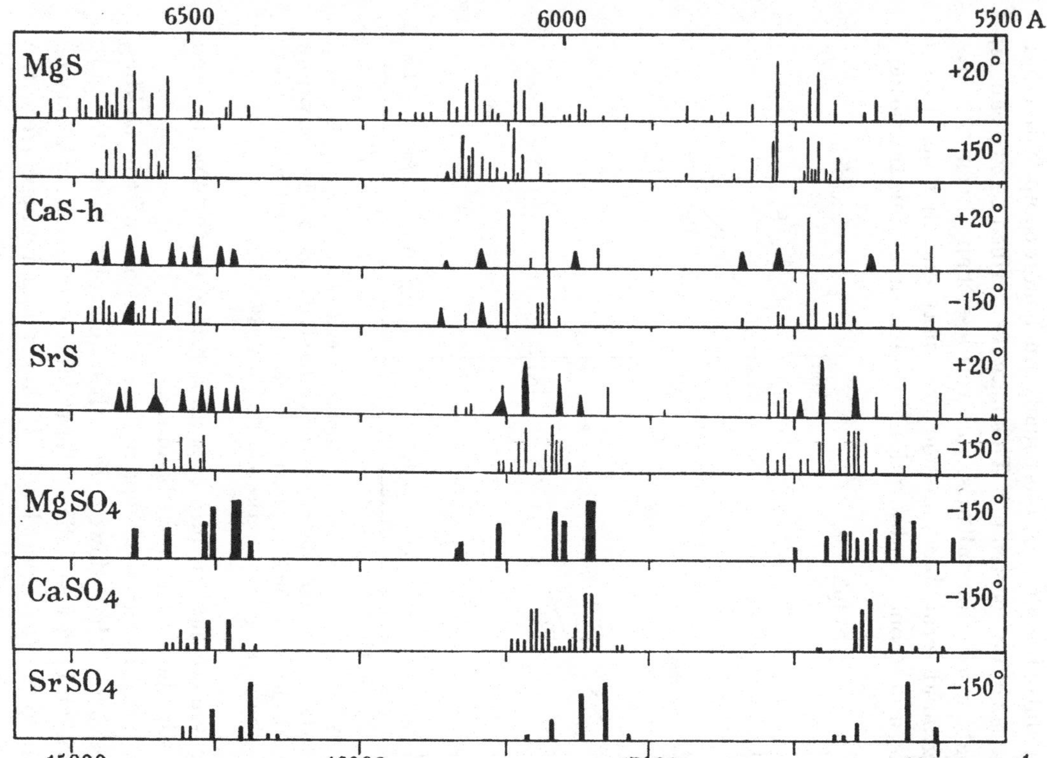

Fig. 22·4. Phosphorescent spectra of samarium bedded in various sulphides and sulphates. (After Tomaschek, *AP*, 1928, **84** 374, 382.)

earth compounds need not be isomorphous with the bedding, and this can very probably be related to the sharpness of the lines. In contrast to the chromium phosphors, however, a phosphorescent spectrum of the rare earths is usually much more complicated than the absorption spectrum.* This may be ascribed to two causes; first absorption lines all arise in levels less than 500 cm.$^{-1}$ above the ground term, but emission lines may end in much higher levels; and second all absorption lines arise in atomic transitions,

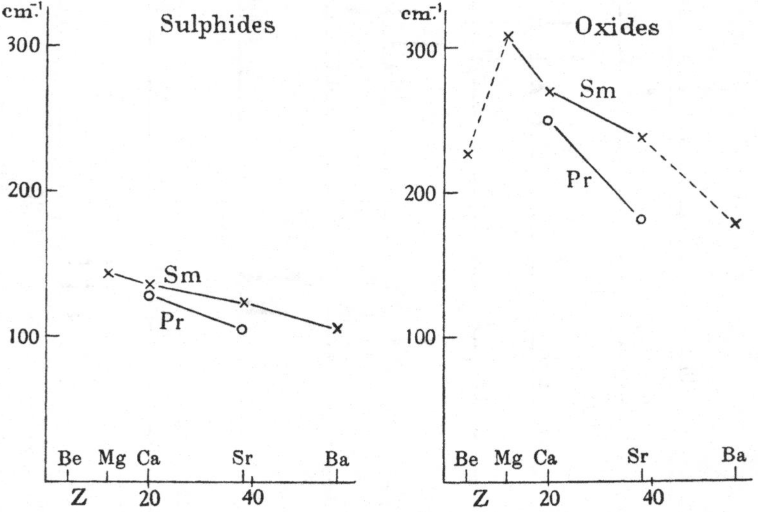

Fig. 22·5. Term displacements of samarium and praseodymium phosphors, when bedded in the sulphides and oxides of the alkaline earths. The heavier the rare earth and the smaller the diameter of the alkaline earth atom, the greater the displacement. (After Tomaschek, *PZ*, 1932, **33** 878.)

while some emission lines arise in the rare earth molecule and others in the lattice of the bedding.

In the rare earth phosphors the vibrations of the bedding lattice are rather slower than those calculated for the pure crystal, and this is significant for the rare earth molecules may be supposed to act as inert loads. If this explanation is correct, the change of frequency should be greater the larger the rare earth and the smaller the bedding molecule; these predictions experi-

* Spedding, *PR*, 1933, **43** 143*a*.

ment confirms. The praseodymium ion has a diameter 3 per cent. greater than the samarium ion, and the change of frequency is slightly greater; barium oxide has a larger diameter than calcium

cm.⁻¹	α	A–X	Type
MgO	4·21	2·10	Sodium chloride
CaO	4·80	2·38	
SrO	5·15	2·59	
BaO	5·50	2·77	
MgS	5·19	2·52	
CaS	5·69	2·80	
SrS	6·01	3·00	
BaS	6·37	3·18	
CaF_2	5·45	2·36	Fluorite
SrF_2	5·78	2·50	
BaF_2	6·19	2·68	
ZrO_2	5·10	2·21	
CeO_2	5·40	2·34	
ThO_2	5·57	2·42	
Al_2O_3	5·15	1·84 1·99	Corundum
Ga_2O_3	5·28		
Li_2SO_4			
$CaSO_4$			
$SrSO_4$			
$BaSO_4$			

Fig. 22·6. The two short wave line groups of some samarium phosphors arranged by the crystal type of bedding. α is the lattice constant and *A–X* the distance between the electropositive and electronegative centres in the lattice; both are in angstroms. (After Tomaschek, *PZ*, 1932, **33** 880.)

oxide, and the change of frequency produced by bedding samarium in it is much less (Fig. 22·5).*

Though in their general features the spectra of samarium in

* Tomaschek, *PZ*, 1932, **33** 878.

different beddings are similar, a closer examination reveals many differences. Thus in Fig. 22·6 a series of spectra are arranged according to their crystal types, the figures alongside being the lattice constant a and the distance between the positive and negative ions, both in angstroms. The crystal type is here revealed as an important influence, but in fact this is clearer still in the original photographs,* for in them a practised eye can recognise the crystal type at a glance.†

With so few principal lines in the samarium emission spectrum, one naturally asks whether they may not be interpreted, like the principal lines of the chromium phosphors, as electronic transitions within the ion. As the spectrum of samarium vapour has not been analysed, the way is by no means clear; but the fact that the rare earths do not have to be isomorphous with their bedding, suggests that the outer electrons can hardly be concerned; rather are the transitions likely to occur between different terms of the $4f^n$ configuration, the brightest lines arising between terms of the same multiplicity. If transitions between the $4f^n$ and $4f^{n-1}.5d$ configurations are responsible for some lines, as Laporte has suggested, they are likely to be much more diffuse than the lines arising within the $4f^n$ configuration; indeed it is tempting to identify a diffuse spectrum, photographed beside the sharp one by Fagerberg in neodymium and by Tomaschek‡ in samarium, with these predicted lines.

Assuming that the principal lines arise within the $4f^n$ configuration, their positions can be calculated by the method which Goudsmit developed.§ This method gives the extreme intervals of the various terms, if the coupling is Russell-Saunders, and the individual terms can then be interpolated with the interval rule. The calculations are laborious, but the results are eminently satisfactory, especially in praseodymium, whose emission spectrum has been recently measured by Evert ||. The lines shown in

* Photographs, *AP*, 1927 **84** Taf. IX–XI. Cf. *Z. Elect.* 1930, **36** 737.

† Tomaschek, *PZ*, 1932, **33** 879.

‡ Tomaschek, *PZ*, 1932, **33** 882.

§ Goudsmit, *PR*, 1928, **31** 948, and chapter XVIII of this book.

|| Evert, *AP*, 1932, **12** 144.

PLATE VIII

1. Fluorescent spectra of samarium bedded in various sulphates at 20 and $-150°$ C. At room temperature the lines are so blurred that the photographs do little more than show that the positions of the multiplets are independent of the bedding, but they sharpen as the temperature is reduced. The samarium lines are much sharper in $La_2(SO_4)_3$ than in any other bedding, probably because lanthanum like samarium is trivalent. (After Tomaschek, *AP*, 1927, **84**, Taf. X, XI.)

2. DF quintet from the iron arc. This multiplet arises as $3d^6 . 4s\,(^4P)\,4p\ ^5F^\circ \rightarrow 3d^6 . 4s^2\ ^5D$. (Lent by Prof. H. Dingle.)

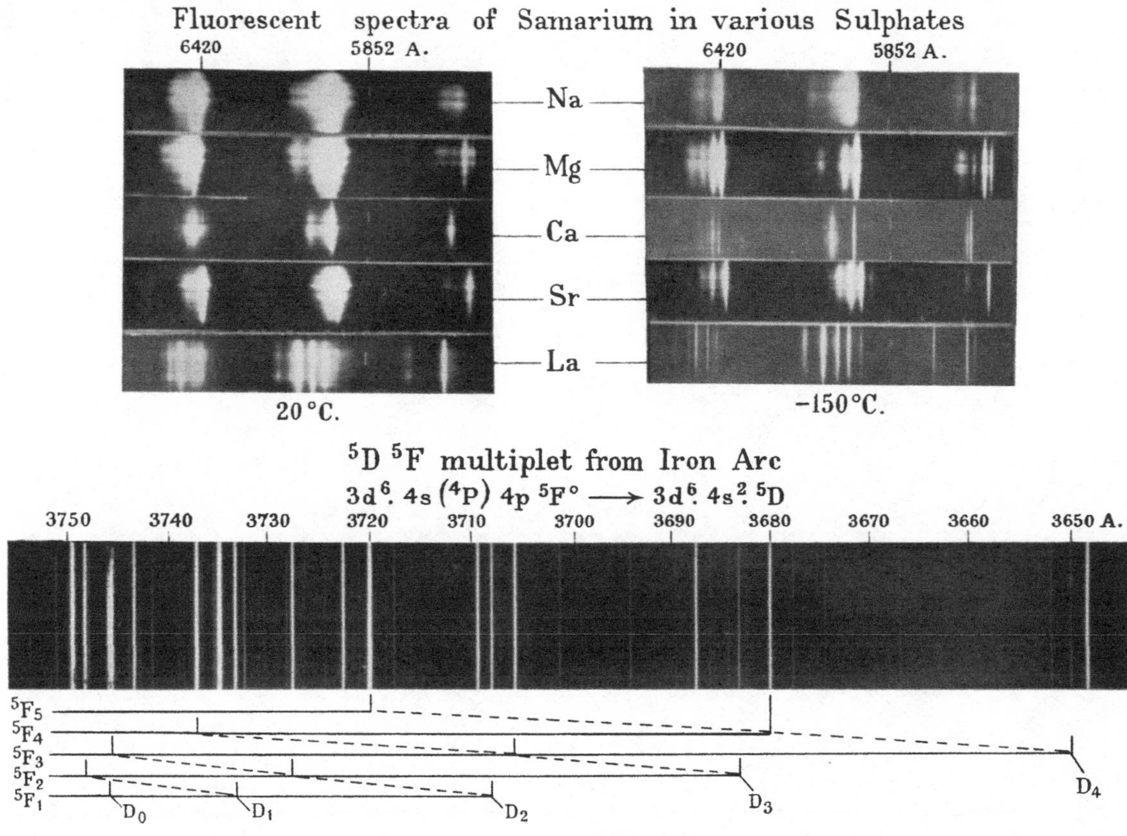

Fluorescent spectra of Samarium in various Sulphates
6420
5852 A.
6420
5852 A.
Na
Mg
Ca
Sr
La
20 °C.
−150 °C.
$^5D\ ^5F$ multiplet from Iron Arc
$3d^6 . 4s\ (^4P)\ 4p\ ^5F° \longrightarrow 3d^6 . 4s^2 . ^5D$
3750
3740
3730
3720
3710
3700
3690
3680
3670
3660
3650 A.
$^5F_5$
$^5F_4$
$^5F_3$
$^5F_2$
$^5F_1$
D_0
D_1
D_2
D_3
D_4

Plate VIII

Fig. 22·7, taken from left to right, arise in transitions between the following values of J:

for $^3F \rightarrow {}^3H$: 4–6, 3–5, 2–4, 4–5, 3–4, 4–4,

for $^1G \rightarrow {}^3H$: 4–5, 4–4.

Since quadripole transitions are probable, J has been allowed to change by two units. These predictions are compared with the phosphorescent spectrum of the metal plotted above the line, and the absorption spectrum plotted below.

Besides praseodymium satisfactory agreement is obtained in neodymium and erbium, spectra arising from 3 and $(14-3)$ electrons respectively. In samarium only the terms of highest multiplicity have yet been calculated, for when there are five electrons the work becomes very heavy, but the $^6F \rightarrow {}^6H$ transition accounts apparently for three strong lines in the infra-red.

5. Uranyl salts*

Like the rare earths the uranyl salts are fluorescent in their own right, and are not dependent on the crystal in which they happen to be embedded; but fluorescence is a property of the uranyl radical UO_2, not of the uranium atom, for salts in which uranium is quadrivalent do not fluoresce. On the other hand the absorption spectra of all uranium compounds are so similar, that the uranyl radical cannot be considered peculiar in the energy it absorbs, but only in re-emitting some of this energy as visible light.

At room temperature and viewed in a spectroscope of low resolving power, both the fluorescent and absorption spectra of uranyl salts consist of bands. In the emission spectra some seven or eight bands appear in the yellow and red, while in absorption a smaller number appear in the green and blue; the emission band of shortest wave-length coincides with the longest absorption band. The emission bands are equally spaced in the scale of frequency; the interval varies a little from salt to salt, but is never far from 830 cm.$^{-1}$ Again, all emission bands show the same variation in intensity, but the intensity distribution in absorption bands is quite different (Fig. 22·8).

* Pringsheim, *Fluorescenz und Phosphorescenz*, 1928, 238; Nichols and Howes, *The fluorescence of the uranyl salts*, Carnegie Institute Publication, 1919, No. 298.

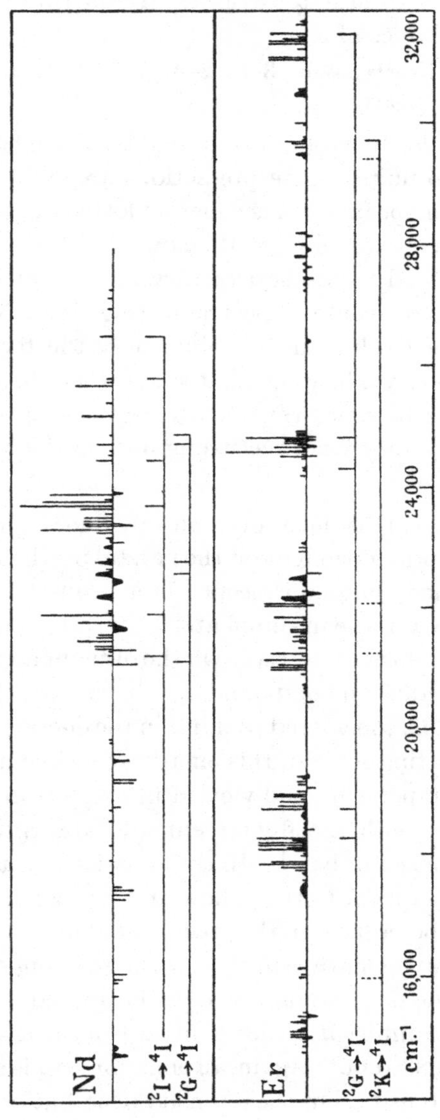
Nd
$^2I \to {}^4I$
$^2G \to {}^4I$
Er
$^2G \to {}^4I$
$^2K \to {}^4I$
cm^{-1}
16,000
20,000
24,000
28,000
32,000

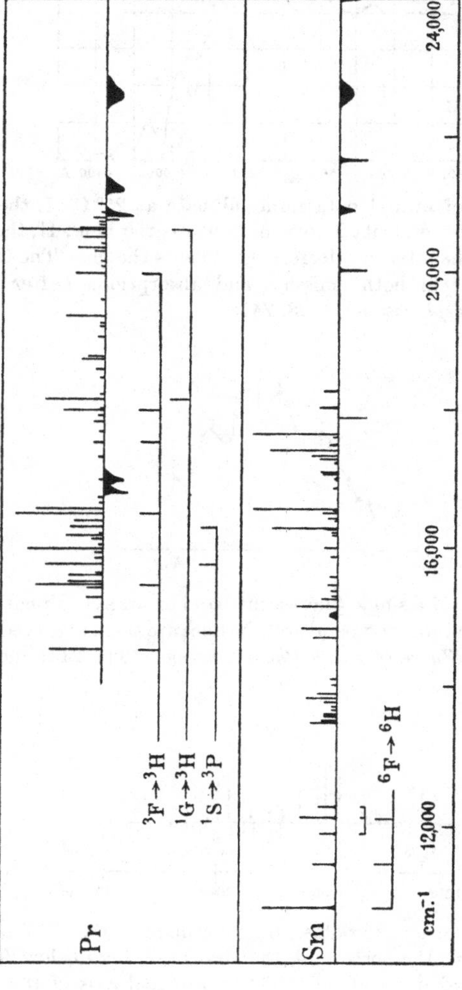

Fig. 22·7. Multiplets of some rare earth phosphors compared with the lines predicted by theory. Above the line is the phosphorescent spectrum when bedded in CaO; below, the absorption of salts in solution. (After Tomaschek, *PZ*, 1932, **33** 883.)

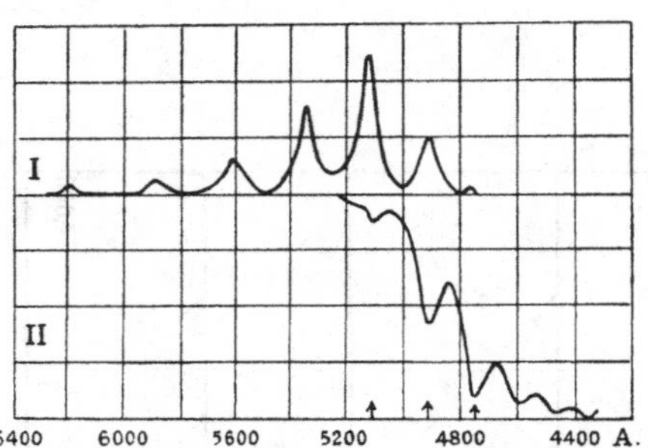

Fig. 22·8. Spectra of uranyl potassium sulphate at 25° C. I, the fluorescent spectrum showing the intensity increasing towards the blue; II, the absorption spectrum showing the intensity decreasing towards the blue. The three marked with arrows appear in both emission and absorption. (After Pringsheim, *Fluorescenz und Phosphorescenz*, 1928, 241.)

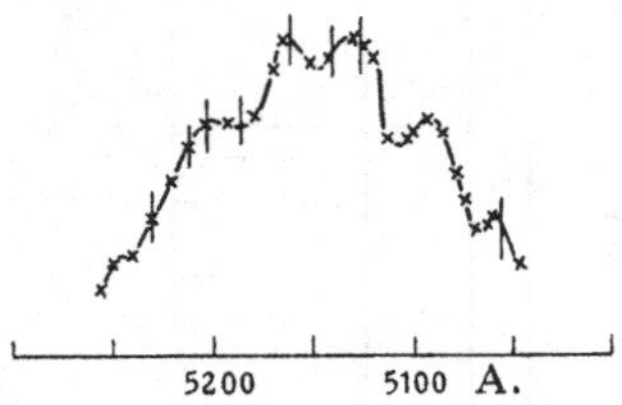

Fig. 22·9. Structure of a single fluorescent band of uranyl sulphate. The curve gives the intensity at room temperature; the lines are those observed at $-185°$ C. (After Pringsheim, *Fluorescenz und Phosphorescenz*, 1928, 243; and Wick, *PR*, 1918, **11** 126.)

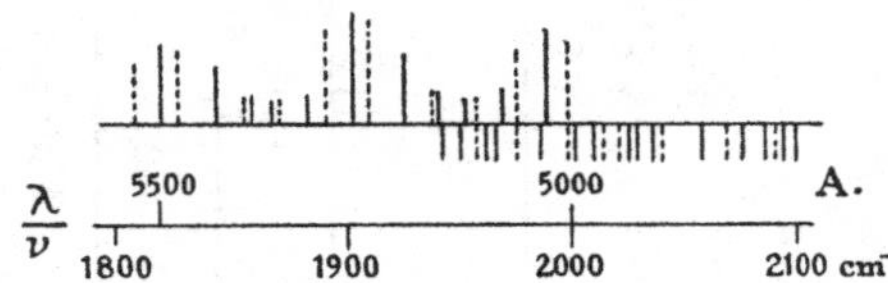

Fig. 22·10. Spectra of uranyl potassium chloride at $-186°$ C. The fluorescent lines are shown above the horizontal, the absorption lines below. The lines are polarised parallel and perpendicular to the principal axis of the crystal; the ordinary spectrum is shown by full and the extraordinary by dotted lines. The lengths of the lines indicate intensities. (After Pringsheim, *Fluorescenz und Phosphorescenz*, 1928, 243.)

At low temperatures high resolution splits these bands into a number of components; indeed in liquid hydrogen, at 20° A., many lines are as fine as the lines of a spark spectrum (Fig. 22.9). But rather curiously even these sharp lines show no trace of broadening in a magnetic field of 2500 gauss. In both emission and absorption spectra many lines are polarised parallel or perpendicular to the principal axis; in Fig. 22·10 these lines are represented by continuous and dotted lines respectively, and their different positions show that when the phosphorescent spectrum is viewed through a nicol prism, two entirely different spectra appear as the nicol is rotated.

The spectra of the uranyl salts resemble those of chromium sufficiently closely to make one wonder whether the frequencies of the strong lines are not equally significant, but as yet none of the higher spark spectra of uranium has been analysed.

6. Lenard phosphors

Many phosphorescent compounds, whether occurring naturally as minerals or artificially produced, consist of a crystalline salt carrying a trace of some other substance which will not fit into the crystal lattice; the intruding substance is commonly a metal, and the greater number of these phosphors may be regarded as members of a family of which the alkaline earth sulphides activated by a heavy metal may be considered the prototype. Having been exhaustively studied by Lenard and his school, they are commonly known as Lenard phosphors.

As the intruding metal seems responsible for the fluorescence, one might have hoped for a line spectrum, instead of the broad bands actually observed, though as many workers have used spectroscopes of low resolving power combined with a slit 1 mm. wide the evidence is not so extensive as could be desired. When a substance exhibits several bands these often behave independently when a change is made in the wave-length of the exciting light or in the temperature; some bands fade and others grow brighter. Further, the positions of the bands change when either the active element or the bedding is changed.*

* Pringsheim, *Hb. d. Phys.* 1929, **21** 600.

Like the Lenard phosphors the platino-cyanides phosphoresce only when bedded in some inert material; neither the pure salts nor their solutions are themselves phosphorescent; but the phosphorescence appears so regularly that it has been attributed to the platino-cyanide molecule in the absence of other evidence. The details of the emission spectra depend largely on the crystal type, and change when water of crystallisation is added. The polarisation depends on the polarisation of the incident light.

At room temperature the emitted light lasts a very short time; but if a body at -250° C. is illuminated, it phosphoresces only when allowed to warm up. In this the platino-cyanides are like the Lenard phosphors.

BIBLIOGRAPHY

The most thorough study of the older work is Lenard, Schmidt u. Tomaschek, "Phosphorescenz und Fluorescenz", *Handbuch der Experimental Physik*, 1928, **23**. This fills two large volumes; briefer summaries have been provided by: Pringsheim, *Fluorescenz und Phosphorescenz*, 1928; Pringsheim, "Luminescenzspektra", *Handbuch der Physik*, 1929, **21** 574.

The newer work of Spedding, Tomaschek and Deutschbein, with which this chapter is chiefly concerned, does not seem to have been written up.

APPENDIX V

KEY TO REFERENCES

The periodicals, in which most papers appear, I have cited by the capitals introduced by Gibbs, the less common by the usual abbreviations. A key to the former is given here; a key to the latter may be found in *Science Abstracts*.

AJ	Astrophysical Journal.
AP	Annalen der Physik.
BSJ	Bureau of Standards, Journal of Research.
EEN	Ergebnisse der Exacten Naturwissenschaften.
JOSA	Journal of the Optical Society of America.
N	Nature.
Nw	Naturwissenschaften.
PM	Philosophical Magazine.
PR	Physical Review.
PRS	Proceedings of the Royal Society, London, series A.
PZ	Physikalische Zeitschrift.
ZP	Zeitschrift für Physik.

An author's initials are not given in the text unless two authors of the same name occur; but all authors are given their initals in the index.

APPENDIX VI

BIBLIOGRAPHY

The bibliography is divided into three sections:

A. Books of reference.

B. Spectra of the elements.

C. Hyperfine structure of the elements.

In the last two sections I follow Bacher and Goudsmit in arranging the elements in the alphabetical order of their chemical symbols.

A. *Books of reference*

The following books are frequently needed for reference, because they contain experimental data conveniently arranged. Books dealing with a particular subject are discussed at the end of the chapter on that subject.

1. Gibbs, J. C. 'A complete bibliography of individual spectra for the years 1920–1931.' *Rev. Mod. Phys.* 1932, **4** 278.

2. Bacher, R. F. and Goudsmit, S. *Atomic energy states*, 1932. The terms of all known spectra, but no wave-lengths.

3. Kayser, H. and Konen, H. *Handbuch der Spektroskopie*, vol. VIII, 1932. A complete review of the first nineteen elements in alphabetical order; that is A, Ag, Al, As, Au, B, Ba, Be, Bi, Br, C, Ca, Cd, Ce, Cl, Co, Cr, Cs, Cu. Wave-lengths, terms, magnetic splitting factors and hyperfine structure are all given with a bibliography.

4. Grotrian, S. *Graphische Darstellung der Spektren von Atomen mit ein, zwei und drei Valenzelektronen*, 1928. The energy diagrams of vol. II show the transitions which produce all the strong lines.

5. Fowler, A. *Report on series in line spectra*, 1922. Paschen, F. and Götze, R. *Seriengesetze der Linienspektren*, 1922. Both books contain lists of terms and lines for the elements of the first three columns.

B. *The spectra of the elements*

These lists are select; where one paper provides a satisfactory summary, earlier papers are not cited. The papers deal chiefly with the analysis into systems of terms, but papers on intensities and the Zeeman effect have been included when they deal with a particular spectrum.

When one of the books of reference, Kayser and Konen, Grotrian and Fowler, deals with a spectrum, I have included the author's name in the bibliography, as these sources are too easily overlooked.

The abbreviations are those used in the text.

A: Argon 18

A Kayser and Konen.

A I Meissner, *ZP*, 1926, **39** 172, **40** 839. Terms and lines.
Dorgelo and Abbink, *ZP*, 1927, **41** 753. Extension in ultra-violet.
Gremmer, *ZP*, 1928, **50** 716. Extension in infra-red.
Rasmussen, *Nw*, 1930, **18** 1112. Bergmann series.
Terrien and Dijkstra, *J. de Phys.* 1934, **5** 443. Zeeman effect.
Pogány, *ZP*, 1935, **93** 364. Zeeman effect.
Boyce, *PR*, 1935, **48** 396. Extreme ultra-violet.

A II Compton, Boyce and Russell, *PR*, 1928, **32** 179. Extreme ultra-violet.
Bakker, de Bruin and Zeeman, *ZP*, 1928, **51** 114, **52** 299; *K. Akad. Amsterdam*, 1928, **31** 780. Zeeman effect.
de Bruin, *ZP*, 1930, **61** 307; *K. Akad. Amsterdam*, 1930, **33**, 198. Complete term scheme and lines.
Boyce, *PR*, 1935, **48** 396. Extreme ultra-violet.

A III Keussler, *ZP*, 1933, **84** 42. Lines and terms.
Boyce, *PR*, 1935, **48** 396. Extreme ultra-violet.
Boyce, *PR*, 1936, **49** 351. Intersystem lines.

A IV Boyce and Compton, *Proc. Nat. Acad. Sci.* 1929, **15** 656. Lines and terms.
Boyce, *PR*, 1935, **48** 396. Extreme ultra-violet.

Ag: Silver 47

Ag Kayser and Konen.

Ag I Fowler. Grotrian.
Blair, *PR*, 1930, **36** 1531. Extension of series.

Ag II Shenstone, *PR*, 1928, **31** 317. Terms and lines.
Blair, *PR*, 1930, **36** 173. Two new terms.
Gilbert, *PR*, 1935, **47** 847. High terms.

Ag III Gilbert, *PR*, 1935, **47** 847. Lines and terms.

Al: Aluminium 13

Al Kayser and Konen.

Al I Fowler. Grotrian.

Al II Sawyer and Paschen, *AP*, 1927, **84** 1. Terms and classified lines.
Ekefors, *ZP*, 1928, **51** 471. New ultra-violet lines.

Al III Paschen, *AP*, 1923, **71** 152. Terms and lines.
Ekefors, *ZP*, 1928, **51** 471. New terms.

Al IV Edlén and Ericson, *Comptes Rendus* 1930, **190** 116, 173. Resonance lines.

Al V *Ibid.*

Al VI *Ibid.*

As: Arsenic 33

As Kayser and Konen.

As I Meggers and de Bruin, *BSJ*, 1929, **3** 765. Lines and terms.
Rao, K. R., *PRS*, 1929, **125** 238. Lines and terms.
Rao, A. S., *Proc. Phys. Soc.* 1932, **44** 243.

As II Rao, A. S., *Proc. Phys. Soc.* 1932, **44** 343.

As III Lang, *PR*, 1928, **32** 737. Series.
Rao, K. R., *Proc. Phys. Soc.* 1931, **43** 68. Series.

As IV Sawyer and Humphreys, *PR*, 1928, **32** 583.

As V *Ibid.*

As: Arsenic 33 (*cont.*)

As VI Borg and Mack, *PR*, 1931, **37** 470. Series.

Au: Gold 79

Au Kayser and Konen.

Au I Grotrian.
McLennan and McLay, *PRS*, 1926, **112** 95. Terms and lines.
Symons and Daley, *Proc. Phys. Soc.* 1929, **41** 431. Zeeman effect.

Au II McLennan and McLay, *T. Roy. Soc. Canada*, 1928, **22** 103. Terms and lines.
Sawyer and Thompson, *PR*, 1931, **38** 2293. Ground term.
Mack and Fromer, *PR*, 1935, **48** 357. Pt I isoelectronic sequence.

B: Boron 5

B Kayser and Konen.

B I Bowen, *PR*, 1927, **29** 231. Terms and lines.
Smith and Sawyer, *JOSA*, 1927, **14** 287. Series.

B II Bacher and Goudsmit, 1932. Terms from unpublished material.

B III *Ibid.*
Edlén, *ZP*, 1931, **72** 763. Ground term.

B IV Edlén, *N*, 1931, **127** 405.

Ba: Barium 56

Ba Kayser and Konen.

Ba I Fowler. Grotrian.
Russell and Saunders, *AJ*, 1925, **61** 38.

Ba II Fowler. Grotrian.
Rasmussen, *ZP*, 1933, **83** 404. New terms.

Be: Beryllium 4

Be Kayser and Konen.

Be I Paschen and Kruger, *AP*, 1931, **8** 1005. Series extended.

Be II *Ibid.*

Be III Edlén and Ericson, *ZP*, 1930, **59** 656. Series.
Edlén, *N*, 1931, **127** 405.

Be IV Edlén and Ericson, *ZP*, 1930, **59** 656.
Edlén, *N*, 1931, **127** 405.

Bi: Bismuth 83

Bi Kayser and Konen.

Bi I Thorsen, *ZP*, 1926, **40** 642. Terms and lines.
Toshniwal, *PM*, 1927, **4** 774. Terms and lines.
Zeeman, Back and Goudsmit, *ZP*, 1930, **66** 1. Interpretation of low terms. Zeeman effect.

Bi II McLennan, McLay and Crawford, *PRS*, 1930, **129** 579. Series.

Bi III Lang, *PR*, 1928, **32** 737. Series.

Bi VI Mack and Fromer, *PR*, 1935, **48** 357. Pt I isoelectronic sequence.

Br: Bromine 35

Br Kayser and Konen.

Br I Kiess, C. C. and de Bruin, *BSJ*, 1930, **4** 667. Lines and terms.

Br II Deb, *PRS*, 1930, **127** 197. Series.

Br III *Ibid.*

Br: Bromine 35 (*cont.*)

Br IV Deb, *PRS*, 1930, **127** 197. Series.
Br V *Ibid.*

C: Carbon 6

C Kayser and Konen.
C I Hopfield, *PR*, 1930, **35** 1586. Ionisation potential.
Paschen and Kruger, *AP*, 1930, **7** 1. Terms and lines.
Birkenbeil, *ZP*, 1934, **88** 1. Extension in infra-red.
C II Fowler and Selwyn, *PRS*, 1928, **120** 312. Terms and lines.
Bowen, *PR*, 1931, **38** 128.
Edlén, *ZP*, 1935, **98** 561. New terms.
C III Bacher and Goudsmit, using unpublished material.
Bowen, *PR*, 1931, **38** 128. Extension and revision of series.
Edlén, *ZP*, 1931, **72** 559. Series.
C IV Edlén and Stenman, *ZP*, 1930, **66** 328. Term system.
Bacher and Goudsmit, using unpublished material.

Ca: Calcium 20

Ca Kayser and Konen.
Ca I Fowler. Grotrian.
Russell and Saunders, *AJ*, 1925, **61** 38. Displaced terms.
Back, *ZP*, 1925, **33** 579. Zeeman effect.
Russell, *AJ*, 1927, **66** 191. New terms.
Ca II Fowler. Grotrian.
Saunders and Russell, *AJ*, 1925, **62** 1. Terms and lines.
Russell, *AJ*, 1927, **66** 283.
Ca III Bowen, *PR*, 1928, **31** 497.
Ca IV *Ibid.*
Ca V *Ibid.*

Cb: Columbium 41

Cb I Meggers, *J. Wash. Acad. Sci.* 1924, **14** 442. Zeeman effect.
Meggers and Kiess, C. C., *JOSA*, 1926, **12** 417. Lines and terms.
King and Meggers, *PR*, 1931, **37** 226*a*. Furnace spectrum.
Meggers and Scribner, *BSJ*, 1935, **14** 629. New terms.
Cb II Meggers and Kiess, C. C., *JOSA*, 1926, **12** 417. Multiplets and Zeeman effect.
King and Meggers, *PR*, 1931, **37** 226*a*. Furnace spectrum.
Meggers and Scribner, *BSJ*, 1935, **14** 629. New terms.
Cb III Gibbs and White, *PR*, 1928, **31** 520. Multiplets.
Eliason, *PR*, 1933, **43** 745. New multiplets.
Cb IV Gibbs and White, *PR*, 1928, **31** 520. Multiplets.

Cd: Cadmium 48

Cd Kayser and Konen.
Cd I Fowler. Grotrian.
Ruark, *JOSA*, 1925, **11** 199. Higher terms added.
Cd II McLennan, McLay and Crawford, *T. Roy. Soc. Canada*, 1928, **22** 45.
Takahashi, *AP*, 1929, **3** 27. Series.
Cd III Gibbs and White, *PR*, 1928, **31** 776. Multiplets.

Ce: Cerium 58

Ce Kayser and Konen.
Ce I Karlson, *ZP*, 1933, **85** 482. Terms and lines.
Ce II Haspas, *ZP*, 1935, **96** 410. Terms and Zeeman effect.
Ce III Kalia, *Indian J. Phys.* 1933, **8** 137. Terms and lines.
Ce IV Gibbs and White, 1929, **33** 157. Doublets of stripped atoms.
Badami, *Proc. Phys. Soc.* 1931, **43** 53. Lines and terms.
Lang, *Can. J. Research*, 1935, A **13** 1. New terms.
Lang, *PR*, 1936, **49** 552 *a*. Ground term.

Cl: Chlorine 17

Cl Kayser and Konen.
Cl I Kiess, C. C. and de Bruin, *BSJ*, 1929, **2** 1117. Lines and terms.
Cl II Bowen, *PR*, 1928, **31** 34; 1934, **45** 401. New lines and terms.
Murakawa, *ZP*, 1931, **69** 507. Terms and lines.
Murakawa, *ZP*, 1935, **96** 117. New terms.
Cl III Bowen, *PR*, 1928, **31** 34; 1934, **45** 401. Lines and terms.
Cl IV *Ibid.*
Cl V *Ibid.*
Cl VI Bowen and Millikan, *PR*, 1925, **25** 591. Lines and terms.
Cl VII Bowen and Millikan, *PR*, 1925, **25** 295. Lines and terms.

Co: Cobalt 27

Co Kayser and Konen.
Co I Catalan, *ZP*, 1928, **47** 89. Terms and classified lines.
Catalan, *An. Soc. fís. y quím. (Madrid)*, 1929, **27** 832.
Co II Findlay, *PR*, 1930, **36** 5. Lines, terms and Zeeman effect.
Co V Gilroy, *PR*, 1931, **38** 2217. V I isoelectronic sequence.

Cr: Chromium 24

Cr Kayser and Konen.
Cr I Kiess, C. C., *BSJ*, 1930, **5** 775. Terms and lines.
Allen and Hesthal, *PR*, 1935, **47** 926. Intensities.
Cr II Krömer, *ZP*, 1928, **52** 531. Zeeman effect.
Kiess, C. C., *BSJ*, 1930, **5** 775. Terms and lines.
Gilroy, *PR*, 1931, **38** 2217. V I isoelectronic sequence.
Cr III White, *PR*, 1929, **33** 914. Ti I isoelectronic sequence.
Cr IV White, *PR*, 1929, **33** 676. Sc I isoelectronic sequence.
Cr V White, *PR*, 1929, **33** 543. Ca I isoelectronic sequence.
Cr VI Gibbs and White, *PR*, 1929, **33** 157. Doublets of stripped atoms.

Cs: Caesium 55

Cs Kayser and Konen.
Cs I Fowler. Grotrian.
Cs II Bacher and Goudsmit, 1932, using unpublished material.
Laporte, Miller and Sawyer, *PR*, 1931, **37** 845; 1932, **39** 458.
Olthoff and Sawyer, *PR*, 1932, **42** 766. Analysis extended.

Cu: Copper 29

Cu Kayser and Konen.
Cu I Sommer, *ZP*, 1926, **39** 711. Zeeman effect.
Bacher and Goudsmit, 1932, using unpublished material.
Cu II Shenstone, *PR*, 1927, **29** 380. Terms and lines.
Bacher and Goudsmit, 1932, using unpublished material.

Eu: Europium 63

Eu I Russell and King, *PR*, 1934, **46** 1023. Low terms.
Eu II Albertson, *PR*, 1934, **45** 499*a*. Low terms.

F: Fluorine 9

F I Bowen, *PR*, 1927, **29** 231. Ground state from ultra-violet lines.
Dingle, *PRS*, 1928, **117** 407. Terms and lines.
Edlén, *ZP*, 1935, **98** 445. New terms.
F II Dingle, *PRS*, 1930, **128** 600. Series.
F III Dingle, *PRS*, 1929, **122** 144. Series.
F IV Bowen, *PR*, 1927, **29** 231.
F VI Edlén, *ZP*, 1934, **89** 179.
F VII Edlén, *ZP*, 1934, **89** 179.

Fe: Iron 26

Fe I Laporte, *ZP*, 1924, **23** 135, **26** 1. Classification and Zeeman effect.
Laporte, *Proc. Nat. Acad. Sci.* 1926, **12** 496. Series.
Moore and Russell, *AJ*, 1928, **68** 151. Terms.
Burns and Walters, *Alleghany Observatory Publications*, 1929, **6** 159.
Fe II Russell, *AJ*, 1926, **64** 194. Terms and classified lines.
Meggers and Walters, *Bur. of Standards, Sci. Pap.* 1927, **22** 205. Low terms.
Dobbie, *PRS*, 1935, **151** 703. New terms.
Fe IV Gilroy, *PR*, 1931, **38** 2217. V I isoelectronic sequences.
Fe V White, *PR*, 1929, **33** 914. Ti I isoelectronic sequences.
Fe VI Bowen, *PR*, 1935, **47** 924.

Ga: Gallium 31

Ga I Grotrian.
Sawyer and Lang, *PR*, 1929, **34** 718. Lines and terms.
Ga II Sawyer and Lang, *PR*, 1929, **34** 712. Series.
Ga III Lang, *PR*, 1927, **30** 762.
Ga IV Mack, Laporte and Lang, *PR*, 1928, **31** 748.

Gd: Gadolinium 64

Gd I Albertson, *PR*, 1935, **47** 370. Ground term.

Ge: Germanium 32

Ge I Gartlein, *PR*, 1928, **31** 782. Lines and terms.
Rao, K. R., *PRS*, 1929, **124** 465. Lines and terms.
Ge II Lang, *PR*, 1929, **34** 697. Lines and terms.
Gartlein, *PR*, 1931, **37** 1704*a*. Series.
Ge III Lang, *PR*, 1929, **34** 697. Series.
Ge IV Lang, *PR*, 1929, **34** 697. Series.
Ge V Mack, Laporte and Lang, *PR*, 1928, **31** 748. Lines and terms.

H: Hydrogen 1

H I Fowler. Grotrian.
Bracket, *AJ*, 1922, **56** 154. A new series.
Pfund, *JOSA*, 1924, **9** 139. A new series.
Hansen, *AP*, 1925, **78** 558. Fine structure.
Houston, *AJ*, 1926, **64** 81. Fine structure.
Kent, Taylor and Pearson, *PR*, 1927, **30** 266. Fine structure.

H: Hydrogen 1 (*cont.*)

H I Houston and Hsieh, *PR*, 1934, **45** 263. Intervals of Balmer lines.
Williams and Gibbs, *PR*, 1934, **45** 491*a*. Correction to R_H.

He: Helium 2

He I Fowler. Grotrian.
Burger, *ZP*, 1929, **54** 643. Intensity.
Hopfield, *AJ*, 1930, **72** 133. Ultra-violet series extended.
Kruger, *PR*, 1930, **36** 855. New lines classified.
Gibbs and Kruger, *PR*, 1931, **37** 1559. Structure of 3888 A. line.

He II Fowler. Grotrian.
Paschen, *AP*, 1927, **82** 689. Fine structure.
Kruger, *PR*, 1930, **36** 855. New lines classified.

Hf: Hafnium 72

Hf I Meggers and Scribner, *BSJ*, 1930, **4** 169. Terms and lines.

Hf II Meggers and Scribner, *JOSA*, 1928, **17** 83. Terms.
Meggers and Scribner, *BSJ*, 1934, **13** 625. Terms, intensities and Zeeman effect.

Hg: Mercury 80

Hg I Fowler. Grotrian.
Takamine and Suga, *Inst. Phys. and Chem. Tokio*, 1930, **13** 1. New series in infra-red.

Hg II Paschen, *Berlin Akad. Sitz.* 1928, **32** 536. Terms and lines.
McLennan, McLay and Crawford, *PRS*, 1931, **134** 41. Terms and lines.

Hg III McLennan, McLay and Crawford, *T. Roy. Soc. Canada*, 1928, **22** 247. Terms and lines.
Mack and Fromer, *PR*, 1935, **48** 357. Pt I isoelectronic sequence.

I: Iodine

I I Evans, *PRS*, 1931, **133** 417. Lines and terms.
Deb, *PRS*, 1933, **139** 380. Lines and terms.

I IV Krishnamurty, *Proc. Phys. Soc.* 1936, **48** 277. Lines and terms.

In: Indium 49

In I Fowler. Grotrian.
Sawyer and Lang, *PR*, 1929, **34** 718. New terms.
Lansing, *PR*, 1929, **34** 597. New terms.
Lang, *PR*, 1930, **35** 126*a*. Ionisation potentials.

In II Bacher and Goudsmit, 1932, using unpublished data.

In III Lang, *Proc. Nat. Acad. Sci.* 1927, **13** 341; 1929, **15** 414.
Rao, K. R., Narayan and Rao, A. S., *Indian J. Phys.* 1928, **2** 482.

In IV Gibbs and White, *PR*, 1928, **31** 776. Some multiplets.

Ir: Iridium 77

Ir I Meggers and Laporte, *PR*, 1926, **28** 642. Low levels.
Albertson, *PR*, 1932, **42** 443*a*. Low terms.

K: Potassium 19

K I Fowler. Grotrian.
Ferschmin and Frisch, *ZP*, 1929, **53** 326. Doublet structure of D terms.
Edlén, *ZP*, 1935, **98** 445. Doublets resolved.

K: Potassium 19 (*cont.*)

K II Bowen, *PR*, 1928, **31** 497. Terms and lines.
Whitford, *PR*, 1932, **39** 898. Zeeman effect.

K III de Bruin, *ZP*, 1929, **53** 658.
Ram, *Indian J. Phys.* 1933, **8** 151.

K IV Bowen, *PR*, 1928, **31** 497. One multiplet.
Ram, *Indian J. Phys.* 1933, **8** 151.

Kr: Krypton 36

Kr I Gremmer, *ZP*, 1929, **54** 215. Lines and terms.
Meggers, de Bruin and Humphreys, *BSJ*, 1929, **3** 129; 1931, **7** 643. Lines and terms.
Pogány, *ZP*, 1933, **86** 729. Zeeman effect.

Kr II Bakker and de Bruin, *ZP*, 1931, **69** 36. Zeeman effect. New terms.
de Bruin, Humphreys and Meggers, *BSJ*, 1933, **11** 409.

Kr III Deb and Dutt, *ZP*, 1931, **67** 138. Series.
Humphreys, *PR*, 1935, **47** 712. Lines and terms.

La: Lanthanum 57

La I Meggers, *BSJ*, 1932, **9** 239. Lines and Zeeman effect.
Russell and Meggers, *BSJ*, 1932, **9** 625. Lines and terms.

La II Russell and Meggers, *BSJ*, 1932, **9** 625. Lines and terms.

La III Russell and Meggers, *BSJ*, 1932, **9** 625. Lines and terms.

Li: Lithium 3

Li I Fowler. Grotrian.

Li II Werner, *N*, 1925, **116** 574; 1926, **118** 154. Series.
Schüler, *ZP*, 1926, **37** 568; 1927, **42** 487. Fine structure.
Ericson and Edlén, *ZP*, 1930, **59** 656. Ground term.

Li III Edlén and Ericson, *ZP*, 1930, **59** 656. Two lines.
Gale and Hoag, *PR*, 1931, **37** 1703*a*. New lines.

Lu: Lutecium 71

Lu I Meggers and Scribner, *BSJ*, 1930, **5** 73. Lines and terms.

Lu II *Ibid.*

Lu III *Ibid.*

Mg: Magnesium 12

Mg I Fowler. Grotrian.
Bowen and Millikan, *PR*, 1925, **26** 150. PP′ multiplets.

Mg II Fowler. Grotrian.

Mg III Mack and Sawyer, *Science*, 1928, **68** 306, 1761. Series.
Edlén and Ericson, *Comptes Rendus*, 1930, **190** 116. Doublet.

Mg IV Edlén and Ericson, *Comptes Rendus*, 1930, **190** 173. Extreme ultra-violet.
Mack and Sawyer, *PR*, 1930, **35** 299. Screening doublets.

Mg V *Ibid.*

Mn: Manganese 25

Mn I McLennan and McLay, *T. Roy. Soc. Canada*, 1926, **20** 89. Lines and terms.
Russell, *AJ*, 1927, **66** 184, 347. Configurations assigned.
Seward, *PR*, 1931, **37** 344. Intensities.

Mn: Manganese 25 (*cont.*)

Mn II Russell, *AJ*, 1927, **66** 233.
Duffendack and Black, *PR*, 1929, **34** 42. Lines and terms.
Seward, *PR*, 1931, **37** 344. Intensities.

Mn III Gibbs and White, *Proc. Nat. Acad. Sci.* 1927, **13** 525. Multiplets.
Gilroy, *PR*, 1931, **38** 2217. Lines and terms.

Mn IV White, *PR*, 1930, **33** 914. Ti I isoelectronic sequence.

Mn V White, *PR*, 1929, **33** 678. Sc I isoelectronic sequence.
Bowen, *PR*, 1935, **47** 924. New terms.

Mn VII Gibbs and White, *Proc. Nat. Acad. Sci.* 1926, **12** 676.

Mo: Molybdenum 42

Mo I Catalan, *An. Soc. fís. y quím.* (*Madrid*), 1923, **21** 213. Lines and terms.
Wilhelmy, *AP*, 1926, **80** 305. Zeeman effect.
Meggers and Kiess, *JOSA*, 1926, **12** 417. Lines and terms.

Mo II Meggers and Kiess, *JOSA*, 1926, **12** 417. Lines and terms.
Wilhelmy, *AP*, 1926, **80** 305. Zeeman effect.

Mo IV Eliason, *PR*, 1933, **43** 745. Multiplets.

Mo V Trawick, *PR*, 1935, **48** 223. Sr I isoelectronic sequence.

N: Nitrogen 7

N I Ingram, *PR*, 1929, **34** 421. Terms and lines.
Hopfield, *PR*, 1930, **36** 789. Adds some higher terms.
Ekefors, *ZP*, 1930, **63** 437. Lines and terms.
Stucklen and Carr, *PR*, 1933, **43** 944*a*. One term.

N II Fowler, A. and Freeman, *PRS*, 1927, **114** 662. Lines and terms.
Bowen, *PR*, 1927, **29** 231. Lines and terms.
Freeman, *PRS*, 1929, **124** 654. New terms.
Bowen, *PR*, 1929, **34** 534. Series.

N III Freeman, *PRS*, 1928, **121** 318. Lines and terms.

N IV Freeman, *PRS*, 1930, **127** 330. Lines and terms.
Edlén, *N*, 1931, **127** 744. Singlets.
Bacher and Goudsmit, using unpublished material.

N V Edlén and Ericson, *ZP*, 1930, **64** 64. Series.

Na: Sodium 11

Na I Fowler. Grotrian.
Ferschmin and Frisch, *ZP*, 1929, **53** 326. Doublets structure of D terms.

Na II Bowen, *PR*, 1928, **31** 967. Lines and terms.
Frisch, *ZP*, 1931, **70** 498. Lines and terms.
Vance, *PR*, 1932, **41** 480. Lines and terms.

Na III Edlén and Ericson, *Comptes Rendus*, 1930, **190** 173. Extreme ultraviolet.
Mack and Sawyer, *PR*, 1930, **35** 299. New lines and levels.
Vance, *PR*, 1932, **41** 480. Lines and terms.
Söderqvist, *ZP*, 1932, **76** 316. Lines and terms.

Na IV Vance, *PR*, 1932, **41** 480.

Ne: Neon 10

Ne I Paschen, *AP*, 1918, **60** 405.
Paschen and Götze.

Ne: Neon 10 (*cont.*)

Ne I Back, *AP*, 1925, **76** 317. Zeeman effect.
Lyman and Saunders, *Proc. Nat. Acad. Sci.* 1926, **12** 92. Low levels.
Gremmer, *ZP*, 1928, **50** 716. New levels from infra-red lines.
Murakawa and Iwama, *Inst. Phys. and Chem. Tokio*, 1930, **13** 283. Zeeman effect.

Ne II Russell, Compton and Boyce, *Proc. Nat. Acad. Sci.* 1928, **14** 280. Lines and terms.
Frisch, *ZP*, 1930, **64** 499. Two new terms.
de Bruin and Bakker, *ZP*, 1931, **69** 19. Zeeman effect.

Ne III Boyce and Compton, *Proc. Nat. Acad. Sci.* 1929, **15** 656. Series.
de Bruin, *ZP*, 1932, **77** 489. New lines and terms.
Keussler, *ZP*, 1933, **85** 1. Ground term.

Ne IV Boyce and Compton, *Proc. Nat. Acad. Sci.* 1929, **15** 656. Series.

Ni: Nickel 28

Ni I Russell, *PR*, 1929, **34** 821. Lines and terms.
Ornstein and Buoma, *PR*, 1930, **36** 679. Intensities.
Marvin and Baragar, *PR*, 1933, **43** 973. Zeeman effect.

Ni II Shenstone, *PR*, 1927, **30** 255. Lines, terms and Zeeman effect.
Menzies, *PRS*, 1929, **122** 134. Ground term.
Ornstein and Buoma, *PR*, 1930, **36** 679. Intensities.

Ni VI Gilroy, *PR*, 1931, **38** 2217. V I isoelectronic sequence.

O: Oxygen 8

O I Hopfield, *PR*, 1931, **37** 160. Lines and terms.
de Bruin, *N*, 1932, **129** 469.

O II Russell, *PR*, 1928, **31** 27. Terms and lines.

O III Fowler, A., *PRS*, 1928, **117** 317. Lines and terms.
Freeman, *PRS*, 1929, **124** 654. Series.

O IV Bowen, *PR*, 1927, **29** 231.
Freeman, *PRS*, 1930, **127** 330.

O V Edlén, *N*, 1931, **127** 744. Singlets.
Bacher and Goudsmit, 1932, using unpublished material.

O VI Edlén and Ericson, *ZP*, 1930, **64** 64. Li-like spectra.

Os: Osmium 76

Os I Meggers and Laporte, *PR*, 1926, **28** 642. Six low levels.

P: Phosphorus 15

P I Kiess, C. C., *BSJ*, 1932, **8** 393. Lines and terms.
Robinson, *PR*, 1936, **49** 297. New terms.

P II Bowen, *PR*, 1927, **29** 510. Lines and terms.
Robinson, *PR*, 1936, **49** 297. New terms.

P III Millikan and Bowen, *PR*, 1925, **25** 600. Lines and terms.
Saltmarsh, *PRS*, 1925, **108** 332. Higher terms.
Bowen, *PR*, 1928, **31** 34. New lines.
Bowen, *PR*, 1932, **39** 8. New terms.

P IV Bowen and Millikan, *PR*, 1925, **25** 591. Terms and lines.
Bowen, *PR*, 1932, **39** 8. New terms.

P V Millikan and Bowen, *PR*, 1925, **25** 295. Lines and terms.

Pb: Lead 82

Pb I Gieseler and Grotrian, *ZP*, 1925, **34** 374; 1926, **39** 377. Series.
Back, *ZP*, 1926, **37** 193. Zeeman effect.
Bacher and Goudsmit, 1932, using unpublished material.

Pb II Gieseler, *ZP*, 1927, **42** 265. Lines and terms.

Pb III Smith, *PR*, 1929, **34** 393; 1930, **36** 1. Terms and lines.
Green and Loring, *PR*, 1932, **41** 389. Zeeman effect.

Pb IV Smith, *PR*, 1930, **36** 1. Lines and terms.

Pb V Mack, *PR*, 1929, **34** 17. Lines and terms.
Mack and Fromer, *PR*, 1935, **48** 357. Pt I isoelectronic sequence.

Pd: Palladium 46

Pd I Shenstone, *PR*, 1930, **36** 669. Lines, terms and Zeeman effect.

Pd II Shenstone, *PR*, 1928, **32** 30. Lines and terms.
Blair, *PR*, 1930, **36** 173. New terms.

Pt: Platinum 78

Pt I Haussmann, *AJ*, 1927, **66** 333; *PR*, 1928, **31** 152. Zeeman effect and terms.
Livingood, *PR*, 1929, **34** 185. Lines and terms.

Ra: Radium 88

Ra I Rasmussen, *ZP*, 1934, **87** 607.

Ra II Fowler.
Rasmussen, *ZP*, 1933, **86** 24.

Rb: Rubidium 37

Rb I Fowler. Grotrian.
Ramb, *AP*, 1931, **10** 311. Doublets resolved.

Rb II Laporte, Miller and Sawyer, *PR*, 1931, **38** 843. Series.

Re: Rhenium 75

Re I Meggers, *BSJ*, 1931, **6** 1027. Lines and terms.

Rh: Rhodium 45

Rh I Sommer, *ZP*, 1927, **45** 147. Lines, terms and Zeeman effect.

Rh II Bacher and Goudsmit, 1932, using unpublished material.

Rn: Radon 86

Rn I Rasmussen, *ZP*, 1933, **80** 726. Lines and terms.

Ru: Ruthenium 44

Ru I Sommer, *ZP*, 1926, **37** 1. Lines, terms and Zeeman effect.

Ru II Meggers and Shenstone, *PR*, 1930, **35** 868.
Bacher and Goudsmit, 1932, using unpublished material.

S: Sulphur 16

S I Frerichs, *ZP*, 1933, **80** 150. Lines and terms.
Meissner, Bartelt and Eckstein, *ZP*, 1933, **86** 54. Additional terms.
Ruedy, *PR*, 1933, **44** 757. New terms.

S II Ingram, *PR*, 1928, **32** 172. Lines and terms.
Gilles, *Ann. de Phys.* 1931, **15** 267. Terms.
Bartelt and Eckstein, *ZP*, 1933 **86** 77. New terms.

S III Ingram, *PR*, 1929, **33** 907. Lines and terms.

S IV Millikan and Bowen, *PR*, 1925, **25** 602. Lines and terms.

S: Sulphur 16 (*cont.*)

S IV Bowen, *PR*, 1928, **31** 38. New terms.

S V Millikan and Bowen, *PR*, 1925, **25** 591, **26** 150. Lines and terms.

S VI Millikan and Bowen, *PR*, 1925, **25** 295. Doublet Laws.

Sb: Antimony 51

Sb I McLennan and McLay, *T. Roy. Soc. Canada*, 1927, **21** 63. Lines and terms.
Löwenthal, *ZP*, 1929, **57** 822. Zeeman effect and new terms.

Sb II Lang and Vestine, *PR*, 1932, **42** 233. Terms.

Sb III Lang, *PR*, 1930, **35** 445. Terms and lines.

Sb IV Green and Lang, *Proc. Nat. Acad. Sci.* 1928, **14** 706. Lines and terms.
Badami, *Proc. Phys. Soc.* 1931, **43** 538. Series.

Sb V Lang, *Proc. Nat. Acad. Sci.* 1927, **13** 341. Series.
Badami, *Proc. Phys. Soc.* 1931, **43** 538. Series.

Sc: Scandium 21

Sc I Russell and Meggers, *Bur. Stand., Sci. Papers*, 1927, **22** 329. Lines and terms.

Sc II Russell and Meggers, *Bur. Stand., Sci. Papers*, 1927, **22** 329. Lines and terms.
Russell and Meggers, *BSJ*, 1929, **2** 733. Comparison with Y.

Sc III Russell and Lang, *AJ*, 1927, **66** 13. Lines and terms.
Smith, *Proc. Nat. Acad. Sci.* 1927, **13** 65. Extension of series.

Sc IV Majumdar and Toshniwal, *N*, 1928, **121** 828. Screening doublets.

Se: Selenium 34

Se I McLennan, McLay and McLeod, *PM*, 1927, **4** 486. Lines and terms.
Gibbs and Ruedy, *PR*, 1932, **40** 204. New terms.

Se II Badami and Rao, K. R., *PRS*, 1933, **140** 387. Lines and terms.
Martin, *PR*, 1935, **48** 938. Lines and terms.

Se IV Pattabhiramayya and Rao, A. S., *Indian J. Phys.* 1929, **3** 531. Series.

Se V Sawyer and Humphreys, *PR*, 1928, **32** 583. Series.

Se VI Sawyer and Humphreys, *PR*, 1928, **32** 583. Series.

Si: Silicon 14

Si I Fowler, A., *PRS*, 1929, **123** 422. Lines and terms.
Kiess, C. C., *BSJ*, 1933, **11** 775. New terms.

Si II Fowler, A., *Phil. Trans. R.S.* 1925, **225** 1. Series.
Bowen and Millikan, *PR*, 1925, **26** 150. PP° multiplets.
Bowen, *PR*, 1928, **31** 34. Series.

Si III Fowler, A., *Phil. Trans. R.S.* 1925, **225** 1. Series.
Sawyer and Paschen, *AP*, 1927, **84** 1. Comparison with Al II.

Si IV Fowler, A., *Phil. Trans. R.S.* 1925, **225** 1. Series.

Si V Edlén and Ericson, *Comptes Rendus*, 1930, **190** 116. Doublets.

Sm: Samarium 62

Sm I Albertson, *PR*, 1935, **47** 370. Ground term.

Sn: Tin 50

Sn I Back, *ZP*, 1927, **43** 309. Lines, terms and Zeeman effect.
Green and Loring, *PR*, 1927, **30** 575. Terms, lines and Zeeman effect.
Bacher and Goudsmit, 1932, using unpublished material.

Sn: Tin 50 (*cont.*)

Sn II Grotrian.
Green and Loring, *PR*, 1927, **30** 574. Lines, terms and Zeeman effect.
Narayan and Rao, K.R., *ZP*, 1927, **45** 350. Terms.
Lang, *PR*, 1930, **35** 445. Lines and terms.

Sn III Gibbs and Vieweg, *PR*, 1929, **34** 400. Cd I isoelectronic sequence.
Green and Loring, *PR*, 1931, **38** 1289. Multiplets and Zeeman effect.

Sn IV Rao, K. R., Narayan and Rao, A. S., *Indian J. Phys.* 1928, **2** 476. Series.

Sn V Gibbs and White, *Proc. Nat. Acad. Sci.* 1928, **14** 345. Series.

Sr: Strontium 38

Sr I Fowler. Grotrian.
Russell and Saunders, *AJ*, 1925, **61** 39. Lines and terms.

Sr II Fowler. Grotrian.

Ta: Tantalum 73

Ta I McLennan and Durnford, *PRS*, 1928, **120** 502. Zeeman effect.
Kiess, C. C. and Kiess, H. K., *BSJ*, 1933, **11** 277. Lines and terms.

Te: Tellurium 52

Te I Bartelt, *ZP*, 1934, **88** 522. Lines and terms.

Te III Krishnamurty, *PRS*, 1935, **151** 178. Lines and terms.

Te IV Rao, K. R., *PRS*, 1931, **133** 220. Lines and terms.

Te V Gibbs and Vieweg, *PR*, 1929, **34** 400. Cd I isoelectronic sequence.

Te VI Lang, *Proc. Nat. Acad. Sci.* 1927, **13** 341. Lines and terms.
Rao, K. R., *PRS*, 1931, **133** 220. Lines and terms.

Ti: Titanium 22

Ti I Russell, *AJ*, 1927, **66** 347. Lines and terms.
Harrison, *JOSA*, 1928, **17** 389. Intensities.
White, *PR*, 1929, **33** 914. Ti I isoelectronic sequence.

Ti II Russell, *AJ*, 1927, **66** 283. Lines and terms.

Ti III Russell and Lang, *AJ*, 1927, **66** 13. Terms.

Ti IV Russell and Lang, *AJ*, 1927, **66** 13. Terms.

Tl: Thallium 81

Tl I Fowler. Grotrian.

Tl II McLennan, McLay and Crawford, *PRS*, 1929, **125** 570. Lines and terms.
Smith, *PR*, 1930, **35** 235. Extension of series.
Ellis and Sawyer, *PR*, 1936, **49** 145. New lines and terms.

Tl III McLennan, McLay and Crawford, *PRS*, 1929, **125** 50. Lines and terms.

Tl IV Rao, K. R., *Proc. Phys. Soc.* 1929, **41** 361. Terms.
Mack and Fromer, *PR*, 1935, **48** 357. Pt I isoelectronic sequence.

V: Vanadium 23

V I Bacher and Goudsmit, 1932, using unpublished material.
Gilroy, *PR*, 1931, **38** 2217. New lines and terms.

V II Meggers, *ZP*, 1925, **33** 509; **39** 114. Lines, terms and Zeeman effect.
Russell, *AJ*, 1927, **66** 184, 194. Terms named.
White, *PR*, 1929, **33** 914. New terms and lines.

V III White, *PR*, 1929, **33** 672. Sc I isoelectronic sequence.

V: Vanadium 23 (*cont.*)

V IV White, *PR*, 1929, **33** 542. Ca I isoelectronic sequence.
V V Gibbs and White, *PR*, 1929, **33** 157. Doublets of stripped atoms.

W: Tungsten 74

W I Beining, *ZP*, 1927, **42** 146. Zeeman effect.
Bacher and Goudsmit, 1932, using unpublished material.
Laun, *PR*, 1935, **48** 572*a*. New terms.
W II Beining, *ZP*, 1927, **42** 146. Zeeman effect.

Xe: Xenon 54

Xe I Rasmussen, *ZP*, 1932, **72** 779.
Humphreys and Meggers, *BSJ*, 1933, **10** 139. Lines and terms.
Pogány, *ZP*, 1935, **93** 364. Zeeman effect.
Xe II Humphreys, de Bruin and Meggers, *BSJ*, 1931, **6** 287. Lines and terms.
Xe III Deb and Dutt, *ZP*, 1931, **67** 138.

Y: Yttrium 39

Y I Meggers and Russell, *BSJ*, 1929, **2** 733. Lines and terms.
Y II Meggers and Russell, *BSJ*, 1929, **2** 733. Lines and terms.
Y III Meggers and Russell, *BSJ*, 1929, **2** 733. Lines and terms.

Zn: Zinc 30

Zn I Fowler. Grotrian.
Sawyer, *JOSA*, 1926, **13** 431. PP° multiplets.
Zn II Lang, *Proc. Nat. Acad. Sci.* 1929, **15** 414. Lines and terms.
Takahashi, *AP*, 1929, **3** 27. Lines and terms.
Zn III Laporte and Lang, *PR*, 1927, **30** 378. Series.

Zr: Zirconium 40

Zr I Kiess, C. C. and Kiess, H. K., *BSJ*, 1931, **6** 621. Lines, terms and Zeeman effect.
Zr II Kiess, C. C. and Kiess, H. K., *BSJ*, 1930, **5** 1205. Lines, terms and Zeeman effect.
Zr III Kiess, C. C. and Lang, *BSJ*, 1930, **5** 305. Lines and terms.
Zr IV Kiess, C. C. and Lang, *BSJ*, 1930, **5** 305. Lines and terms.

C. *The hyperfine structure of the elements*

Ag Tolansky, *PPS*, 1933, **45** 559. No hyperfine structure in Ag I.
Williams and Middleton, *N*, 1933, **131** 691. No hyperfine structure found.
Jackson, *N*, 1933, **131** 691. No hyperfine structure found.
Hill, *PR*, 1934, **46** 536. No hyperfine structure found.

Al Gibbs and Kruger, *PR*, 1931, **37** 656*a*. No hyperfine structure in 4 lines.
White, *PR*, 1931, **37** 1175*a*. Interpretation of Gibbs and Kruger.
Tolansky, *ZP*, 1932, **74** 336. Narrow lines suggests g (I) small.
Williams and Sabine, *PR*, 1933, **43** 362*a*. No hyperfine structure in Al III.
Paschen and Ritschl, *AP*, 1933, **18** 867. Hyperfine structure of Al II shows $I=\frac{1}{2}$.
Brown and Cook, *PR*, 1934, **45** 731*a*. Work of Paschen shows $\mu=2{\cdot}4$.

As Tolansky, *PRS*, 1932, **137** 541; *ZP*, 1933, **87** 210. $I=1\frac{1}{2}$ in As II.
Rao, A. S., *ZP*, 1933, **84** 236. I is $1\frac{1}{2}$ in As II.
Crawford and Crooker, *N*, 1933, **131** 655. As IV confirms.
Tolansky and Heard, *PRS*, 1934, **146** 818. Intensities of As II confirm.

Ba Ritschl and Sawyer, *ZP*, 1931, **72** 36. Resonance lines of Ba II.
Kruger, Gibbs and Williams, *PR*, 1932, **41** 322. $I=2\frac{1}{2}$ in 135 and 137.

Be Kruger and Wagner, *PR*, 1932, **41** 373*a*. Three lines of Be I and two lines of Be II appear sharp.
Parker, *PR*, 1933, **43** 1035*a*. I is probably $\frac{1}{2}$.

Bi Goudsmit and Back, *ZP*, 1927, **43** 321; 1928, **47** 174. Hyperfine structure in magnetic field shows $I=4\frac{1}{2}$.
McLennan, McLay and Crawford, *PRS*, 1930, **129** 579. Bi II and Bi III.
Zeeman, Back and Goudsmit, *ZP*, 1931, **66** 1. Magnetic analysis of Bi I.
Fisher and Goudsmit, *PR*, 1931, **37** 1057. Analysis of partially resolved patterns.

Br de Bruin, *N*, 1930, **125** 414. Hyperfine structure shows $I=1\frac{1}{2}$.
Tolansky, *PRS*, 1932, **136** 585. Confirms de Bruin.
Brown, *PR*, 1932, **39** 777. Band spectrum shows $I=1\frac{1}{2}$ or $2\frac{1}{2}$ in Br^{79} and Br^{81}.

Cb Ballard, *PR*, 1934, **46** 806. $I=4\frac{1}{2}$; $\mu=3{\cdot}7$.

Cd Schüler and Bruck, *ZP*, 1929, **56** 291. Hyperfine structure shows $I=0$ in even and $I=\frac{1}{2}$ in odd isotopes.
Schüler and Keyston, *ZP*, 1931, **71** 413. Abnormal intensities.
Jones, *Proc. Phys. Soc.*, 1933, **45** 625. In Cd II μ is $-0{\cdot}625$.
Schüler and Westmeyer, *ZP*, 1933, **82** 685. Isotope displacement in Cd II.

Cl Elliott, *PRS*, 1930, **127** 638. Band spectrum shows $I=2\frac{1}{2}$.
Tolansky, *ZP*, 1931, **73** 470; 1932, **74** 336. Hyperfine structure shows μ is small; isotope displacement.

Co Grace, *PR*, 1933, **43** 762*a*. I is probably $3\frac{1}{2}$.
More, *PR*, 1934, **46** 470. $I=3\frac{1}{2}$; μ is between 2 and 3.
Kopfermann and Rasmussen, *ZP*, 1935, **94** 58. $I=3\frac{1}{2}$ confirmed.

Cs Jackson, *PRS*, 1934, **143** 455. Hyperfine intensity ratio shows $I=3\frac{1}{2}$.
Granath and Stranathan, *PR*, 1934, **46** 317. μ is $2{\cdot}45$.
Kopfermann, *ZP*, 1932, **73** 437. $I=3\frac{1}{2}$ in Cs II.
Cohen, *PR*, 1934, **46** 713. Atomic ray shows $I=3\frac{1}{2}$.
Heydenburg, *PR*, 1934, **46** 802. Polarisation of resonance radiation shows μ between $2{\cdot}40$ and $2{\cdot}52$.
Jackson, *ZP*, 1935, **93** 809. Intensity ratios in 4555 A line.
Granath and Stranathan, *PR*, 1935, **48** 725. Hyperfine structure shows μ between $2{\cdot}4$ and $3{\cdot}0$.

Cu Ritschl, *ZP*, 1932, **79** 1. I is $1\frac{1}{2}$ in both isotopes. Cu I. Isotope displacement.
Schüler and Schmidt, *ZP*, 1936, **100** 113. $\mu=2{\cdot}5$ in 63 and $\mu=2{\cdot}6$ in 65 Electric quadripole moment.

Eu Schüler and Schmidt, *ZP*, 1935, **94** 457. $I=2\frac{1}{2}$ in 151 and 153; both μ's are positive; $\mu_{151}:\mu_{153}=2{\cdot}2$. Abnormal intervals.

F Gale and Monk, *PR*, 1929, **33** 114. Band spectrum shows $I=\frac{1}{2}$.
Campbell, *ZP*, 1933, **84** 393. In F I, $I=\frac{1}{2}$.
Brown and Bartlett, *PR*, 1934, **45** 527. μ is about $3\mu_N$.

Ga Jackson, *ZP*, 1932, **74** 291. Resonance lines of Ga I.
Campbell, *N*, 1933, **131** 204. $I=1\frac{1}{2}$ in 69 and 71 from Ga II.

H¹ Kapuściński and Eymers, *PRS*, 1929, **122** 58. Band spectrum shows $I=\frac{1}{2}$.
Rabi, Kellogg and Zacharias, *PR*, 1934, **46** 157. μ for proton is 3·25.

H² Urey, Brickwedde and Murphy, *PR*, 1932, **40** 1. Mass effect in Balmer lines.
Murphy and Johnston, *PR*, 1934, **46** 95. Band spectrum shows $I=1$.
Estermann and Stern, *N*, 1934, **133** 911. μ is $0{\cdot}7\mu_N$.
Rabi, Kellogg and Zacharias, *PR*, 1934, **46** 163. μ for deuton is $0{\cdot}75 \pm 0{\cdot}2$.

Hg Schüler and Keyston, *ZP*, 1931, **72** 423. Hg I. Isotope displacement.
Schüler and Jones, *ZP*, 1932, **74** 631. Hg II. Isotope displacement.
Schüler and Jones, *ZP*, 1932, **77** 801. Irregularities.
Mrozowski, *Acad. Pol. Sci. et Lettres*, 1931, **9** 464. 2537 A. line.
Inglis, *ZP*, 1933, **84** 466. Magnetic wandering of 2537 A. line.
Schüler and Schmidt, *ZP*, 1935, **98** 239. Asymmetry of electric field of Hg^{201}.
Venkatesachar and Sibaiya, *Indian Acad. of Sci. Proc.* 1934, **1** 8. Hyperfine structure of Hg II.

I Loomis, *PR*, 1927, **29** 112. Band spectrum shows I is large.
Tolansky, *PRS*, 1935, **149** 269. Hyperfine structure of I II shows $I=2\frac{1}{2}$.
Tolansky, *PRS*, 1935, **152** 663. I I confirms. A perturbed term.
Strait and Jenkins, *PR*, 1936, **49** 635*a*. Alternating intensities confirm $I=2\frac{1}{2}$.

In Jackson, *ZP*, 1933, **80** 59. Hyperfine intensity ratio shows $I=4\frac{1}{2}$.
Paschen, *Sitz. d. Preusz. Akad. d. Wiss. Phys.-Math.* 1934, 456. Departures from interval rule in In II.

Ir Schüler and Schmidt, *ZP*, 1935, **94** 460. I is probably $\frac{1}{2}$ in 191 and 193.

K Loomis and Wood, *PR*, 1931, **38** 854. Band spectrum shows $I>\frac{1}{2}$.
Schüler, *ZP*, 1932, **76** 14. Isotope displacement.
Fermi and Segré, *ZP*, 1933, **82** 749. Hyperfine structure cannot be resolved.
Frisch, *P.Z. Sowjetunion*, 1933, **4** 557. 30 lines of K II are single.
Jackson and Kuhn, *PRS*, 1934, **148** 335. Hyperfine structure of resonance lines shows $I \geqslant 2\frac{1}{2}$.
Millman, Fox and Rabi, *PR*, 1934, **46** 320. Splitting of atomic rays gives $I=1\frac{1}{2}$ and $\mu=0{\cdot}38\mu_N$.
Millman, *PR*, 1935, **47** 739. $I>\frac{1}{2}$ in K^{41}.
Fox and Rabi, *PR*, 1935, **48** 746. Atomic ray shows $\mu=0{\cdot}379$ in K^{39} and $I \leqslant 2\frac{1}{2}$ in K^{41}.

Kr Kopfermann and Wieth-Knudsen, *ZP*, 1933, **85** 353. Work on Kr I shows in Kr^{83} I is $\geqslant 3\frac{1}{2}$ and μ probably negative.

La Anderson, *PR*, 1934, **45** 685; **46** 473. Hyperfine structure shows $I=3\frac{1}{2}$ and $\mu=2{\cdot}5$.

Crawford, *PR*, 1935, **47** 768. Hyperfine structure of La I confirms Anderson.

Li Schüler, *AP*, 1925, **76** 292; *ZP*, 1927, **42** 487. Hyperfine structure of Li II.

Harvey and Jenkins, *PR*, 1930, **35** 789. Band spectrum shows $I=1\frac{1}{2}$ in Li^7.

Hughes and Eckart, *PR*, 1930, **36** 694. Mass effect.

van Wijk and van Koeveringe, *PRS*, 1931, **132** 98. Band spectrum shows $I=1\frac{1}{2}$ in Li^7.

Güttinger and Pauli, *ZP*, 1931, **67** 743. Perturbations in Li II.

Goudsmit and Inglis, *PR*, 1931, **37** 328*a*. Hyperfine structure of Li II.

Granath, *PR*, 1932, **42** 44. From 5485 A. of Li II, $I=1\frac{1}{2}$ and $\mu=3{\cdot}29\mu_N$.

Gray, *PR*, 1933, **44** 570. $I=1\frac{1}{2}$ confirmed.

Ladenburg and Levy, *ZP*, 1934, **88** 449. Alternating intensities suggest $I=2\frac{1}{2}$ in Li.

Fox and Rabi, *PR*, 1935, **48** 746. Atomic ray shows $I \geqslant 1$ in Li^6 and $\mu=3{\cdot}20$ in Li.

Fock and Petrashen, *P.Z. Sowjetunion*, 1935, **8** 547. Theory suggests $\mu=4{\cdot}57$.

Bartlett and Gibbons, *PR*, 1936, **49** 552*a*. Comment on Fock and Petrashen.

Schüler and Schmidt, *ZP*, 1936, **99** 285. If $I=1$ in Li^6, then $\mu=0{\cdot}6$.

Lu Schüler and Schmidt, *ZP*, 1935, **95** 265. $I=3\frac{1}{2}$, deviations from interval rule.

Mg Murakawa, *ZP*, 1931, **72** 793. Absence of hyperfine structure in Mg I suggests $I=0$.

Jackson and Kuhn, *PRS*, 1936, **154** 679. Isotope displacement in resonance line.

Mn White, *PR*, 1929, **34** 1404. Hyperfine structure shows $I = 2\frac{1}{2}$.

White and Ritschl, 1930, **35** 1146. Vector coupling in Mn I.

N Ornstein and van Wijk, *ZP*, 1928, **49** 315. Band spectrum shows $I=1$.

Bacher, *PR*, 1933, **43** 1001. Magnetic moment of nucleus $\geqslant 0{\cdot}2\mu_N$ since lines very fine.

Na Frisch, *P.Z. Sowjetunion*, 1933, **4** 559. Lines of Na II show $I=1\frac{1}{2}$.

Rabi and Cohen, *PR*, 1933, **43** 582. Atomic rays show $I=1\frac{1}{2}$.

van Atta and Granath, *PR*, 1933, **44** 61*a*. $\mu=2{\cdot}6\mu_N$.

Joffe, *PR*, 1934, **45** 468. Band spectrum shows $I=1\frac{1}{2}$.

Rabi and Cohen, *PR*, 1934, **46** 707. Atomic ray shows $I=1\frac{1}{2}$.

Larrick, *PR*, 1934, **46** 581. Polarisation of resonance radiation explained by $I=1\frac{1}{2}$.

Fox and Rabi, *PR*, 1935, **48** 746. Method of "zero moments" shows $\mu=2{\cdot}08$.

Ne Nagaoka and Mishima, *P. Imp. Acad. Tokio*, 1929, **5** 200; 1930, **6** 143. Mass effect and magnetic splitting.

Bartlett and Gibbons, *PR*, 1933, **44** 538. Isotope displacement theory.

P Herzberg, *PR*, 1932, **40** 313. Band spectrum.
Ashley, *PR*, 1933, **44** 919. Band spectrum shows $I=\frac{1}{2}$.
Jenkins, *PR*, 1935, **47** 783*a*. Alternating intensities confirm Ashley.

Pb Murakawa, *ZP*, 1931, **72** 793. $I=0$ for 206 and 208, but $\frac{1}{2}$ for 207.
Kopfermann, *ZP*, 1932, **75** 363. Isotope displacements.
Schüler and Jones, *ZP*, 1932, **75** 563. Hyperfine structure shows new isotope.
Rose and Granath, *PR*, 1932, **40** 760. Isotope displacements.
Dickinson, *PR*, 1934, **46** 498. Isotope displacements.
Rose, *PR*, 1935, **47** 122. Isotope displacements reviewed.

Pr White, *PR*, 1929, **34** 1397. In Pr II, $I=2\frac{1}{2}$.

Pt Fuchs and Kopfermann, *Nw*, 1935 ,**23** 372.
Jaeckel and Kopfermann, *ZP*, 1936, **99** 492. $I=\frac{1}{2}$ in Pt^{195}; isotope displacements.
Jaeckel, *ZP*, 1936, **100** 513. Confirms $I=\frac{1}{2}$.
Kopfermann and Krebs, *ZP*, 1936, **101** 193. Isotope proportions present.

Rb Kopfermann, *ZP*, 1933, **83** 417. Rb II shows $I=1\frac{1}{2}$ in 87 and $2\frac{1}{2}$ in 85. μ_{87} is twice μ_{85}. No isotope displacement.
Jackson, *ZP*, 1933, **86** 131. Confirms Kopfermann.

Re Zeeman, Gisolf and de Bruin, *N*, 1931, **128** 637. Zeeman effect of hyperfine structure shows $I=2\frac{1}{2}$.
Meggers, King and Bacher, *PR*, 1931, **38** 1258. Hyperfine structure shows $I=2\frac{1}{2}$ in both isotopes.
Sommer and Karlson, *Nw*, 1931, **19** 1021. Confirms Meggers.

Sb Tolansky, *PRS*, 1934, **146** 182. $I=2\frac{1}{2}$ in 121 and 123; $\mu_{121}=2{\cdot}75$, $\mu_{123}=2{\cdot}0$.
Crawford and Bateson, *Can. J. Research*, 1934, **10** 693. Sb III shows $I=2\frac{1}{2}$ in 121 and $3\frac{1}{2}$ in 123; $\mu_{121}=4{\cdot}0$, $\mu_{123}=3{\cdot}2$.

Sc Kopfermann and Rasmussen, *ZP*, 1934, **92** 82. $I=3\frac{1}{2}$ and $\mu=3{\cdot}6$.

Se Rafalowski, *Acta Phys. Polonica*, 1933, **2** 119. Lines are single.
Olsson, *ZP*, 1934, **90** 134. Band spectrum shows $I=0$ in Se^{80}.

Sm Schüler and Schmidt, *ZP*, 1934, **92** 148. Isotope displacements.

Sn Tolansky, *PRS*, 1934, **144** 574. Sn II shows $I=\frac{1}{2}$ in 117 and 119; μ is $0{\cdot}89\mu_N$ in both. One line of Sn I confirms *I*. Isotope displacement.

Sr Benson and Sawyer, *PR*, 1933, **43** 766*a*. No hyperfine structure wider than 0·05 cm.$^{-1}$

Ta Gisolf and Zeeman, *N*, 1933, **132** 566. Hyperfine structure shows $I=3\frac{1}{2}$.
McMillan and Grace, *PR*, 1933, **44** 949. Hyperfine structure shows $I=3\frac{1}{2}$.

Te Rafalowski, *Acta Phys. Polonica*, 1933, **2** 119. Lines are single.

Tl Schüler and Keyston, *ZP*, 1931, **70** 1. Isotope displacement. Tl I, Tl II.
Crooker, *PM*, 1933, **16** 994. Paschen-Back effect.
Wills, *PR*, 1934, **45** 883. $\mu=2{\cdot}7$. A perturbed term.

V Kopfermann and Rasmussen, *ZP*, 1935, **98** 624. Hyperfine structure shows $I=3\frac{1}{2}$.

W Grace and White, *PR*, 1933, **43** 1039*a*. No value of *I*. Lines complex.

Xe Kopfermann and Rindal, *ZP*, 1933, **87** 460. $I=\frac{1}{2}$ in 129 and $1\frac{1}{2}$ in 131. μ_{129} is negative, μ_{131} is positive. μ_{129}/μ_{131} is 1·1. No isotope displacement.

Gwynne-Jones, *PRS*, 1934, **144** 587. Xe I confirms.

Y Kruger and Challacombe, *PR*, 1935, **48** 111*a*. Earlier report incorrect.

Zn Schüler and Bruck, *ZP*, 1929, **56** 291. No hyperfine structure found. Zn I.

Murakawa, *ZP*, 1931, **72** 793. Absence of hyperfine structure in Zn I suggests $I=0$.

Schüler and Westmeyer, *ZP*, 1933, **81** 565. Zn II shows isotope displacement. In Zn 67, $I=1\frac{1}{2}$.

Billeter, *Helv. Phys. Acta*, 1934, **7** 413. No hyperfine structure found in resonance line.

SUBJECT INDEX

Numbers in Clarendon refer to the second volume.

AUTHOR INDEX

Numbers in Clarendon refer to the second volume.